Oncogenesis of Lymphoma

Oncogenesis of Lymphoma

Guest Editors

Annunziata Gloghini
Giancarlo Pruneri

MDPI

Basel • Beijing • Wuhan • Barcelona • Belgrade • Novi Sad • Cluj • Manchester

Guest Editors
Annunziata Gloghini
Department of Diagnostic Innovation
Fondazione IRCCS Istituto Nazionale dei Tumori
Milan
Italy

Giancarlo Pruneri
Department of Diagnostic Innovation
Fondazione IRCCS Istituto Nazionale dei Tumori
Milan
Italy

Editorial Office
MDPI AG
Grosspeteranlage 5
4052 Basel, Switzerland

This is a reprint of the Special Issue, published open access by the journal *Cancers* (ISSN 2072-6694), freely accessible at: https://www.mdpi.com/journal/cancers/special_issues/GN76QL2O0V.

For citation purposes, cite each article independently as indicated on the article page online and as indicated below:

Lastname, A.A.; Lastname, B.B. Article Title. *Journal Name* **Year**, *Volume Number*, Page Range.

ISBN 978-3-7258-6684-7 (Hbk)
ISBN 978-3-7258-6685-4 (PDF)
https://doi.org/10.3390/books978-3-7258-6685-4

Contents

About the Editors

Annunziata Gloghini

Annunziata Gloghini is a Senior Scientist at the Department of Advanced Pathology at the Fondazione IRCCS, Istituto Nazionale dei Tumori in Milan, Italy, where she has been engaged since 2021 after a long tenure as senior staff researcher at the same department. Previously, she occupied key positions at the Centro di Riferimento Oncologico in Aviano, including as Head of the Diagnostic Immunohistochemistry and Molecular Pathology Unit, Deputy Director of the Division of Pathology, and Assistant in the same division. As a biologist with postgraduate training in General Pathology from the University of Genova and additional research experience at the University of Geneva, she has authored 380 peer-reviewed publications on the pathogenesis, classification, and diagnosis of human cancers, achieving an h-index of 72 (Scopus). Her research focuses on Hodgkin and non-Hodgkin lymphomas, particularly HIV-associated lymphomas, to which she has contributed by refining their morphologic classification, elucidating viral associations, and exploring the impact of the tumor microenvironment on lymphomagenesis. She is an expert in bright-field in situ molecular pathology and in the application of advanced molecular and virologic techniques to the study of lymphomas and solid tumors, including those of the breast, gastrointestinal tract, and head and neck.

Giancarlo Pruneri

Giancarlo Pruneri is Full Professor in Pathology and Director of the School of Specialization in Pathological Anatomy at the University of Milan, School of Medicine. He is also Chair of the Department of Advanced Diagnostics, National Cancer Institute of Milan, a comprehensive cancer center fully equipped for translational research and innovative care, including a molecular lab for next-generation sequencing. He is devoted to translational research programs, currently focused on lymphoma and breast pathology, sustained by several national and international peer-reviewed grants, resulting in 400 scientific papers in international journals and gaining an H-index of 78. He is involved in several international networks, including the International Breast Cancer Study Group (IBCSG, where he was appointed chair of the Translational Research Working Group) and the International Immuno-Oncology Working Group, focused on the validation of biomarkers for immune therapy. Prof. Pruneri is Editor-in-Chief of *TJ Tumori* Journal, Specialty Editor for *The Breast*, and in charge of the Precision Medicine program at the National Cancer Institute of Milan.

cancers

MDPI

Review

Immune Deficiency/Dysregulation-Associated EBV-Positive Classic Hodgkin Lymphoma

Mohamed Nazem Alibrahim [1,*], Annunziata Gloghini [2] and Antonino Carbone [3,*]

1 Faculty of Medicine, Zagazig University, Zagazig 44519, Egypt
2 Department of Avanced Pathology, Fondazione Istituto di Ricovero e Cura a Carattere Scientifico IRCCS, Istituto Nazionale dei Tumori Milano, 20133 Milano, Italy; annunziata.gloghini@istitutotumori.mi.it
3 Centro di Riferimento Oncologico, Istituto di Ricovero e Cura a Carattere Scientifico, National Cancer Institute, 33081 Aviano, Italy
* Correspondence: 20512020101186@medicine.zu.edu.eg (M.N.A.); acarbone@cro.it (A.C.); Tel.: +39-0434-659085 (A.C.)

Simple Summary: Classic Hodgkin lymphoma (cHL) associated with immune deficiency or dysregulation, particularly involving Epstein–Barr virus (EBV), represents a distinct and challenging clinical entity. EBV plays a crucial role by promoting cancer cell survival, altering immune responses, and fostering immune escape mechanisms. This subtype of lymphoma frequently arises in immunocompromised individuals, such as HIV-positive patients, transplant recipients, and those with primary immunodeficiencies, and is often resistant to standard treatments. Current classifications recognize the importance of EBV status and immune dysfunction in diagnosis and therapy planning. New strategies, including immune checkpoint inhibitors, EBV-targeted T-cell therapies, and novel molecular approaches, show promise in improving outcomes. Understanding the complex interplay between viral infections, immune status, and genetics is essential for developing targeted treatments and personalized patient care.

Abstract: Classic Hodgkin lymphoma (cHL) in patients with immune deficiency/dysregulation represents a critical unmet need in hematology, demanding the appropriate revision of classification and therapeutic paradigms. Epstein–Barr virus (EBV) is a pivotal driver of lymphomagenesis in this high-risk subset, where viral oncoproteins (e.g., LMP1/2A) exploit immune vulnerabilities to activate NF-κB, rewire tumor microenvironments (TME), and evade immune surveillance. EBV-positive cHL, prevalent in immunosuppressed populations, exhibits distinct molecular hallmarks, including reduced somatic mutations, unique HLA associations, and profound PD-L1-mediated immune suppression, that diverge from EBV-negative cases reliant on genetic aberrations. Despite advances in combined antiretroviral therapy, HIV co-infection exacerbates pathogenesis, M2 macrophage dominance, and T-cell exhaustion, while links to other viruses remain ambiguous. Current therapies fail to adequately target these viral and immune complexities, leaving patients with poorer outcomes. This review synthesizes insights into EBV's etiological role, immune contexture disparities, and the genetic–environmental interplay shaping cHL heterogeneity. The WHO classification highlights the need to reclassify EBV-associated cHL as a distinct subset, integrating viral status and immune biomarkers into diagnostic frameworks. Urgent priorities include global epidemiological studies to clarify causal mechanisms, development of virus-targeted therapies (e.g., EBV-specific T-cell strategies, PD-1/CTLA-4 blockade), and personalized regimens for immune-dysregulated cohorts.

Cancers **2025**, *17*, 1433 https://doi.org/10.3390/cancers17091433

Keywords: classic Hodgkin lymphoma; EBV+ cHL; WHO classification; immune deficiency; immune dysregulation; management; diagnosis

1. Introduction

Hodgkin lymphoma is categorized into two primary forms, each with unique pathological and clinical characteristics: classic Hodgkin lymphoma (cHL) and nodular lymphocyte-predominant Hodgkin lymphoma (NLPHL). cHL is further divided into four subtypes, nodular sclerosis (NSHL), lymphocyte-rich (LRHL), mixed cellularity (MCHL), and lymphocyte-depleted (LDHL) [1]. cHL is defined by the presence of Reed–Sternberg (RS) cells, large, often multinucleated malignant cells with characteristic "owl-eye" nuclei, originating from germinal center (GC) B cells with immunoglobulin gene rearrangements. These cells are typically positive for CD30 and CD15, but negative for B cell markers CD20 and CD45, and are often accompanied by a rich inflammatory background shaped by cytokines secreted by RS cells. Variants of RS cells include lacunar cells, mummified cells, and mononuclear Hodgkin cells, which are seen in specific cHL subtypes such as nodular sclerosis and mixed cellularity (Figure 1) [1,2]. In contrast, NLPHL lacks RS cells and instead features lymphocyte-predominant (LP) or "popcorn" cells, which display multilobulated, folded nuclei with smaller nucleoli. LP cells are positive for CD20, CD45, EMA, BCL6, and other B cell markers, reflecting their preserved B cell phenotype [3]. Morphologically, NLPHL shows a nodular architecture with a background rich in small B lymphocytes and follicular dendritic cells, differing from the mixed inflammatory milieu and architectural patterns seen in cHL subtypes [1,4]. cHL is a rare type of hematological cancer and has good prognosis; cHL accounts for about 95% of all cases of Hodgkin lymphoma (HL). HL accounted for 0.4% of new cancer cases and 0.2% of cancer-related deaths worldwide in 2020. In the United States, the disease had an estimated incidence rate of 0.26 cases per 100,000 people that year, representing 10% of all lymphoma diagnoses. cHL exhibits a bimodal age distribution, with peak incidence observed in young adults and those aged 55 and older [5–8].

The precise etiology of cHL remains unclear. However, several studies have identified an elevated risk in individuals with Epstein–Barr virus (EBV) infection. In resource-rich countries, EBV positivity in cHL ranges from 20% to 50%, with higher prevalence observed among older adults, children, and males. In contrast, resource-poor countries exhibit significantly higher EBV positivity rates, with the highest prevalence in Africa (74%) and lower rates in Europe (36%) and North America (32%), highlighting geographic differences in EBV-related HL cases [9,10]. This association is particularly notable in mixed cellularity and lymphocyte depletion subtypes, which is more common in the lower human development index and certain ethnic groups, even after accounting for socioeconomic factors [11,12].

EBV plays a significant role in cHL pathogenesis, with evidence suggesting it acts as an early pathogenic event. Monoclonal EBV genomes found in Hodgkin and Reed–Sternberg (HRS) cells indicate infection precedes clonal expansion. The virus employs type II latency, expressing proteins such as EBNA-1, LMP-1, and LMP-2, which promote tumorigenesis and create an immunosuppressive tumor microenvironment (TME) through cytokine induction and PD-L1 upregulation (Figure 2). This viral involvement is further highlighted in HIV-associated cHL, where nearly all cases are EBV-positive. Similarly, human immunodeficiency virus (HIV) infection increases the risk of cHL by 6–15 fold [9], alongside autoimmune disorders and immunosuppressive conditions. Moreover, familial predisposition has been identified as a contributing factor in Hodgkin lymphoma [1,13–18].

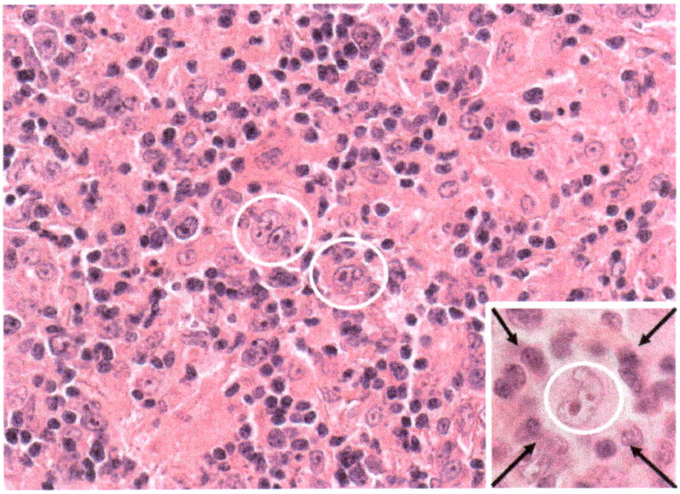

Figure 1. Immune deficiency/dysregulation-associated classic Hodgkin lymphoma (cHL) (HIV-related subset). EBV-positive cHL from an HIV-infected patient. Two Hodgkin–Reed–Sternberg (HRS) cells (within white circle) are recognizable as large multinucleated cells within inflammatory cellularity showing "mixed cellularity" pattern. In inset, HRS cell (within white circle) is "rosetted" by small lymphocytes (arrows). Haematoxliyn and Eosin stain; original magnification 40×, inset 100×.

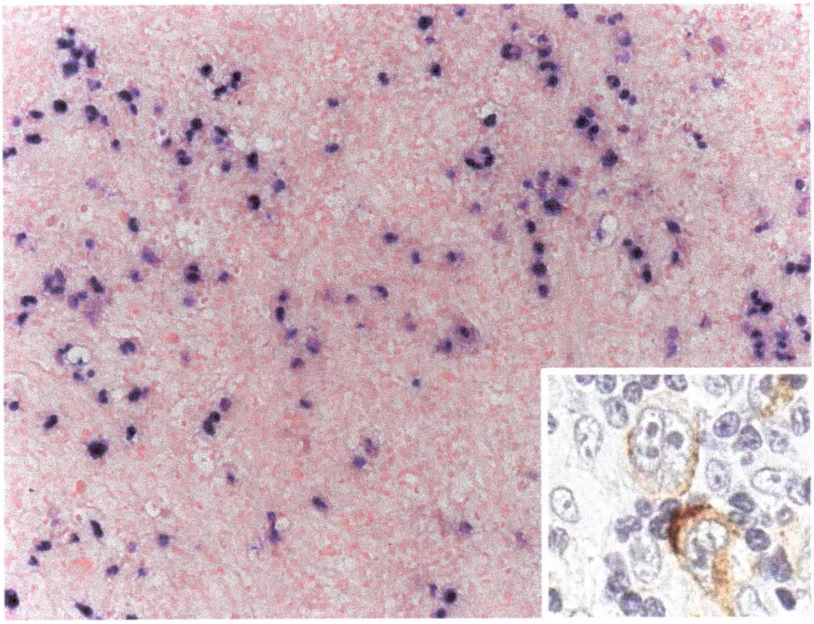

Figure 2. Immune deficiency/dysregulation-associated classic Hodgkin lymphoma (cHL) (HIV-related subset). EBV-positive cHL from HIV-infected patient. Hodgkin Reed–Sternberg cells (HRS) are EBV-positive, as demonstrated by in situ hybridization, for EBERs. Positivity is nuclear. Inset: EBV-infected HRS cells express, as demonstrated by immunohistochemistry, latent membrane protein 1. Staining is cytoplasmic. Original magnification 20×; inset: 100×.

Additionally, cHL arises from complex genetic alterations primarily involving HRS cells, characterized by mutations in key driver genes affecting signaling pathways such as NF-κB and JAK/STAT. Additional genetic aberrations include alterations in SOCS1, PTPN1, TNFAIP3, and B2M, contributing to immune escape and cytokine dysregulation [19–22].

The recently updated 2023 WHO classification has introduced refined diagnostic criteria for immune deficiency/dysregulation-associated lymphoproliferative disorders (IDD-LPDs), integrating histologic subtype (hyperplasia, polymorphic disorder, or lymphoma), viral associations (particularly EBV latency states), and underlying immune dysfunction contexts (e.g., primary immunodeficiency, post-transplant, HIV). This standardized diagnostic framework, as proposed by the WHO, underscores the importance of accurately characterizing the immune environment and EBV latency patterns (latency I or II), thus enabling improved prognostication, targeted therapeutic strategies, and a deeper understanding of the interplay between genetic alterations, viral pathogenesis, and immune dysregulation in cHL (Table 1) [23,24].

Table 1. Morphologic characteristics and EBV infection in immune deficiency/dysregulation-associated classic Hodgkin lymphoma.

Immunodeficiency-Associated cHL	Hodgkin Lymphoma Subtype	EBV Infection
Immune senescence-associated cHL	cHL, nodular sclerosis	Usually Neg
	cHL, mixed cellularity	Usually Pos
	Rare Types	
	cHL, lymphocyte rich	Variably Pos
	cHL, lymphocyte depleted	Variably Pos
HIV-associated cHL	cHL, nodular sclerosis	Pos
	cHL, lymphocyte depleted	Pos
	cHL, mixed cellularity	Pos
	cHL, lymphohistiocyoid	Pos
Post-transplant (cHL type PTLD)	Similar subtypes as above	Pos
Other iatrogenic immune deficiency-associated cHL	cHL, mixed cellularity	Usually Pos

Abbreviations: cHL, classic Hodgkin lymphoma; PTLD, post-transplant lymphoproliferative disorder; Neg, negative; Pos, positive.

The scope of this paper is to explore the potential causal link between viral infections, specifically EBV infection, and the development of cHL. It will examine the molecular mechanisms underlying viral contributions to cHL, including immune evasion, oncogenic signaling, and tumor microenvironment modulation, while addressing epidemiological disparities. The objective is to synthesize current evidence, highlight unresolved questions, and propose future research directions to deepen the understanding of viral-associated cHL and improve strategies for diagnosis and treatment.

2. Methodology

This narrative review aims to explore the potential causal link between viral infections, specifically EBV, HIV, hepatitis viruses, and the development of cHL. The review synthesizes findings from a wide range of primary studies, reviews, and expert opinions to provide a comprehensive understanding of the role of EBV, HIV, HCV, HBV, and their contribution in cHL pathogenesis.

The relevant literature was identified through searches of academic databases, including PubMed, Scopus, and Google Scholar, using search terms such as "classical Hodgkin lymphoma", "viral oncogenesis", "HCV AND HBV", "Human herpesvirus", "Sars-Cov-2", "Covid-19", "EBV and Hodgkin lymphoma", "immunodeficiency", "PTLD", "immunosuppression", and "HIV-associated lymphoma". References from key articles were also reviewed to identify additional sources of relevance. The focus was placed on studies published in English that discuss viral involvement in cHL, its epidemiology, molecular mechanisms, immune modulation, and implications for treatment. The exclusion criteria omitted studies with limited data on viral mechanisms, non-peer review, non-English studies, or unrelated to cHL pathogenesis.

3. Causal Link Between EBV Viral Infection and cHL Development

3.1. Background on EBV and HL

EBV has long been suspected of playing a role in HL, supported by elevated antibody titers against EBV proteins in some patients. Research in the 1970s by IARC showed that EBV infection is nearly universal, with over 90% of adults worldwide carrying the virus. The presence of EBV in every tumor cell of EBV-positive cHL strongly supports its role in the disease's pathogenesis. In Western populations, EBV is found in about one-third of cHL cases, with higher prevalence in other regions. If EBV were incidental, its frequency in cHL would reflect the rare occurrence of EBV-infected B cells (1 in 10^6).

However, since cHL arises from GC B cells, the observed frequency aligns with the higher rate of EBV infection in GC B cells (4.6 per 10^4) [25], reinforcing its etiological significance. EBV normally infects naive B cells in the Waldeyer ring, where it can differentiate into memory cells and exit the cell cycle without causing harm [26]. Over a 54-year period in the Brazilian state of São Paulo, the frequency of EBV-associated cHL showed a significant decline. Among the 817 cases studied, EBV positivity dropped from 87% in the period 1954–1979 to 46% in 2000–2008. This decline was most pronounced in children, adolescents, and young adults, while the rate remained relatively stable in patients over 45 years. These trends reflect a shift toward the epidemiological pattern seen in developed countries, likely influenced by socioeconomic improvements, which are associated with reduced early EBV exposure and enhanced immune control over the virus [27]. However, if EBV infects other types of cells, they may express the growth program and continue proliferating, a rare event that is usually controlled by cytotoxic T lymphocytes (CTLs). If the CTL response is suppressed, these cells can develop into post-transplant lymphoproliferative disorder (PTLD) or HIV-associated lymphomas [28,29].

PTLDs represent a group of lymphoid proliferations occurring due to therapeutic immunosuppression administered to prevent graft rejection after solid organ or hematopoietic stem cell transplantation [30]. The degree, duration, and type of immunosuppression significantly influence the development of PTLD, with potent immunosuppressants such as T-cell-depleting antibodies and calcineurin inhibitors associated with higher risks [29,31,32]. Immunosuppression impairs EBV-specific T-cell immune surveillance, enabling EBV-infected B cells to proliferate unchecked, particularly in patients who are EBV-seronegative prior to transplantation [33]. This immunologic dysfunction leads to a microenvironment enriched with exhausted and ineffective CD8+ T-cells overexpressing inhibitory markers like PD-1, which further supports malignant transformation and the progression of EBV-driven lymphoproliferative disease [29,32,34].

Hodgkin lymphoma arises from the default program (Latency II) of EBV infection, while Burkitt lymphoma develops when a germinal center cell entering the memory compartment becomes stuck in proliferation, expressing only EBNA-1 [17,35–38]. However, given that over 90% of humans harbor latent EBV infections, initial findings were inconclusive. Significant advances occurred in 1985 and 1989 when the EBNA protein (Epstein–Barr nuclear antigen, now known to be EBNA1) was detected in HRS cells, indicating clonal viral infection and suggesting that HRS cells are monoclonal [39–42].

3.2. EBV-Mediated Oncogenesis

EBV plays a critical role in the pathogenesis of certain lymphomas, including cHL, Burkitt lymphoma (BL), and diffuse large B cell lymphoma (DLBCL) [43,44] (Table 2). EBV infection can be lytic, involving widespread viral gene expression, virus production, and cell death [45], or latent, observed in most tumor cells, with limited gene expression, persistent infection, and host cell transformation. EBV infects HRS cells clonally, promoting their survival by preventing apoptosis and upregulating PDL1 expression, mediated by latent membrane protein 1 (LMP1), which is a viral protein [14].

Table 2. EBV-associated lymphomas.

Lymphoma Type	Approximate Percentage of EBV Positivity	EBV Latency
Burkitt lymphoma		
Endemic	>95	I
Sporadic	20–80	I
HIV-associated	30–50	I
HIV-associated DLBCL		
Immunoblastic	70–100	II/III
Non-immunoblastic	10–30	I
CNS lymphomas	>95	II/III
Plasmablastic	60–75	I/II
Primary effusion lymphoma	70–90	I
Post-transplant LPD	>90	II/III
EBV-associated DLBCL (of the elderly)	100	II
Classic Hodgkin lymphoma *	20–90	II
Extranodal T/NK cell lymphoma nasal type	100	II

* Classic HL of the general population include the following subtypes: nodular sclerosis (usually EBV-negative), mixed cellularity (usually EBV-positive), lymphocyte rich (variably EBV-positive), and lymphocyte depleted (variably EBV-positive). Association with EBV is less frequent in nodular sclerosis (10–40%) than in mixed cellularity cHL (approximately 75%). For immune deficiency/dysregulation-associated classic Hodgkin lymphoma, see Table 1. Abbreviations: CNS, central nervous system; DLBCL, diffuse large B cell lymphoma; lLPD, lymphoproliferative disorder.

HRS cells express EBNA1, LMP1, LMP2A, and LMP2B, while the other EBNAs are downregulated [46], HRS cells bypass normal GC B cell survival mechanisms, relying on nonphysiological pathways. LMP1 acts as an oncogene by mimicking CD40 signaling [47,48], activating AP-1, JAK\STAT, NF-κB, Jun N-terminal kinase (JNK), and p38 mitogen-activated protein kinase and other pathways through its cytoplasmic domain [49–54].

The NF-κB pathway is a cornerstone of HRS cell survival, driving the upregulation of anti-apoptotic genes that protect these malignant cells from programmed cell death [55–57]. LMP1 further amplifies this effect in GC B cells by inducing Id2 expression [58], which suppresses the E2A transcription factor and silences B cell signature genes, facilitating extensive transcriptional reprogramming [58]. Beyond its role in cellular survival, NF-κB activation orchestrates a broad spectrum of critical processes, including immune system regulation, cell proliferation, viral replication, tumor metastasis, and inflammation [56,57,59]. This multifaceted activity positions NF-κB as a central player in tumor progression, immune evasion, and the pathogenesis of cHL; however, these pathways may also be activated through genetic mutations within the cells [56,57,59].

Meanwhile, LMP2a resembles components of the B cell receptor (BCR) and sustains survival signaling [48], LMP2A contributes to HRS cell survival and transcriptional reprogramming by activating Notch1, inhibiting E2A, and downregulating EBF, key transcription factors for B cell development [26,60,61]. However, the exact role of LMP2A in cHL remains unclear due to the downregulation of many BCR signaling molecules in HRS cells, although its effects extend beyond classical BCR pathways. Furthermore, latently infected cells transcribe noncoding viral RNAs, including two small non-polyadenylated transcripts called EBERs and more than 44 viral microRNAs (miRNAs). These miRNAs are either embedded within the introns of BARTs or located near the coding region of the BHRF1 gene; in cHL, cells express BART miRNAs but lack BHRF1 miRNAs which is linked to latency type III. The simultaneous expression of all latent genes characterizes what is known as latency III [36–38,40,62–65].

Certain miRNAs such as BART2-5p are expressed in cHL and help suppress the lytic cycle by downregulating the viral DNA polymerase BALF5, supporting the maintenance of latency [66,67]. BART2-5p can also impair B cell receptor signaling. By aligning with the characteristic loss of BCR function in cHL, it can also inhibit the expression of the viral DNA polymerase BALF5, thereby suppressing the lytic cycle and promoting latency [66,68]. Additionally, BART13-3p is highly expressed and secreted via exosomes, where it can alter macrophage behavior to promote a tumor-supportive microenvironment by increasing cytokines like TNF-α and IL-10 [69,70]. The expression patterns of EBV miRNAs also shift across stages of latency, with different subsets activated during transitions from latency III to latency II and 0 [71]. However, detecting EBV miRNAs in cHL is challenging due to the low number of malignant HRS cells and contamination by surrounding non-neoplastic immune cells [72]. Additionally, EBV can influence host miRNA expression; in EBV-positive cHL, miR-96, miR-128a, and miR-128b are notably downregulated [73]. These findings underline the multifaceted role of EBV miRNAs in shaping the TME, maintaining latency, and modulating immune escape mechanisms in cHL. Recent findings emphasize the importance of both host and EBV-encoded microRNAs as potential circulating biomarkers in EBV-associated lymphomas, including cHL [74–76]. While EBV miRNAs such as BART13-3p have been detected in exosomes and linked to immune modulation, several host miRNAs (e.g., miR-21, miR-155) are also differentially expressed and could serve as minimally invasive tools for diagnosis or prognosis [70,72].

3.3. *Differences Between EBV-Positive and EBV-Negative cHL*

EBV-positive cHL is typically associated with MCHL and LDHL and shows a dense inflammatory background composed of numerous CD68+ and CD163+ macrophages, cytotoxic T lymphocytes, and frequent necrotic areas [35,77]. The RS cells are often large, with prominent nucleoli, and test positive for EBV markers such as LMP1 and EBER [78]. In contrast, EBV-negative cHL, commonly seen in nodular sclerosis and lymphocyte-rich subtypes, displays a more fibrotic architecture with lacunar-type HRS cells, fewer macrophages, and a less pronounced inflammatory infiltrate, reflecting fundamental histopathological differences between the two variants [13].

In EBV-negative cHL, oncogenesis is driven by mutations in genes like C-REL and tyrosine kinase receptors [79–82]. Additionally, hypothetical cause of cHL in EBV-negative cases is alternative pathogenic mechanisms unrelated to viral influence, which include genetic abnormalities, aberrant signaling pathways, and interactions with the TME. Specifically, EBV-negative cHL often exhibits dysregulation of immune checkpoints, such as PD-1/PD-L1, and a distinct inflammatory microenvironment characterized by a Th17 profile, which differs from the Th1-dominant environment in EBV-positive cases [13,44,83,84].

EBV-infected HRS cells also foster an immunosuppressive environment by interacting with immune cells and secreting suppressive mediators [85,86]. EBV causes the overexpression of PD-L1, PD-L2, and CTLA-4 [87,88], further amplifying immune evasion mechanisms through inhibiting T-cell activation. PD-1 (CD279) inhibits T-cell activity upon engagement with its ligands PD-L1 and PD-L2. CTLA-4 (CD152) competes with CD28 for binding to CD80/CD86, suppressing T-cell activation, leading to immune exhaustion or apoptosis. T-cell activation depends on TCR signaling and co-stimulatory signals, regulated by immune checkpoints like PD-1 and CTLA-4, which maintain immune tolerance and prevent overactivation. In EBV-positive lymphomas, the virus enhances immune evasion by upregulating PD-L1 (Figure 3) and PD-L2 through the NF-κB and JAK/STAT pathways, with additional contributions from 9p24.1 genetic alterations. These mechanisms impair T-cell function, leading to immune suppression and facilitating lymphoma progression. Additionally, EBV-positive HL is linked to the suppression of the gene p21^cip1/waf1, which plays a key role in cell cycle regulation and apoptosis. The EBV-encoded RNA EBER1 was shown to inhibit p21^cip1/waf1 transcription by downregulating its regulators, p53, EGR1, and STAT1, making cells more resistant to apoptosis induced by common lymphoma treatments. Clinically, EBV+ HL cases exhibited lower p21^cip1/waf1 expression and had significantly worse two-year survival outcomes compared to EBV- cases, suggesting that EBER1′s anti-apoptotic effect contributes to treatment resistance and poorer prognosis [13,40,64,65,86].

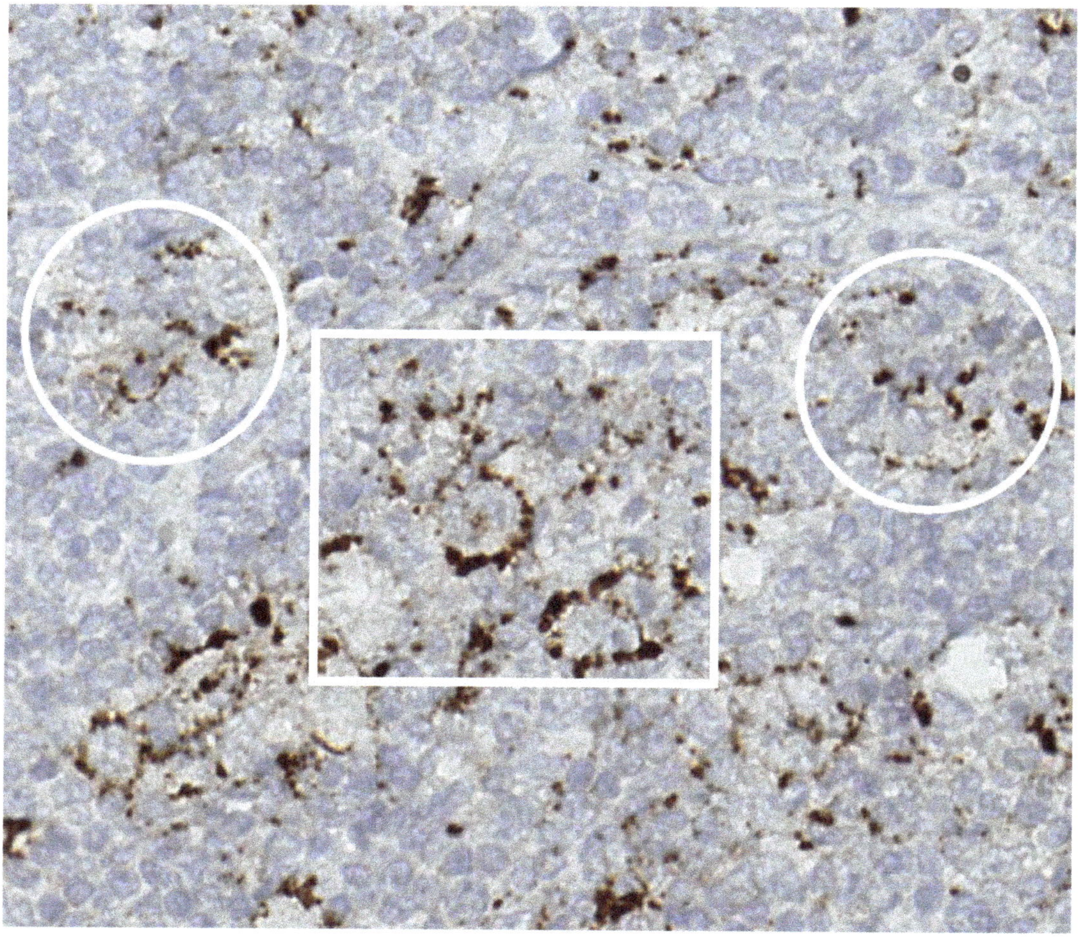

Figure 3. Immune deficiency/dysregulation-associated cHL (HIV-related subset). EBV-positive classic Hodgkin lymphoma (cHL) from HIV-infected patient. PD-L1 expression in Hodgkin Reed–Sternberg (HRS) cells (within white rectangle) and macrophages (areas highlighted by white circles) in cHL. Immunohistochemical staining; Haematoxylin counterstain; original magnification 40×.

Studies have demonstrated that immune checkpoint blockade therapies can effectively restore T-cell activity, reduce tumor burden, and counteract the immune evasion caused by EBV-driven PD-L1 expression. The expression occurs through the activation of the AP1 transcription factor pathway, which upregulates c-Jun and JunB. Notably, PD-L1 expression is observed in 89.2% of HRS cells and extensively in the TME of cHL patients. Despite this high prevalence, studies have shown no significant correlation between PD-L1 expression, EBV status, clinical parameters, or prognosis. This pathway, a hallmark of cHL, functions independently of 9p24.1 copy number variations, underscoring its potential as a target for emerging immunotherapies. EBV's latency states influence its role in lymphomas. In BL and related B cell malignancies, EBV employs latency type 1, expressing EBNA1, EBERs, and BARF0 to evade immune detection. In cHL and NK/T-cell lymphomas, latency type 2 predominates, involving LMP1, BARF0, EBERs, EBNA-1, and LMP2 expression but not EBNA2. Latency type 2 is a target for PD1/PDL1-blocking therapies. In immunosuppressed individuals, latency type 3, expressing a broader range of EBV proteins, emerges. EBV-positive and EBV-negative tumors exhibit distinct pathogenic mechanisms and prognoses [13,35,40,64,65,86,89–93].

In EBV-positive cHL cases, LMP1 and LMP2a provide critical survival signals to GC B cells, compensating for the absence of functional BCRs due to destructive IgV mutations. Notably, EBV-positive HL cases often harbor fewer mutations in genes regulating the NF-κB pathway, such as TNFAIP3 and NFKBIA, due to the strong NF-κB activation. Additionally, EBV-positive cases show lower overall mutation loads in oncogenes and tumor suppressor genes compared to EBV-negative cases, suggesting a pathogenetic role for EBV; viral genes alone can drive abnormal signaling pathway activation, reducing the need for cellular mutations. This suggests that EBV and cellular mutations represent distinct, mutually exclusive mechanisms leading to the same disease outcome. Beyond protein expression, EBV-infected cells produce noncoding RNAs, such as miR-BHRF1-2-5p and miR-BARTs. These ncRNAs target and modulate PD-L1/PD-L2 expression, further contributing to immune suppression. Collectively, these findings underscore EBV's significant role in the survival and transformation of pre-apoptotic GC B cells, particularly in EBV-positive HL cases [37,40,89,94].

4. Genetic Associations

A study in 2015 examined HLA associations with EBV-positive and EBV-negative cHL using allele selection modeling. For EBV-positive cHL, HLA-A01:01 and B37:01 were associated with increased risk, while DRB115:01, E*01:01 Is, and DPB101:01 were linked to decreased risk. In EBV-negative cHL, the strongest predictor was SNP rs6903608 in the class II region, with B15:01, DRB103:01, and DQB1*03:03 associated with increased risk. These findings reveal distinct HLA associations for EBV-positive and EBV-negative cHL, suggesting different mechanisms in disease pathogenesis and highlighting the importance of analyzing EBV-stratified subgroups in cHL research [95–97].

Additionally, a genetic variant, rs6457715, located near the HLA-DPB1 gene in the MHC region, was identified as strongly associated with EBV-positive cHL. This variant confers a 2.3-fold increased risk specifically for EBV-positive cHL, with no significant association for EBV-negative cHL. The effect is consistent across histological subtypes, indicating that EBV status is the primary driver of the association. rs6457715 represents the third independent genetic locus in the MHC region exclusively associated with EBV-positive cHL. These findings reinforce the critical role of EBV in cHL etiology and the importance of considering tumor EBV status in genetic susceptibility studies [98–100].

In summary, EBV plays a significant causal role in cHL, particularly in immune-suppressed settings, where it provides survival signals to B cells through latency proteins like LMP1 and LMP2a. EBV-positive cHL cases rely less on cellular mutations for NF-κB activation due to strong viral signaling, while EBV-negative cases involve alternative mechanisms, including genetic mutations, socioeconomic interactions, and distinct HLA associations. These findings highlight the importance of EBV status in understanding cHL pathogenesis and its implications for targeted therapies.

5. Beyond EBV: HIV Infections in CHL Development

5.1. Background

Classic HL is the most prevalent non-AIDS defining tumor. HIV-positive individuals have a significantly elevated risk of developing cHL, with incidence rates 5–26 times higher than those observed in the general population [101]. This increased vulnerability is attributed to the combined effects of HIV-induced immune suppression and frequent co-infection with EBV. Studies show that nearly 90–100% of cHL cases in HIV-infected individuals are EBV-positive, compared to only 30–40% in HIV-negative cases, highlighting the virus's central role in lymphomagenesis. HIV-related cHL predominantly presents as the mixed cellularity subtype, followed by nodular sclerosis. Pathologically, it shares similarities with cHL in HIV-negative individuals, characterized by HRS cells in a background of lymphocytes, eosinophils, neutrophils, macrophages, and plasma cells. However, HIV-related cHL features a higher number of HRS cells, frequent co-infection with EBV, a "sarcomatoid" pattern, and distinct microenvironmental changes, including reduced CD4+ T-cell counts, an inverted CD4/CD8 T-cell ratio, a reduction in infiltrating GrB+ cells (activated cytotoxic cells), and an increase in infiltrating TIA+ T-cells (primarily non-activated cytotoxic cells) were noted.

Adding to this complexity, a recent genetic study has revealed significant heterogeneity in HIV-associated lymphomas. Next-generation sequencing of diagnostic biopsies has identified 213 mutations across 42 genes, with frequent mutations in KMT2D (67%), TP53 (61%), and TNFAIP3 (50%). Most mutations are unique to individual patients, reflecting diverse genetic and pathophysiological mechanisms. This genetic diversity underscores the multifaceted drivers of lymphomagenesis in HIV-positive individuals, further compounded by the immune dysregulation caused by HIV.

Chronic antigenic stimulation by HIV leads to B cell hyperactivation, increasing the likelihood of genetic mutations and chromosomal translocations that predispose cells to malignant transformation. Furthermore, the depletion of CD4+ T-cells impairs the immune system's ability to surveil and eliminate aberrant cells. Interestingly, a moderately suppressed immune state appears more conducive to cHL development than severe immunosuppression, as some level of immune activity is necessary to create the pro-inflammatory microenvironment that supports the growth and survival of HRS cells [93,101–107].

In summary, the development of cHL in HIV-positive individuals results from a complex interplay between immune suppression, EBV-driven oncogenesis, and chronic immune activation. These factors collectively create a permissive environment for the transformation and proliferation of malignant B cells, contributing to the disproportionately high incidence of cHL in this population.

5.2. TME in HIV-Associated and EBV-Positive cHL

TME in HIV-associated cHL is shaped by a complex interplay of HIV-induced immune suppression and EBV-driven oncogenesis. HIV-associated cHL predominantly features unfavorable histological subtypes, such as mixed cellularity and lymphocyte-depleted cHL, characterized by extensive necrosis and sarcomatoid macrophage infiltration. EBV universally infects HRS cells in this context, with its LMP1 driving NF-κB activation, chronic inflammation, immune evasion, and tumor cell survival. Compared to non-HIV cHL, the TME in HIV-associated cHL is enriched with CD68+ and CD163+ spindle-shaped M2 macrophages, PD-L1-expressing cells, and Galectin-1 and significantly lower levels of CD4+ T-cells, CD56+ NK cells, CD57+ cells, CD123+ dendritic cells, and B cells in the TME of HIV-positive cHL cases compared to HIV-negative ones, fostering a highly immunosuppressive environment that shields HRS cells from immune attack [101,108–111].

HIV/EBV co-infection further alters the TME by disrupting immune dynamics. While EBV-positive cHL typically displays an inflammatory profile with increased CD8+ T-cell activity, HIV co-infection reduces CD8+ T-cell density and functionality, downregulating key T-cell receptor (TCR) signaling mediators such as VAV1, FYN, and AKT1. Additionally, the spatial organization of effector cells is disrupted, with CD8+ T-cells and GrzB+ T-cells found farther from neoplastic cells. These changes are accompanied by extracellular matrix (ECM) remodeling, including disrupted CXCL13 networks and reduced heparan sulfate levels, impairing T-cell trafficking and antigen presentation. The immune evasion mechanisms are further enhanced by the increased expression of coinhibitory ligands like CD155 and nectin-3, promoting interactions with inhibitory receptors on T-cells, such as TIGIT [112,113].

These dynamic interactions create a permissive environment for tumor progression and immune escape. Immune checkpoint inhibitors, such as pembrolizumab and nivolumab, show promise in reactivating suppressed immune responses, while ART may partially reduce inflammation and regulatory T-cell activity. However, its effects on TME immune activation remain incomplete, underscoring the need for combinatorial therapies targeting these specific molecular alterations in HIV/EBV-associated cHL [108,112,113].

5.3. Management

The treatment of EBV-positive cHL and HIV-associated cHL in immunosuppressed individuals aligns with protocols for the general population but requires the integration of combination antiretroviral therapy (cART) or reducing immunosuppression. However, HIV-positive individuals on cART have a high risk of developing HL, particularly in the early months of therapy [23]. HIV-HL presents with a more aggressive phenotype, advanced-stage disease, and atypical tumor spread. Most patients have moderate immunosuppression, and a declining CD4 count can be an early HL marker [106–108].

Historically, outcomes were poor, but the combination of cART, chemotherapy, and improved supportive care has significantly enhanced survival [114]. However, disparities in treatment access continue to impact prognosis, emphasizing the need for broader implementation of effective management strategies [115].

For HIV-related cHL, standard treatment includes ABVD (doxorubicin, bleomycin, vinblastine, dacarbazine) combined with cART, achieving complete response (CR) rates of 74% and 5-year overall survival (OS) of 81%. In limited-stage HIV–HL, the recommended approach is two cycles of ABVD followed by 20 Gy involved-site radiation therapy (ISRT) [114,116]. Data support using lower radiation doses (20 Gy instead of 30 Gy) and smaller ISRT fields because of good disease control and reduced toxicity [116]. BEACOPP (bleomycin, etoposide, doxorubicin, cyclophosphamide, vincristine, procarbazine) is an alternative for advanced stages with consolidation radiation therapy reserved for FDG-PET-CT-positive residual disease but is limited by toxicity; brentuximab vedotin (BV)-AVD may be considered for stage IV disease in select cases despite limited data [116]. Interim PET-CT guides risk-adapted therapy, with negative scans predicting improved progression-free survival (PFS) but this still needs confirmation [116]. Novel therapies like brentuximab vedotin (anti-CD30 antibody-drug conjugate) combined with AVD achieved a 2-year PFS of 86% and a 2-year overall survival rate of 92%, with manageable toxicity [9]. Immune checkpoint inhibitors (ICIs) have transformed cancer treatment, including cHL, with pembrolizumab and nivolumab showing high efficacy (65–87% response rates) in relapsed/refractory HIV-negative cHL. However, data on ICIs in HIV-associated cHL are limited, as PLWH have been excluded from most trials. A phase I study found pembrolizumab safe in PLWH with advanced cancers, warranting further research [9,117]. Only two trials confirm ICIs' safety in PLWH, with no impact on viral suppression or CD4 counts [23].

For relapsed/refractory cases, high-dose chemotherapy (ESHAP, DHAP, ICE) followed by autologous stem cell transplantation (ASCT) is the standard [118,119], with real-world data indicating a higher infection risk in lymphoma patients with HIV compared to those without, emphasizing the need for infection prevention and supportive management [23]. Chemotherapy with cART poses drug interaction risks, particularly with CYP3A4 inhibitors, though integrase inhibitors offer safer alternatives. In EBV-associated PTLD, rituximab with chemotherapy and EBV-specific cytotoxic T-cell (EBV-CTL) therapy are effective with overall response rates of 68% following HSCT and 54% after SOT, and no significant toxicity observed [29,32,120].

Those who relapse post-ASCT can receive PD-1 inhibitors such as nivolumab or pembrolizumab. Patients ineligible for HDCT should follow HIV-negative HL recommendations, with a preference for pembrolizumab given data from a prospective phase III trial [23,106–108,116,120–123].

Recent evidence highlights the significance of LAG-3 as an immune checkpoint in cHL, particularly in the context of EBV-associated cases. A study by Jimenez et al. demonstrated that LAG-3 was expressed in 57% of pediatric cHL cases, with 70% of these being EBV-positive. Notably, a strong correlation between LAG-3 and PD-1 expression was observed exclusively in EBV+ cases, suggesting that EBV may promote a synergistic immune exhaustion phenotype [124]. This dual expression was associated with significantly worse event-free survival (54% vs. 100% in other subgroups), underscoring the potential role of LAG-3 and PD-1 in fostering a tolerogenic microenvironment that facilitates lymphomagenesis [124]. These findings support the rationale for dual checkpoint inhibition strategies targeting both PD-1 and LAG-3 in EBV+ cHL, an approach currently being explored in clinical trials. More recent clinical data further supports the therapeutic relevance of targeting LAG-3 in cHL, particularly in patients with relapsed R/R disease following anti-PD-1 therapy. In an open-label phase 1/2 trial (NCT03598608), the combination of favezelimab (an anti-LAG-3 antibody) and pembrolizumab demonstrated promising antitumor activity in heavily pretreated R/R cHL patients who had progressed after prior PD-1 blockade. The objective response rate was 31%, with 79% achieving some degree of tumor reduction. Notably, dual LAG-3/PD-1 inhibition showed a tolerable safety profile and yielded a 12-month PFS rate of 39%, with a 91% OS rate. These findings underscore the potential of LAG-3/PD-1 co-blockade to re-induce responses in patients who have otherwise exhausted standard immunotherapeutic options, reinforcing the importance of incorporating LAG-3 in the immune checkpoint landscape of cHL [125].

Summing up, both HIV-associated and EBV-driven lymphomas treatment combine the control of the underlying immunosuppression and targeted oncologic therapy (Table 3) to improve survival and reduce relapses. In HIV-positive cases, cART is continued throughout chemotherapy to restore immune function, maintain virologic suppression, and diminish opportunistic infections. HL in immunodeficient populations, typically EBV-positive, generally responds to front-line ABVD chemotherapy, sometimes adapted by stage or combined with risk-based radiotherapy. In relapsed disease, high-dose chemotherapy and autologous stem cell transplantation can be considered, and agents such as brentuximab vedotin and ICIs may offer benefits for refractory cases; however, the efficacy of ICIs requires further validation through larger cohorts and additional clinical trials. PTLD involving EBV often regress with immunosuppression reduction, but persistent or polymorphic lesions may require rituximab or chemoimmunotherapy, similar to HIV lymphomas but adjusted for transplant organ function. Across all these settings, reconstitution or modulation of immune status is fundamental, cART in HIV, immunosuppression tapering in transplant recipients, or novel T-cell-directed strategies, because restoring or optimizing antiviral immune surveillance synergizes with standard lymphoma-directed regimens to achieve better control of these virally driven malignancies. Additionally, novel EBV-targeted therapies focus on neutralizing or disrupting tumor-promoting miRNAs. Tools such as miRNA sponges, anti-miRNA oligonucleotides, and CRISPR/Cas9 are rapidly advancing, offering promising, more precise strategies to counteract EBV-driven malignancies by specifically blocking or editing viral miRNAs [23,29,32,106–108,116,120–123,126] (Table 3).

Table 3. Current and novel therapies for immune deficiency/dysregulation-associated classic Hodgkin lymphoma.

Category	Treatment Strategy	Key Outcomes	Reference
General treatment approach	Follows general population protocols but integrates cART or reduces immunosuppression.	Survival improvement but disparities in treatment access persist.	[23,106–108,116,120–123]
HIV-HL standard treatment	ABVD + cART.	High response rates with ABVD; 74% CR rate, 5-year OS of 81%. Integration with cART essential.	[114,116]
Limited-stage HIV-HL	Two cycles of ABVD + 20 Gy ISRT; lower doses (20 Gy) and smaller fields reduce toxicity.	CR: 96%, 2-yr OS: 95.7%. Reduced radiation dose is effective; lower toxicity, good disease control.	[114,116]
Advanced-stage HIV-HL	ABVD/BEACOPP baseline; BV-AVD OR nivolumab-AVD for stage IV in select patients; PET-positive residual disease requires consolidation radiation.	ABVD/BEACOPP baseline: 2-yr PFS: 87.5%, 2-yr OS: 86.8%, 4 toxicity-related deaths with >6 BEACOPP cycles; max 6 recommended.Effective for aggressive disease; PET-CT guides therapy for improved outcomes.	[108,113,116,127]
Relapsed/refractory cases	High-dose chemotherapy (ESHAP, DHAP, ICE) followed by ASCT; higher infection risk in HIV patients.	Higher infection risk in HIV patients undergoing ASCT; supportive care is key.	[118,119]
Novel therapies	BV-AVD, ICIs (pembrolizumab, nivolumab) show high efficacy but need validation in HIV-HL.	BV-AVD: 2-year PFS 86%, OS 92%. Strong potential, but ICIs require larger trials in HIV-HL patients.	[9,117,120,128,129]
Chemotherapy and cART considerations	Chemotherapy with cART poses drug interaction risks; integrase inhibitors are safer; ICIs safe in PLWH with no impact on viral suppression.	Drug interactions manageable with proper selection of ART regimens.	[120]
PTLD management	Rituximab + chemotherapy or EBV-CTL for PTLD.	ORR of 68% post-HSCT, 54% post-SOT. Effective in PTLD; good response rates with acceptable toxicity. PTLD is treated like cHL but requires careful adjustments due to poorer tolerance, limited efficacy of immunosuppression reduction.	[29,32,120]
Emerging EBV-targeted therapies	miRNA sponges, anti-miRNA oligonucleotides, CRISPR/Cas9 emerging as precise EBV-targeted strategies.	Targeted therapies improving precision in EBV-related malignancies.	[126]

5.4. Preventive Measures

In people living with HIV (PLWH), EBV plays a central role in the pathogenesis of lymphoproliferative disorders, including cHL, which is frequently EBV-positive (up to 90% in HIV-associated cases). Chronic immune dysregulation caused by HIV, marked by CD4+ T-cell depletion, cytokine dysregulation, and persistent inflammation, creates an environment permissive for EBV reactivation and lymphomagenesis. The article by [23], 2024 underscores that initiating cART early in PLWH reduces HIV viremia, restores immune function, and lowers the risk of EBV-driven malignancies by mitigating immune activation and improving viral control. Persistent HIV replication (≥100,000 copies/mL) and low CD4 counts (<50/µL) correlate with heightened lymphoma risk, emphasizing the need for strict cART adherence. While no direct EBV-targeted prevention exists for PLWH, lifestyle modifications (e.g., smoking cessation), vaccinations (e.g., HBV, HPV), and avoidance of co-infections (e.g., KSHV/HHV8) are recommended to minimize immune dysregulation and secondary cancer risks. The use of colony-stimulating factors like G-CSF after each chemotherapy cycle is strongly recommended for primary infection prophylaxis. Additionally, all PLWH undergoing chemotherapy or radiotherapy should receive prophylaxis against *Pneumocystis jirovecii*, as these treatments reduce CD4+ lymphocyte counts. Cotrimoxazole is the preferred option, offering added protection against bacterial infections and toxoplasmosis. For patients with CD4+ counts below 50/µL, oral azithromycin is advised to prevent *Mycobacterium avium* complex infections. The article also notes that EBV-positive cHL in PLWH often presents with advanced-stage disease and extranodal involvement, necessitating extensive monitoring for lymphoproliferative symptoms in this population [23,106,130].

6. Beyond EBV: Other Viruses

A recent 2021 study evaluated the association between hepatitis B virus (HBV) and hepatitis C virus (HCV) infections and the development of HL. A systematic review and meta-analysis found inconclusive evidence for a link between these hepatitis infections and HL. For HBV, pooled data suggested a relative risk (RR) of 1.39, but this was not statistically significant. Similarly, HCV studies showed an RR of 1.09, also lacking significance. Overall, while some patterns suggest a possible association between hepatitis infections and HL under specific conditions, the evidence remains insufficient to establish a direct causative link. Further research is needed to explore these relationships in diverse populations, particularly considering factors like immune status and Epstein–Barr virus co-infection [131].

A study from 2019 investigated Human herpesvirus-6 (HHV-6) for its potential role in HL using IARC criteria for oncogenicity, but the evidence remains inconclusive. Studies have detected HHV-6 DNA in some HL cases, but it is not consistently present in all tumor cells, failing a key criterion for establishing causality. Elevated HHV-6 antibody levels have been reported in some HL patients, but these findings are inconsistent across studies. Unlike EBV, which is firmly linked to HL, HHV-6′s association with the disease appears limited or secondary. Current research suggests that HHV-6 may contribute indirectly or be a passenger virus rather than a direct cause of HL [132].

7. Challenges

The heterogeneity of cHL reflects significant challenges and controversies, especially regarding EBV prevalence and impact. EBV positivity varies widely across cHL subtypes and populations. For instance, in developing regions, mixed cellularity subtype is frequently EBV-positive, particularly in children and older adults, while the nodular sclerosis subtype, which is more common in industrialized nations, tends to be EBV-negative.

Further complicating the picture is the variability in EBV protein expression. Proteins like LMP1 and LMP2A, expressed in EBV-positive cases, contribute to immune evasion and oncogenesis by promoting NF-κB activation and altering the TME. However, this pathogenic impact differs among cases, influenced by genetic predispositions, immune responses, and geographic or socioeconomic factors. These disparities not only challenge uniform diagnostic and therapeutic strategies but also highlight the complex interplay of viral, host, and environmental factors in cHL [14,27].

Establishing causality in viral associations with cHL presents significant challenges due to the lack of direct evidence linking some viruses to the disease. While EBV is frequently detected in tumor cells, the precise mechanisms by which it contributes to oncogenesis remain incompletely understood. Confounding factors, such as genetic susceptibility and environmental exposures, further complicate this relationship. Genetic predispositions, including variations in HLA alleles, may influence immune responses to EBV, while environmental conditions, such as socioeconomic status and early-life infections, can alter disease patterns, making it difficult to isolate the virus's role in cHL pathogenesis.

8. Future Directions

Future research on immune deficiency/dysregulation-associated EBV-positive cHL should focus on integrating large-scale, longitudinal, and collaborative epidemiological efforts to clarify incidence, patterns of immunosuppression, and viral cofactors across diverse geographical and socioeconomic settings. Comprehensive genomic and environmental data, coupled with HLA typing and immune profiling, can illuminate gene or environment interactions and refine risk stratification. Advances in virus-targeted therapeutics, ranging from EBV-specific T-cell infusions and CRISPR/Cas9-based strategies to prophylactic or therapeutic EBV vaccines, hold promise for more precise interventions. Recent studies have successfully identified and characterized the first functional HLA-A01:01-restricted T-cell receptors (TCRs) specific for EBV LMP2, demonstrating their ability to recognize and kill target cells in vitro. These TCRs, when engineered into CD8+ and CD4+ T-cells, restored cytotoxic activity against EBV-infected cells and may serve as a foundation for adoptive T-cell therapies in EBV-driven lymphomas, including cHL. This approach highlights the potential of precision immunotherapy that directly targets viral antigens involved in lymphomagenesis [133].

Emerging classification frameworks that incorporate histopathologic features, underlying immune context, and advanced molecular diagnostics will facilitate more accurate diagnosis and prognostication, especially in complex settings like HIV infection or post-transplant immunosuppression. Equitable access to these diagnostic and therapeutic advances depends on addressing global disparities in healthcare infrastructure, training, and policy, underscoring the importance of capacity-building programs and real-world implementation science to evaluate outcomes in heterogenous populations. A multi-disciplinary approach, often involving oncology, infectious disease, immunology, and transplant specialists, is critical for managing opportunistic infections, optimizing immune reconstitution (through antiretroviral therapy or immunosuppression tapering), and coordinating supportive care measures.

While EBV remains the principal pathogen of interest, further research into co-infection scenarios, such as hepatitis viruses, KSHV, or even SARS-CoV-2, could clarify additional mechanisms of viral-driven lymphomagenesis and pave the way for multi-targeted prophylaxis or intervention strategies. Ultimately, by embracing precision oncology, strengthening evidence through robust clinical trials, and ensuring global accessibility to advanced diagnostics and treatments, future efforts aim to transform EBV-positive cHL into a more preventable and precisely managed malignancy for both immunocompetent and immunocompromised populations worldwide.

9. Conclusions

cHL is a biologically diverse disease, heavily influenced by immune status and viral infections, notably EBV. In immunocompromised individuals, EBV serves as a key driver of lymphomagenesis through viral oncoproteins (LMP1 and LMP2A), which activate essential survival signaling pathways such as NF-κB, reshape the tumor microenvironment, and facilitate immune evasion. These molecular and clinical characteristics clearly distinguish EBV-positive from EBV-negative cHL, highlighting the necessity for revised diagnostic frameworks, as recently emphasized by the WHO 2023 classification.

Clinically, this distinction guides the implementation of targeted treatments, including immune checkpoint inhibitors, EBV-specific T-cell therapies, and emerging miRNA-based interventions, which offer the potential for improved therapeutic outcomes. Moving forward, concerted efforts in epidemiological research, molecular diagnostics, and precision medicine strategies, particularly in underserved populations, are critical. These steps will ensure better disease management and ultimately reduce disparities in clinical outcomes globally.

Author Contributions: M.N.A. selected the literature, wrote the manuscript, created Table 3. A.G. selected the literature, created Tables 1 and 2 all the figures. A.C. was responsible for the design of the review article. All authors have read and agreed to the published version of the manuscript.

Funding: This research received no external funding.

Conflicts of Interest: The authors declare no conflicts of interest.

List of Abbreviations

ABVD	adriamycin (doxorubicin), bleomycin, vinblastine, dacarbazine
AIDS	acquired immunodeficiency syndrome
AMC	AIDS malignancy consortium
ART	antiretroviral therapy
ASCT	autologous stem cell transplantation
AVD	adriamycin (Doxorubicin), vinblastine, dacarbazine
BART	BamHI a rightward transcript (EBV miRNAs)
BCR	B cell receptor
BEACOPP	bleomycin, etoposide, doxorubicin, cyclophosphamide, vincristine, procarbazine, prednisone
BL	Burkitt lymphoma
BV	brentuximab vedotin
CHL	classic Hodgkin lymphoma
CNS	central vervous system
CR	complete response
CRISPR	clustered regularly interspaced short palindromic repeats
CTL	cytotoxic T lymphocyte
CTLA-4	cytotoxic T-lymphocyte antigen 4
DHAP	dexamethasone, cytarabine, cisplatin
DLBCL	diffuse large B cell lymphoma
DNA	deoxyribonucleic acid
EBER	Epstein–Barr virus-encoded RNA
EBNA	Epstein–Barr nuclear antigen

EBV	Epstein–Barr virus
FDG	fluorodeoxyglucose
GC	germinal center
HBV	hepatitis B virus
HCV	hepatitis C virus
HDCT	high-dose chemotherapy
HIV	human immunodeficiency virus
HL	Hodgkin lymphoma
HLA	human leukocyte antigen
HPV	human papillomavirus
HRS	Hodgkin and Reed–Sternberg cells
HSCT	hematopoietic stem cell transplantation
ISRT	involved site radiation therapy
ICIs	immune checkpoint inhibitors
NF-κB	nuclear factor kappa-light-chain-enhancer of activated B cells
PD-1/PD-L1	programmed cell death protein 1/ligand 1
PTLD	post-transplant lymphoproliferative disorder
TME	tumor microenvironment
WHO	world health organization
RS	Reed–Sternberg cells

References

1. Kaseb, H.; Babiker, H.M. *Hodgkin Lymphoma*; StatPearls Publishing: Treasure Island, FL, USA, 2018.
2. Brice, P.; de Kerviler, E.; Friedberg, J.W. Classical hodgkin lymphoma. *Lancet* **2021**, *398*, 1518–1527. [CrossRef] [PubMed]
3. Eichenauer, D.A.; Engert, A. VII. Management of nodular lymphocyte-predominant Hodgkin lymphoma. *Hematol. Oncol.* **2013**, *31*, 47–50. [CrossRef]
4. Eichenauer, D.A.; Engert, A. Nodular lymphocyte-predominant Hodgkin lymphoma: A unique disease deserving unique management. In *Hematology 2014, the American Society of Hematology Education Program Book 2017*; Volume 2017, pp. 324–328. Available online: https://europepmc.org/article/NBK/nbk499969 (accessed on 21 April 2025).
5. Alibrahim, M.N.; Gloghini, A.; Carbone, A. Classic Hodgkin lymphoma: Pathobiological features that impact emerging therapies. *Blood Rev.* 2025; *online ahead of print*.
6. Aslani, A.; Morsali, S.; Mousavi, S.E.; Choupani, S.; Yekta, Z.; Nejadghaderi, S.A. Adult Hodgkin lymphoma incidence trends in the United States from 2000 to 2020. *Sci. Rep.* **2024**, *14*, 20500. [CrossRef] [PubMed]
7. Lee, J.-H.; Kim, Y.; Choi, J.-W.; Kim, Y.-S. Prevalence and prognostic significance of Epstein–Barr virus infection in classical Hodgkin's lymphoma: A meta-analysis. *Arch. Med. Res.* **2014**, *45*, 417–431. [CrossRef] [PubMed]
8. Munir, F.; Hardit, V.; Sheikh, I.N.; AlQahtani, S.; He, J.; Cuglievan, B.; Hosing, C.; Tewari, P.; Khazal, S. Classical Hodgkin Lymphoma: From Past to Future—A Comprehensive Review of Pathophysiology and Therapeutic Advances. *Int. J. Mol. Sci.* **2023**, *24*, 10095. [CrossRef]
9. Rubinstein, P.G.; Moore, P.C.; Bimali, M.; Lee, J.Y.; Rudek, M.A.; Chadburn, A.; Ratner, L.; Henry, D.H.; Cesarman, E.; DeMarco, C.E. Brentuximab vedotin with AVD for stage II–IV HIV-related Hodgkin lymphoma (AMC 085): Phase 2 results from an open-label, single arm, multicentre phase 1/2 trial. *Lancet Haematol.* **2023**, *10*, e624–e632. [CrossRef]
10. Wong, Y.; Meehan, M.T.; Burrows, S.R.; Doolan, D.L.; Miles, J.J. Estimating the global burden of Epstein–Barr virus-related cancers. *J. Cancer Res. Clin. Oncol.* **2022**, *148*, 31–46. [CrossRef]
11. Moscona-Nissan, A.; Mancilla-Osuna, M.F.; Bardán-Duarte, A.; Rendón-Macías, M.E. Classical Hodgkin lymphoma histologic subtypes distribution among geographical regions and correlation with Human Development Index. *Health Sci. Rev.* **2023**, *9*, 100117. [CrossRef]
12. Huang, X.; Nolte, I.; Gao, Z.; Vos, H.; Hepkema, B.; Poppema, S.; van den Berg, A.; Diepstra, A. Epidemiology of classical Hodgkin lymphoma and its association with Epstein Barr virus in Northern China. *PLoS ONE* **2011**, *6*, e21152. [CrossRef]
13. Carbone, A.; Gloghini, A. Epstein Barr virus-associated Hodgkin lymphoma. *Cancers* **2018**, *10*, 163. [CrossRef]
14. Massini, G.; Siemer, D.; Hohaus, S. EBV in Hodgkin lymphoma. *Mediterr. J. Hematol. Infect. Dis.* **2009**, *1*, e2009013. [CrossRef] [PubMed]
15. Mundo, L.; Del Porro, L.; Granai, M.; Siciliano, M.C.; Mancini, V.; Santi, R.; Marcar, L.; Vrzalikova, K.; Vergoni, F.; Di Stefano, G. Frequent traces of EBV infection in Hodgkin and non-Hodgkin lymphomas classified as EBV-negative by routine methods: Expanding the landscape of EBV-related lymphomas. *Mod. Pathol.* **2020**, *33*, 2407–2421. [CrossRef] [PubMed]

16. Rubinstein, P.G.; Moore, P.C.; Bimali, M.; Chadburn, A.; Cesarman, E.; Rudek, M.; Ratner, L.; Ambinder, R.F.; Henry, D.H.; Barta, S.K. Safety and efficacy of brentuximab vedotin in combination with avd in stage ii-iv hiv-associated classical hodgkin lymphoma: Results of the phase 2 study, amc 085. *Blood* **2019**, *134*, 130. [CrossRef]
17. Vrzalikova, K.; Pugh, M.; Mundo, L.; Murray, P. The Contribution of Ebv to the Pathogenesis of Classical Hodgkin Lymphoma. 2021. Available online: https://doi.org/10.34961/researchrepository-ul.23702151.v1 (accessed on 21 April 2025).
18. Mack, T.M.; Cozen, W.; Shibata, D.K.; Weiss, L.M.; Nathwani, B.N.; Hernandez, A.M.; Taylor, C.R.; Hamilton, A.S.; Deapen, D.M.; Rappaport, E.B. Concordance for Hodgkin's disease in identical twins suggesting genetic susceptibility to the young-adult form of the disease. *N. Engl. J. Med.* **1995**, *332*, 413–419. [CrossRef]
19. Alibrahim, M.N.; Gloghini, A.; Carbone, A. Pathobiological Features and Therapeutic Opportunities Linked to TNF Family Member Expression in Classic Hodgkin Lymphoma. *Cancers* **2024**, *16*, 4070. [CrossRef]
20. Camus, V.; Viennot, M.; Lequesne, J.; Viailly, P.-J.; Bohers, E.; Bessi, L.; Marcq, B.; Etancelin, P.; Dubois, S.; Picquenot, J.M. Targeted genotyping of circulating tumor DNA for classical Hodgkin lymphoma monitoring: A prospective study. *Haematologica* **2020**, *106*, 154. [CrossRef] [PubMed]
21. Tanwar, P.; Naagar, M.; Maity, M.K. A review on Hodgkin lymphoma: Etiology, epidemiology, pathophysiology, differential diagnosis, management strategies and treatment modalities. *Int. J. Pharm. Res. Dev.* **2022**, *6*, 35–38. [CrossRef]
22. Tiacci, E.; Ladewig, E.; Schiavoni, G.; Penson, A.; Fortini, E.; Pettirossi, V.; Wang, Y.; Rosseto, A.; Venanzi, A.; Vlasevska, S. Pervasive mutations of JAK-STAT pathway genes in classical Hodgkin lymphoma. *Blood J. Am. Soc. Hematol.* **2018**, *131*, 2454–2465. [CrossRef]
23. Carbone, A.; Chadburn, A.; Gloghini, A.; Vaccher, E.; Bower, M. Immune deficiency/dysregulation-associated lymphoproliferative disorders. Revised classification and management. *Blood Rev.* **2024**, *64*, 101167. [CrossRef]
24. Alaggio, R.; Amador, C.; Anagnostopoulos, I.; Attygalle, A.D.; Araujo, I.B.O.; Berti, E.; Bhagat, G.; Borges, A.M.; Boyer, D.; Calaminici, M.; et al. The 5th edition of the World Health Organization Classification of Haematolymphoid Tumours: Lymphoid Neoplasms. *Leukemia* **2022**, *36*, 1720–1748. [CrossRef]
25. Roughan, J.E.; Torgbor, C.; Thorley-Lawson, D.A. Germinal center B cells latently infected with Epstein-Barr virus proliferate extensively but do not increase in number. *J. Virol.* **2010**, *84*, 1158–1168. [CrossRef] [PubMed]
26. Thorley-Lawson, D.A. Epstein-Barr virus: Exploiting the immune system. *Nat. Rev. Immunol.* **2001**, *1*, 75–82. [CrossRef] [PubMed]
27. Campos, A.H.J.F.M.; Moreira, A.; Ribeiro, K.B.; Paes, R.P.; Zerbini, M.C.; Aldred, V.; De Souza, C.A.; Neto, C.S.; Soares, F.A.; Vassallo, J. Frequency of EBV associated classical Hodgkin lymphoma decreases over a 54-year period in a Brazilian population. *Sci. Rep.* **2018**, *8*, 1849. [CrossRef] [PubMed]
28. Cameron, J.E.; Fewell, C.; Yin, Q.; McBride, J.; Wang, X.; Lin, Z.; Flemington, E.K. Epstein–Barr virus growth/latency III program alters cellular microRNA expression. *Virology* **2008**, *382*, 257–266. [CrossRef]
29. De Jong, D.; Roemer, M.G.; Chan, J.K.; Goodlad, J.; Gratzinger, D.; Chadburn, A.; Jaffe, E.S.; Said, J.; Natkunam, Y. B-cell and classical Hodgkin lymphomas associated with immunodeficiency: 2015 SH/EAHP workshop report—Part 2. *Am. J. Clin. Pathol.* **2017**, *147*, 153–170. [CrossRef]
30. Nourse, J.; Jones, K.; Gandhi, M. Epstein–Barr virus-related post-transplant lymphoproliferative disorders: Pathogenetic insights for targeted therapy. *Am. J. Transplant.* **2011**, *11*, 888–895. [CrossRef]
31. Dharnidharka, V.R.; Ho, P.L.; Stablein, D.M.; Harmon, W.E.; Tejani, A.H. Mycophenolate, tacrolimus and post-transplant lymphoproliferative disorder: A report of the North American Pediatric Renal Transplant Cooperative Study. *Pediatr. Transplant.* **2002**, *6*, 396–399. [CrossRef]
32. Adams, H.; Campidelli, C.; Dirnhofer, S.; Pileri, S.A.; Tzankov, A. Clinical, phenotypic and genetic similarities and disparities between post-transplant and classical Hodgkin lymphomas with respect to therapeutic targets. *Expert Opin. Ther. Targets* **2009**, *13*, 1137–1145. [CrossRef]
33. Salmerón-Villalobos, J.; Castrejón-de-Anta, N.; Guerra-García, P.; Ramis-Zaldivar, J.E.; López-Guerra, M.; Mato, S.; Colomer, D.; Diaz-Crespo, F.; Menarguez, J.; Garrido-Pontnou, M. Decoding the molecular heterogeneity of pediatric monomorphic post–solid organ transplant lymphoproliferative disorders. *Blood* **2023**, *142*, 434–445. [CrossRef]
34. Nakid-Cordero, C.; Choquet, S.; Gauthier, N.; Balegroune, N.; Tarantino, N.; Morel, V.; Arzouk, N.; Burrel, S.; Rousseau, G.; Charlotte, F. Distinct immunopathological mechanisms of EBV-positive and EBV-negative posttransplant lymphoproliferative disorders. *Am. J. Transplant.* **2021**, *21*, 2846–2863. [CrossRef]
35. IARC, I. *Monographs vol. 100B Evaluation of Carcinogenic Risks to Humans*; International Agency for Research on Cancer: Lyon, France, 2012.
36. Jarrett, R.F.; Hjalgrim, H.; Murray, P.G. The role of viruses in the genesis of Hodgkin lymphoma. In *Hodgkin Lymphoma: A Comprehensive Overview*; 2020; pp. 25–45. Available online: https://link.springer.com/chapter/10.1007/978-3-030-32482-7_2 (accessed on 21 April 2025).
37. Murray, P.G.; Young, L.S. An etiological role for the Epstein-Barr virus in the pathogenesis of classical Hodgkin lymphoma. *Blood J. Am. Soc. Hematol.* **2019**, *134*, 591–596. [CrossRef]

38. Vockerodt, M.; Cader, F.Z.; Shannon-Lowe, C.; Murray, P. Epstein-Barr virus and the origin of Hodgkin lymphoma. *Chin. J. Cancer* **2014**, *33*, 591. [CrossRef] [PubMed]
39. Anagnostopoulos, I.; Herbst, H.; Niedobitek, G.; Stein, H. Demonstration of monoclonal EBV genomes in Hodgkin's disease and Ki-l-positive anaplastic large cell lymphoma by combined Southern blot and in situ hybridization. *Blood* **1989**, *74*, 810–816. [CrossRef] [PubMed]
40. Küppers, R. Advances in Hodgkin lymphoma research. *Trends Mol. Med.* **2024**, *31*, 326–343. [CrossRef] [PubMed]
41. Poppema, S.; van Imhoff, G.; Torensma, R.; Smit, J. Lymphadenopathy morphologically consistent with Hodgkin's disease associated with Epstein-Barr virus infection. *Am. J. Clin. Pathol.* **1985**, *84*, 385–390. [CrossRef]
42. Weiss, L.M.; Movahed, L.A.; Warnke, R.A.; Sklar, J. Detection of Epstein–Barr viral genomes in Reed–Sternberg cells of Hodgkin's disease. *N. Engl. J. Med.* **1989**, *320*, 502–506. [CrossRef]
43. Saha, A.; Robertson, E.S. Epstein-Barr virus–associated B-cell lymphomas: Pathogenesis and clinical outcomes. *Clin. Cancer Res.* **2011**, *17*, 3056–3063. [CrossRef]
44. Shannon-Lowe, C.; Rickinson, A.B.; Bell, A.I. Epstein–Barr virus-associated lymphomas. *Philos. Trans. R. Soc. B Biol. Sci.* **2017**, *372*, 20160271. [CrossRef]
45. Kenney, S.C. Reactivation and lytic replication of EBV. In *Human Herpesviruses: Biology, Therapy, and Immunoprophylaxis*; Cambridge University Press: Cambridge, UK, 2007.
46. Babcock, G.J.; Hochberg, D.; Thorley-Lawson, D.A. The expression pattern of Epstein-Barr virus latent genes in vivo is dependent upon the differentiation stage of the infected B cell. *Immunity* **2000**, *13*, 497–506. [CrossRef]
47. Gires, O.; Zimber-Strobl, U.; Gonnella, R.; Ueffing, M.; Marschall, G.; Zeidler, R.; Pich, D.; Hammerschmidt, W. Latent membrane protein 1 of Epstein–Barr virus mimics a constitutively active receptor molecule. *EMBO J.* **1997**, *16*, 6131–6140. [CrossRef]
48. Caldwell, R.G.; Wilson, J.B.; Anderson, S.J.; Longnecker, R. Epstein-Barr virus LMP2A drives B cell development and survival in the absence of normal B cell receptor signals. *Immunity* **1998**, *9*, 405–411. [CrossRef] [PubMed]
49. Gires, O.; Kohlhuber, F.; Kilger, E.; Baumann, M.; Kieser, A.; Kaiser, C.; Zeidler, R.; Scheffer, B.; Ueffing, M.; Hammerschmidt, W. Latent membrane protein 1 of Epstein–Barr virus interacts with JAK3 and activates STAT proteins. *EMBO J.* **1999**, *18*, 3064–3073. [CrossRef] [PubMed]
50. Eliopoulos, A.G.; Young, L.S. Activation of the cJun N-terminal kinase (JNK) pathway by the Epstein-Barr virus-encoded latent membrane protein 1 (LMP1). *Oncogene* **1998**, *16*, 1731–1742. [CrossRef]
51. Laherty, C.; Hu, H.M.; Opipari, A.; Wang, F.; Dixit, V. The Epstein-Barr virus LMP1 gene product induces A20 zinc finger protein expression by activating nuclear factor kappa B. *J. Biol. Chem.* **1992**, *267*, 24157–24160. [CrossRef]
52. Huen, D.; Henderson, S.; Croom-Carter, D.; Rowe, M. The Epstein-Barr virus latent membrane protein-1 (LMP1) mediates activation of NF-kappa B and cell surface phenotype via two effector regions in its carboxy-terminal cytoplasmic domain. *Oncogene* **1995**, *10*, 549–560. [PubMed]
53. Eliopoulos, A.G.; Gallagher, N.J.; Blake, S.M.; Dawson, C.W.; Young, L.S. Activation of the p38 mitogen-activated protein kinase pathway by Epstein-Barr virus-encoded latent membrane protein 1 coregulates interleukin-6 and interleukin-8 production. *J. Biol. Chem.* **1999**, *274*, 16085–16096. [CrossRef]
54. Cheerathodi, M.R.; Meckes Jr, D.G. The Epstein–Barr virus LMP1 interactome: Biological implications and therapeutic targets. *Future Virol.* **2018**, *13*, 863–887. [CrossRef]
55. Izban, K.F.; Ergin, M.; Huang, Q.; Qin, J.-z.; Martinez, R.L.; Schnitzer, B.; Ni, H.; Nickoloff, B.J.; Alkan, S. Characterization of NF-κB expression in Hodgkin's disease: Inhibition of constitutively expressed NF-κB results in spontaneous caspase-independent apoptosis in Hodgkin and Reed-Sternberg cells. *Mod. Pathol.* **2001**, *14*, 297–310. [CrossRef]
56. Bargou, R.C.; Emmerich, F.; Krappmann, D.; Bommert, K.; Mapara, M.Y.; Arnold, W.; Royer, H.D.; Grinstein, E.; Greiner, A.; Scheidereit, C. Constitutive nuclear factor-kappaB-RelA activation is required for proliferation and survival of Hodgkin's disease tumor cells. *J. Clin. Investig.* **1997**, *100*, 2961–2969. [CrossRef]
57. Jardin, F. NFkB pathway and Hodgkin lymphoma. *Biomedicines* **2022**, *10*, 2153. [CrossRef]
58. Zhao, P.; Lu, Y.; Liu, L.; Zhong, M. Aberrant Expression of ID2 protein and its correlation with EBV-LMP1 and P16 (INK4A) in Classical Hodgkin Lymphoma in China. *BMC Cancer* **2008**, *8*, 379. [CrossRef] [PubMed]
59. Toscano, M.A.; Campagna, L.; Molinero, L.L.; Cerliani, J.P.; Croci, D.O.; Ilarregui, J.M.; Fuertes, M.B.; Nojek, I.M.; Fededa, J.P.; Zwirner, N.W. Nuclear factor (NF)-κB controls expression of the immunoregulatory glycan-binding protein galectin-1. *Mol. Immunol.* **2011**, *48*, 1940–1949. [CrossRef]
60. Bell, A.I.; Groves, K.; Kelly, G.L.; Croom-Carter, D.; Hui, E.; Chan, A.T.; Rickinson, A.B. Analysis of Epstein–Barr virus latent gene expression in endemic Burkitt's lymphoma and nasopharyngeal carcinoma tumour cells by using quantitative real-time PCR assays. *J. Gen. Virol.* **2006**, *87*, 2885–2890. [CrossRef]
61. Price, A.M.; Luftig, M.A. To be or not IIb: A multi-step process for Epstein-Barr virus latency establishment and consequences for B cell tumorigenesis. *PLoS Pathog.* **2015**, *11*, e1004656. [CrossRef]

62. Crombie, J.L.; LaCasce, A.S. Epstein Barr virus associated B-cell lymphomas and iatrogenic lymphoproliferative disorders. *Front. Oncol.* **2019**, *9*, 109. [CrossRef]
63. Donzel, M.; Bonjour, M.; Combes, J.-D.; Broussais, F.; Sesques, P.; Traverse-Glehen, A.; de Martel, C. Lymphomas associated with Epstein-Barr virus infection in 2020: Results from a large, unselected case series in France. *Eclinicalmedicine* **2022**, *54*, 101674. [CrossRef]
64. Kennedy-Nasser, A.A.; Hanley, P.; Bollard, C.M. Hodgkin disease and the role of the immune system. *Pediatr. Hematol. Oncol.* **2011**, *28*, 176–186. [CrossRef] [PubMed]
65. Liu, T.-Y.; Wu, S.-J.; Huang, M.-H.; Lo, F.-Y.; Tsai, M.-H.; Tsai, C.-H.; Hsu, S.-M.; Lin, C.-W. EBV-positive Hodgkin lymphoma is associated with suppression of p21 cip1/waf1 and a worse prognosis. *Mol. Cancer* **2010**, *9*, 32. [CrossRef] [PubMed]
66. Barth, S.; Pfuhl, T.; Mamiani, A.; Ehses, C.; Roemer, K.; Kremmer, E.; Jäker, C.; Höck, J.; Meister, G.; Grässer, F.A. Epstein–Barr virus-encoded microRNA miR-BART2 down-regulates the viral DNA polymerase BALF5. *Nucleic Acids Res.* **2008**, *36*, 666–675. [CrossRef]
67. Vrzalikova, K.; Vockerodt, M.; Leonard, S.; Bell, A.; Wei, W.; Schrader, A.; Wright, K.L.; Kube, D.; Rowe, M.; Woodman, C.B. Down-regulation of BLIMP1α by the EBV oncogene, LMP-1, disrupts the plasma cell differentiation program and prevents viral replication in B cells: Implications for the pathogenesis of EBV-associated B-cell lymphomas. *Blood J. Am. Soc. Hematol.* **2011**, *117*, 5907–5917. [CrossRef]
68. Chen, Y.; Fachko, D.; Ivanov, N.S.; Skinner, C.M.; Skalsky, R.L. Epstein-Barr virus microRNAs regulate B cell receptor signal transduction and lytic reactivation. *PLoS Pathog.* **2019**, *15*, e1007535. [CrossRef] [PubMed]
69. Sakamoto, K.; Sekizuka, T.; Uehara, T.; Hishima, T.; Mine, S.; Fukumoto, H.; Sato, Y.; Hasegawa, H.; Kuroda, M.; Katano, H. Next-generation sequencing of miRNAs in clinical samples of Epstein–Barr virus-associated B-cell lymphomas. *Cancer Med.* **2017**, *6*, 605–618. [CrossRef] [PubMed]
70. Higuchi, H.; Yamakawa, N.; Imadome, K.-I.; Yahata, T.; Kotaki, R.; Ogata, J.; Kakizaki, M.; Fujita, K.; Lu, J.; Yokoyama, K. Role of exosomes as a proinflammatory mediator in the development of EBV-associated lymphoma. *Blood J. Am. Soc. Hematol.* **2018**, *131*, 2552–2567. [CrossRef]
71. Qiu, J.; Cosmopoulos, K.; Pegtel, M.; Hopmans, E.; Murray, P.; Middeldorp, J.; Shapiro, M.; Thorley-Lawson, D.A. A novel persistence associated EBV miRNA expression profile is disrupted in neoplasia. *PLoS Pathog.* **2011**, *7*, e1002193. [CrossRef] [PubMed]
72. Notarte, K.I.; Senanayake, S.; Macaranas, I.; Albano, P.M.; Mundo, L.; Fennell, E.; Leoncini, L.; Murray, P. MicroRNA and other non-coding RNAs in Epstein–Barr virus-associated cancers. *Cancers* **2021**, *13*, 3909. [CrossRef]
73. Navarro, A.; Gaya, A.; Martinez, A.; Urbano-Ispizua, A.; Pons, A.; Balagué, O.; Gel, B.; Abrisqueta, P.; Lopez-Guillermo, A.; Artells, R. MicroRNA expression profiling in classic Hodgkin lymphoma. *Blood J. Am. Soc. Hematol.* **2008**, *111*, 2825–2832. [CrossRef]
74. Di Leva, G.; Croce, C.M. miRNA profiling of cancer. *Curr. Opin. Genet. Dev.* **2013**, *23*, 3–11. [CrossRef]
75. Reddy, K.B. MicroRNA (miRNA) in cancer. *Cancer Cell Int.* **2015**, *15*, 38. [CrossRef]
76. Kosaka, N.; Iguchi, H.; Ochiya, T. Circulating microRNA in body fluid: A new potential biomarker for cancer diagnosis and prognosis. *Cancer Sci.* **2010**, *101*, 2087–2092. [CrossRef]
77. Chetaille, B.; Bertucci, F.; Finetti, P.; Esterni, B.; Stamatoullas, A.; Picquenot, J.M.; Copin, M.C.; Morschhauser, F.; Casasnovas, O.; Petrella, T. Molecular profiling of classical Hodgkin lymphoma tissues uncovers variations in the tumor microenvironment and correlations with EBV infection and outcome. *Blood J. Am. Soc. Hematol.* **2009**, *113*, 2765–3775. [CrossRef] [PubMed]
78. Grywalska, E.; Rolinski, J. Epstein-barr virus–associated lymphomas. *Semin. Oncol.* **2015**, *42*, 291–303. [CrossRef] [PubMed]
79. Renné, C.; Willenbrock, K.; Küppers, R.; Hansmann, M.-L.; Bräuninger, A. Autocrine-and paracrine-activated receptor tyrosine kinases in classic Hodgkin lymphoma. *Blood* **2005**, *105*, 4051–4059. [CrossRef]
80. Teofili, L.; Di Febo, A.L.; Pierconti, F.; Maggiano, N.; Bendandi, M.; Rutella, S.; Cingolani, A.; Di Renzo, N.; Musto, P.; Pileri, S. Expression of the c-met proto-oncogene and its ligand, hepatocyte growth factor, in Hodgkin disease. *Blood J. Am. Soc. Hematol.* **2001**, *97*, 1063–1069. [CrossRef] [PubMed]
81. Lamprecht, B.; Walter, K.; Kreher, S.; Kumar, R.; Hummel, M.; Lenze, D.; Köchert, K.; Bouhlel, M.A.; Richter, J.; Soler, E. Derepression of an endogenous long terminal repeat activates the CSF1R proto-oncogene in human lymphoma. *Nat. Med.* **2010**, *16*, 571–579. [CrossRef]
82. Cader, F.Z.; Vockerodt, M.; Bose, S.; Nagy, E.; Brundler, M.-A.; Kearns, P.; Murray, P.G. The EBV oncogene LMP1 protects lymphoma cells from cell death through the collagen-mediated activation of DDR1. *Blood J. Am. Soc. Hematol.* **2013**, *122*, 4237–4245. [CrossRef]
83. Meti, N.; Esfahani, K.; Johnson, N.A. The role of immune checkpoint inhibitors in classical Hodgkin lymphoma. *Cancers* **2018**, *10*, 204. [CrossRef]
84. Pasqualucci, L.; Klein, U. NF-κB mutations in germinal center B-cell lymphomas: Relation to NF-κB function in normal B cells. *Biomedicines* **2022**, *10*, 2450. [CrossRef]

85. Skinnider, B.F.; Mak, T.W. The role of cytokines in classical Hodgkin lymphoma. *Blood J. Am. Soc. Hematol.* **2002**, *99*, 4283–4297. [CrossRef] [PubMed]
86. Liu, W.R.; Shipp, M.A. Signaling pathways and immune evasion mechanisms in classical Hodgkin lymphoma. *Blood J. Am. Soc. Hematol.* **2017**, *130*, 2265–2270.
87. Green, M.R.; Rodig, S.; Juszczynski, P.; Ouyang, J.; Sinha, P.; O'Donnell, E.; Neuberg, D.; Shipp, M.A. Constitutive AP-1 activity and EBV infection induce PD-L1 in Hodgkin lymphomas and posttransplant lymphoproliferative disorders: Implications for targeted therapy. *Clin. Cancer Res.* **2012**, *18*, 1611–1618. [CrossRef]
88. Ma, S.-D.; Xu, X.; Jones, R.; Delecluse, H.-J.; Zumwalde, N.A.; Sharma, A.; Gumperz, J.E.; Kenney, S.C. PD-1/CTLA-4 blockade inhibits Epstein-Barr virus-induced lymphoma growth in a cord blood humanized-mouse model. *PLoS Pathog.* **2016**, *12*, e1005642. [CrossRef] [PubMed]
89. Biggi, A.F.B.; Elgui de Oliveira, D. The Epstein-Barr virus hacks immune checkpoints: Evidence and consequences for lymphoproliferative disorders and cancers. *Biomolecules* **2022**, *12*, 397. [CrossRef] [PubMed]
90. Menter, T.; Tzankov, A. Mechanisms of immune evasion and immune modulation by lymphoma cells. *Front. Oncol.* **2018**, *8*, 54. [CrossRef] [PubMed]
91. Özdemir, S.; Ton, Ö.; Kabukcuoğlu, F. PD-L1 and EBV LMP1 expressions in classic Hodgkin lymphomas and its correlation with clinicopathological parameters and prognosis. *Turk. J. Med. Sci.* **2022**, *52*, 1013–1021. [CrossRef]
92. Volk, V.; Theobald, S.J.; Danisch, S.; Khailaie, S.; Kalbarczyk, M.; Schneider, A.; Bialek-Waldmann, J.; Krönke, N.; Deng, Y.; Eiz-Vesper, B. PD-1 blockade aggravates Epstein–Barr virus+ post-transplant lymphoproliferative disorder in humanized mice resulting in central nervous system involvement and CD4+ T cell dysregulations. *Front. Oncol.* **2021**, *10*, 614876. [CrossRef]
93. Beck, S.; Roush, S.M.; Mponda, M.; Kasonkanji, E.; Tomoka, T.; Painschab, M.S.; Galeotti, J.; Westmoreland, K.D.; Fedoriw, Y.D. Immunogenomic Profiles of EBV-Associated Classic Hodgkin Lymphoma in an HIV-Inclusive Cohort. *Blood* **2024**, *144*, 4344. [CrossRef]
94. Chabay, P.; Preciado, M.V. Epidemiology of Epstein-Barr virus-associated pediatric lymphomas from Argentina. *Boletín Médico Hosp. Infant. México* **2016**, *73*, 47–54. [CrossRef]
95. Huang, X.; Kushekhar, K.; Nolte, I.; Kooistra, W.; Visser, L.; Bouwman, I.; Kouprie, N.; Veenstra, R.; van Imhoff, G.; Olver, B. HLA associations in classical Hodgkin lymphoma: EBV status matters. *PLoS ONE* **2012**, *7*, e39986. [CrossRef]
96. Urayama, K.Y.; Jarrett, R.F.; Hjalgrim, H.; Diepstra, A.; Kamatani, Y.; Chabrier, A.; Gaborieau, V.; Boland, A.; Nieters, A.; Becker, N.; et al. Genome-Wide Association Study of Classical Hodgkin Lymphoma and Epstein–Barr Virus Status–Defined Subgroups. *JNCI J. Natl. Cancer Inst.* **2012**, *104*, 240–253. [CrossRef]
97. Martín, P.; Krsnik, I.; Navarro, B.; Provencio, M.; García, J.F.; Bellas, C.; Vilches, C.; Gomez-Lozano, N. HLA allele E* 01: 01 is associated with a reduced risk of EBV-related classical Hodgkin lymphoma independently of HLA-A* 01/* 02. *PLoS ONE* **2015**, *10*, e0135512. [CrossRef]
98. Delahaye-Sourdeix, M.; Urayama, K.Y.; Gaborieau, V.; Veenstra, R.; Foll, M.; Chabrier, A.; Benavente, Y.; Nieters, A.; Becker, N.; Foretova, L. A Novel Risk Locus at 6p21. 3 for Epstein–Barr Virus-Positive Hodgkin Lymphoma. *Cancer Epidemiol. Biomark. Prev.* **2015**, *24*, 1838–1843. [CrossRef]
99. Bakkalci, D.; Jia, Y.; Winter, J.R.; Lewis, J.E.; Taylor, G.S.; Stagg, H.R. Risk factors for Epstein Barr virus-associated cancers: A systematic review, critical appraisal, and mapping of the epidemiological evidence. *J. Glob. Health* **2020**, *10*, 010405. [CrossRef] [PubMed]
100. Broccia, G.; Carter, J.; Ozsin-Ozler, C.; De Matteis, S.; Cocco, P. Time and space co-ordinates of Hodgkin's lymphoma in Sardinia, Italy. *J. Natl. Cancer Cent.* **2024**, *5*, 50–56. [CrossRef] [PubMed]
101. Carbone, A.; Vaccher, E.; Gloghini, A.; Pantanowitz, L.; Abayomi, A.; De Paoli, P.; Franceschi, S. Diagnosis and management of lymphomas and other cancers in HIV-infected patients. *Nat. Rev. Clin. Oncol.* **2014**, *11*, 223–238. [CrossRef]
102. Besson, C.; Lancar, R.; Prevot, S.; Brice, P.; Meyohas, M.-C.; Marchou, B.; Gabarre, J.; Bonnet, F.; Goujard, C.; Lambotte, O. Characteristics and outcome of HIV associated classical Hodgkin's lymphoma among 68 patients included in the French ANRS CO16 lymphovir cohort study. *Blood* **2014**, *124*, 2954. [CrossRef]
103. Bibas, M.; Antinori, A. EBV and HIV-related lymphoma. *Mediterr. J. Hematol. Infect. Dis.* **2009**, *1*, e2009032. [CrossRef]
104. Carbone, A.; Gloghini, A.; Caruso, A.; De Paoli, P.; Dolcetti, R. The impact of EBV and HIV infection on the microenvironmental niche underlying H odgkin lymphoma pathogenesis. *Int. J. Cancer* **2017**, *140*, 1233–1245. [CrossRef]
105. Hybel, T.E.; Sørensen, E.F.; Enemark, M.H.; Hemmingsen, J.K.; Simonsen, A.T.; Lauridsen, K.L.; Møller, M.B.; Pedersen, C.; Pedersen, G.; Obel, N.; et al. Characterization of the genomic landscape of HIV-associated lymphoma reveals heterogeneity across histological subtypes. *AIDS* **2024**, *38*, 1897–1906. [CrossRef]
106. Navarro, J.-T.; Moltó, J.; Tapia, G.; Ribera, J.-M. Hodgkin lymphoma in people living with HIV. *Cancers* **2021**, *13*, 4366. [CrossRef]
107. Verdu-Bou, M.; Tapia, G.; Hernandez-Rodriguez, A.; Navarro, J.-T. Clinical and therapeutic implications of Epstein–Barr virus in HIV-related lymphomas. *Cancers* **2021**, *13*, 5534. [CrossRef]

108. Carbone, A.; Gloghini, A.; Serraino, D.; Spina, M.; Tirelli, U.; Vaccher, E. Immunodeficiency-associated Hodgkin lymphoma. *Expert Rev. Hematol.* **2021**, *14*, 547–559. [CrossRef]
109. Hartmann, S.; Jakobus, C.; Rengstl, B.; Döring, C.; Newrzela, S.; Brodt, H.-R.; Wolf, T.; Hansmann, M.-L. Spindle-shaped CD163+ rosetting macrophages replace CD4+ T-cells in HIV-related classical Hodgkin lymphoma. *Mod. Pathol.* **2013**, *26*, 648–657. [CrossRef]
110. Kamper, P.; Bendix, K.; Hamilton-Dutoit, S.; Honoré, B.; Nyengaard, J.R.; d'Amore, F. Tumor-infiltrating macrophages correlate with adverse prognosis and Epstein-Barr virus status in classical Hodgkin's lymphoma. *Haematologica* **2010**, *96*, 269. [CrossRef]
111. Koulis, A.; Trivedi, P.; Ibrahim, H.; Bower, M.; Naresh, K.N. The role of the microenvironment in human immunodeficiency virus-associated classical Hodgkin lymphoma. *Histopathology* **2014**, *65*, 749–756. [CrossRef]
112. Chantziou, A.; Brenna, C.; Ioannidou, K.; Chen, O.Y.; Korkolopoulou, P.; Antoniadou, A.; Psichogiou, M.; Papaioannou, M.; Tsirigotis, P.; Foukas, P.G. HIV infection is associated with compromised tumor microenvironment adaptive immune reactivity in Hodgkin lymphoma. *Blood Adv.* **2024**, *8*, 6215–6231. [CrossRef]
113. Liu, Y.; Xie, X.; Li, J.; Xiao, Q.; He, S.; Fu, H.; Zhang, X.; Liu, Y. Immune Characteristics and Immunotherapy of HIV-Associated Lymphoma. *Curr. Issues Mol. Biol.* **2024**, *46*, 9984–9997. [CrossRef]
114. Jacobson, C.A.; Abramson, J.S. HIV-Associated Hodgkin′ s Lymphoma: Prognosis and Therapy in the Era of cART. *Adv. Hematol.* **2012**, *2012*, 507257. [CrossRef]
115. Suneja, G.; Shiels, M.S.; Angulo, R.; Copeland, G.E.; Gonsalves, L.; Hakenewerth, A.M.; Macomber, K.E.; Melville, S.K.; Engels, E.A. Cancer treatment disparities in HIV-infected individuals in the United States. *J. Clin. Oncol.* **2014**, *32*, 2344–2350. [CrossRef]
116. Hübel, K.; Bower, M.; Aurer, I.; Bastos-Oreiro, M.; Besson, C.; Brunnberg, U.; Cattaneo, C.; Collins, S.; Cwynarski, K.; Pria, A.D. Human immunodeficiency virus-associated Lymphomas: EHA–ESMO Clinical Practice Guideline for diagnosis, treatment and follow-up. *HemaSphere* **2024**, *8*, e150. [CrossRef]
117. Uldrick, T.S.; Gonçalves, P.H.; Abdul-Hay, M.; Claeys, A.J.; Emu, B.; Ernstoff, M.S.; Fling, S.P.; Fong, L.; Kaiser, J.C.; Lacroix, A.M. Assessment of the safety of pembrolizumab in patients with HIV and advanced cancer—A phase 1 study. *JAMA Oncol.* **2019**, *5*, 1332–1339. [CrossRef]
118. Krishnan, A.; Molina, A.; Zaia, J.; Smith, D.; Vasquez, D.; Kogut, N.; Falk, P.M.; Rosenthal, J.; Alvarnas, J.; Forman, S.J. Durable remissions with autologous stem cell transplantation for high-risk HIV-associated lymphomas. *Blood* **2005**, *105*, 874–878. [CrossRef] [PubMed]
119. Balsalobre, P.; Díez-Martín, J.L.; Re, A.; Michieli, M.; Ribera, J.M.; Canals, C.; Rosselet, A.; Conde, E.; Varela, R.; Cwynarski, K. Autologous stem-cell transplantation in patients with HIV-related lymphoma. *J. Clin. Oncol.* **2009**, *27*, 2192–2198. [CrossRef]
120. Carbone, A.; Vaccher, E.; Gloghini, A. Hematologic cancers in individuals infected by HIV. *Blood J. Am. Soc. Hematol.* **2022**, *139*, 995–1012. [CrossRef] [PubMed]
121. Berhan, A.; Bayleyegn, B.; Getaneh, Z. HIV/AIDS associated lymphoma. *Blood Lymphat. Cancer Targets Ther.* **2022**, *12*, 31–45. [CrossRef]
122. Khwaja, J.; Burns, J.E.; Ahmed, N.; Cwynarski, K. HIV-associated lymphoma—advances in clinical management. *Ann. Lymphoma* **2021**, *5*, 26. [CrossRef]
123. Vaccher, E.; Gloghini, A.; Carbone, A. HIV-related lymphomas. *Curr. Opin. Oncol.* **2022**, *34*, 439–445. [CrossRef]
124. Jimenez, O.; Mangiaterra, T.; Colli, S.; García Lombardi, M.; Preciado, M.V.; De Matteo, E.; Chabay, P. PD-1 and LAG-3 expression in EBV-associated pediatric Hodgkin lymphoma has influence on survival. *Front. Oncol.* **2022**, *12*, 957208. [CrossRef]
125. Timmerman, J.; Lavie, D.; Johnson, N.A.; Avigdor, A.; Borchmann, P.; Andreadis, C.; Bazargan, A.; Gregory, G.; Keane, C.; Inna, T. Favezelimab (anti–LAG-3) plus pembrolizumab in patients with relapsed or refractory (R/R) classical Hodgkin lymphoma (cHL) after anti–PD-1 treatment: An open-label phase 1/2 study. *Hemasphere* **2022**, *40*, 16. [CrossRef]
126. Awasthi, P.; Kohli, A.S.; Dwivedi, M.; Hasan, S. Implications of EBV-Encoded and EBV-Related miRNAs in Tumors. *Curr. Gene Ther.* 2025; *online ahead of print*.
127. Hentrich, M.; Berger, M.; Wyen, C.; Siehl, J.; Rockstroh, J.K.; Müller, M.; Fätkenheuer, G.; Seidel, E.; Nickelsen, M.; Wolf, T. Stage-adapted treatment of HIV-associated Hodgkin lymphoma: Results of a prospective multicenter study. *J. Clin. Oncol.* **2012**, *30*, 4117–4123. [CrossRef]
128. Graeter, A.; Bowles, K.E.; Coghill, A. Safety and efficacy of immune checkpoint inhibitors in the treatment of hodgkin lymphoma in people living with human immunodeficiency virus: A case series. *Blood* **2023**, *142*, 6176. [CrossRef]
129. Lurain, K.; El Zarif, T.; Ramaswami, R.; Nassar, A.H.; Adib, E.; Abdel-Wahab, N.; Chintapally, N.; Drolen, C.E.; Feldman, T.; Haykal, T. Real-World Multicenter Study of PD-1 Blockade in HIV-Associated Classical Hodgkin Lymphoma Across the United States. *Clin. Lymphoma Myeloma Leuk.* **2024**, *24*, 523–530. [CrossRef] [PubMed]
130. Bower, M.; Carbone, A. KSHV/HHV8-associated lymphoproliferative disorders: Lessons learnt from people living with HIV. *Hemato* **2021**, *2*, 703–712. [CrossRef]
131. Mullen, C.J.; Volesky, K.D.; Greenwald, Z.R.; El-Zein, M.; Franco, E.L. Is Hodgkin Lymphoma associated with Hepatitis B and C viruses? A systematic review and meta-analysis. *Cancer Epidemiol. Biomark. Prev.* **2021**, *30*, 2167–2175. [CrossRef] [PubMed]

132. Wells, M.J.; Jacobson, S.; Levine, P.H. An evaluation of HHV-6 as an etiologic agent in Hodgkin lymphoma and brain cancer using IARC criteria for oncogenicity. *Infect. Agents Cancer* **2019**, *14*, 31. [CrossRef]
133. Huisman, W.; Gille, I.; van der Maarel, L.E.; Hageman, L.; Morton, L.T.; de Jong, R.C.; Heemskerk, M.H.; Amsen, D.; Falkenburg, J.F.; Jedema, I. Identification of functional HLA-A* 01: 01–restricted Epstein-Barr latent membrane protein 2–specific T-cell receptors. *J. Infect. Dis.* **2022**, *226*, 833–842. [CrossRef]

cancers

MDPI

Review

The Functional Interaction Between Epstein–Barr Virus and MYC in the Pathogenesis of Burkitt Lymphoma

Sandra Solares, Javier León and Lucía García-Gutiérrez *

Instituto de Biomedicina y Biotecnología de Cantabria, Departamento de Biología Molecular, Universidad de Cantabria-CSIC, Albert Einstein 22, 39011 Cantabria, Spain; sandra.solares@alumnos.unican.es (S.S.); javier.leon@unican.es (J.L.)
* Correspondence: lucia.garcia@unican.es

Simple Summary: The Epstein–Barr virus (EBV) infection does not induce any apparent pathology in most people but it has been associated with an increased risk of developing a number of non-malignant diseases (e.g., infectious mononucleosis and multiple sclerosis) and some cancers. Among these, the association between EBV and Burkitt lymphoma (BL) is striking, involving a tumor where *MYC* is deregulated by translocation in all cases. BL is more prevalent in children from equatorial Africa (>90% of the cases) whereas the association of EBV with BL is much lower (25–40%) in other regions. This high association suggests that EBV is a driving mechanism, but whether it is sufficient to trigger lymphomagenesis or it is a cooperative factor is under debate. Indeed, the precise molecular mechanisms underlying the virus activity in infected B cells in collaboration with MYC is still unclear. The molecular mechanisms by which EBV operates in tumor B cells will be discussed.

Abstract: The Epstein–Barr virus (EBV) is associated with a wide range of diseases, malignant and non-malignant. EBV was, in fact, the first virus described with cell transformation capacity, discovered by Epstein in 1964 in lymphoma samples from African children. Since then, EBV has been associated with several human tumors including nasopharyngeal carcinoma, gastric carcinoma, T-cell lymphoma, Hodgkin lymphoma, diffuse large B cell lymphoma, and Burkitt lymphoma among others. The molecular hallmark of Burkitt lymphoma (BL) is a chromosomal translocation that involves the *MYC* gene and immunoglobulin loci, resulting in the deregulated expression of *MYC*, an oncogenic transcription factor that appears deregulated in about half of human tumors. The role of MYC in lymphoma is well established, as MYC overexpression drives B cell proliferation through multiple mechanisms, foremost, the stimulation of the cell cycle. Indeed, MYC is found overexpressed or deregulated in several non-Hodgkin lymphomas. Most endemic and many sporadic BLs are associated with EBV infection. While some mechanisms by which EBV can contribute to BL have been reported, the mechanism that links *MYC* translocation and EBV infection in BL is still under debate. Here, we review the main EBV-associated diseases, with a special focus on BL, and we discuss the interaction of EBV and *MYC* translocation during B cell malignant transformation in BL.

Keywords: MYC; Epstein–Barr virus; Burkitt lymphoma

1. The Epstein Barr Virus, General Introduction, and Incidence

The Epstein–Barr virus (EBV) is a human herpesvirus (known as Human Herpesvirus 4 or HHV-4) that belongs to the gamma-herpesvirus subfamily. It was identified by Epstein in 1964 in samples of Burkitt lymphoma (BL) from African children [1]. The majority of higher primates have their own EBV-like virus, but humans serve as the only natural host for EBV. Two major types of EBV have been reported: EBV1 and EBV2, also known as A and B, respectively, being the sequence of the genes that encode for their nuclear antigens considering the main difference between them [2]. While EBV is detected worldwide, EBV1 and EBV2 seem to have different prevalences depending on the geographical location [3,4].

EBV1 is the most frequent subtype globally. However, in specific regions such as Central Africa, Papua New Guinea, or Alaska, EBV2 is the most prevalent one. In the rest of the world, EBV2 is very rare and only found in immunosuppressed individuals [3].

EBV infection is associated with a wide range of diseases, either malignant or non-malignant, which can be classified into (i) non-malignant: infectious mononucleosis and autoimmune diseases (multiple sclerosis, rheumatoid arthritis, and lupus erythematosus); (ii) non-lymphoid cancers: nasopharyngeal carcinoma, gastric cancer, and lymphoepithelial-like cancer; and (iii) lymphomas: BL, T cells and NK lymphoma, Hodgkin lymphoma, DLBCL, and lymphomas of immunosuppressed individuals. The incidence of EBV infection is very high. More than 90% of the human population become infected at some point in their lives with EBV, a fact that was already reported during the 70s and continues to be reported [5,6]. Thus, EBV is likely the most ubiquitous virus infecting humans. EBV is mainly spread from saliva-containing virus-infected epithelial cells, but it can be also transmitted through blood transfusions or organ transplants [7]. The primary contact with EBV usually happens during childhood and its first and most common form of presentation is infectious mononucleosis [8]. However, it is important to highlight that, despite EBV prevalence, only a low percentage of people among the infected population develop any EBV-associated disease, as discussed later.

2. EBV Genome and Latency Programs

The EBV is a 122–180 nm diameter virus and consists of a DNA core inside a 162 capsomers-nucleocapsid surrounded by an envelope. The inner part between the nucleocapsid and the envelope is composed of a protein viral tegument, while the outer envelope carries the external virus-encoded glycoprotein spikes [9]. Among these, gp350/220 is the most abundant glycoprotein, together with gH, gL, and gp42, which comprise the machinery that mediate the virus entry into the host cell [10] (Figure 1a).

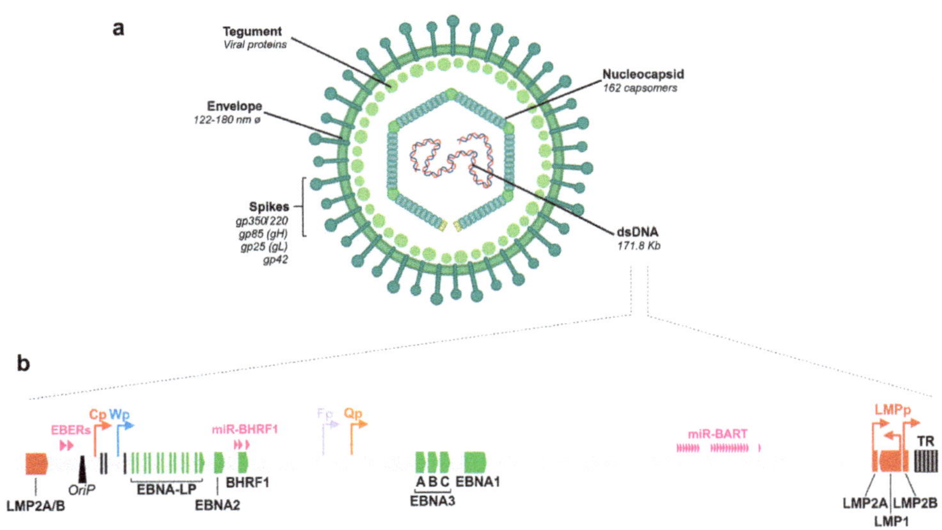

Figure 1. Epstein–Barr virus (EBV) structure and genome. (**a**) Scheme depicting the most relevant features of the EBV structure. Ø: virion diameter. (**b**) Schematic representation of the EBV linearized genome. Green and red boxes represent the EBNAs and LMPs latent genes, respectively. Violet arrows represent ncRNAs. The four alternative promoters from which EBNA genes are transcribed are represented by colored arrows (Cp, Wp, Fp, Qp). *OriP*: Replication origin. TR: Terminal Repeats. Created in BioRender. Garcia, L. (2024) BioRender.com/y04j261.

The EBV genome consists of a linear double-stranded DNA of 171.8 Kb in length, which encodes for more than 94 protein-coding genes (NCBI Reference Sequence: NC_007605.1) (Figure 1b). The nomenclature of the open reading frames (ORFs) is determined by their location with respect to the BamHI fragment. These ORFs are grouped into latent or lytic, and the lytic group is also subdivided into immediate early, early, and late. EBV also encodes non-coding RNAs (ncRNAs), including EBV-encoded RNAs or EBERs [11] and 44 mature miRNAs [12] (from 25 precursor miRNAs, according to https://www.mirbase.org/ accessed on 12/08/2024). The miRNAs are grouped into two clusters, depending on whether they are encoded within the BHRF1 or BART regions, known as ebv-miR-BHRF1 or ebv-miR-BART, respectively. The viral genome also contains 0.5 Kb terminal repeat sequences (TRs) located either at both termini [13,14] or within internal regions, which divide the genome into short and long unique sequences [15,16]. The repeats at the terminal regions serve as good markers to identify if specific EBV-infected cells derive from a common progenitor.

EBV has the ability to infect B cells, transforming them into Lymphoblastoid Cell Lines (LCLs) in vitro. After the primary infection in vivo, the virus silences almost completely its genome to avoid the host immune surveillance, allowing the establishment of a persistent infection. Thus, EBV is found in a dormant state or latent infection for most of its life within the host. This latency can be classified into four different programs (0, I, II, III) depending on the expression of the EBV latency genes (reviewed in [17–19]). Interestingly, each EBV-associated cancer shows one specific EBV latency program, suggesting an important role of the different latency genes in promoting particular tumors (Figure 2). The different transcriptional program activated is based on complex epigenetic regulations selecting viral promoters (reviewed in [20]). Genes expressed along the different latency programs are transcribed from specific promoters within the EBV genome (Figure 1b). In latency 0 (also known as the germinal center model of EBV persistence), only EBERs and miRNAs are expressed. Latency I is characterized by the expression of EBNA1 along with EBERs and BamH1A transcripts, and it is characteristic of BL. In latency II (or the default program), EBNA1, LMP1, LMP2A/B, EBERs, and BamH1A transcripts are expressed and are typically found in nasopharyngeal carcinoma and Hodgkin lymphoma. The promoter usage in latency 0, I, and II is the same, given that EBNA1 is transcribed from the Qp promoter, while Wp and Cp are silenced [21–23]. Latency III (also known as growth program) expresses all latent proteins: six EBNAs (EBNA1, EBNA2, EBNA3A/B/C, and EBNA-LP) and three LMPs (LMP1, LMP2A/B), together with EBER 1 and 2 and BARTs. In this case, promoters Wp and Cp drive the expression of all EBNA genes [24,25]. This type of latency is found in LCLs and in the majority of post-transplant lymphoproliferative disorders [26]. The model of EBV persistent infection states that EBV uses its growth program to activate naïve B cells, forcing them to undergo a germinal center reaction, in the absence of a foreign antigen, leading to a long-lived memory B cell in which EBV latently persists (reviewed in [27,28]).

The latent state of the EBV is concurrent with the establishment of a persistent infection as a mechanism to avoid host immunity, with the limited expression of specific gene sets. However, upon still poorly understood mechanisms, a switch from the latent infection to the lytic replication may occur, leading to the formation of new virion particles. While this is mainly characteristic of epithelial cells and salivary glands, allowing the spread of viral particles toward new cells [29,30], B cells also undergo the EBV lytic replication cycle. This switch from EBV latent to lytic cycle has been related to the cell differentiation processes of both epithelial and B cells [27]. The lytic cycle is divided into three sequential phases, which also classify the gene sets expressed within each of them: (i) immediate-early genes: mainly transcription factors in charge of triggering the lytic cascade [31–33]. *BZLF1* and *BRLF1* are the two determining immediate-early genes, simultaneously expressed and needed for the transition from latent to lytic infection [34]; (ii) early genes: genes encoding nucleotide metabolism-related enzymes (dispensable due to redundancy with host's enzymes), the vPIC (viral preinitiation complex) composed of BcRF1 and another five cofactors that interact with the cellular RNA pol II [35], and DNA replication proteins, essential for the proper replication of the viral DNA and the early to late phase transition [36]; and (iii) late genes: structural proteins for the assembly of the new virions, allowing the packaging of the viral DNA and further release and spread of the new particles from productively infected cells. This process has been extensively reviewed elsewhere [37].

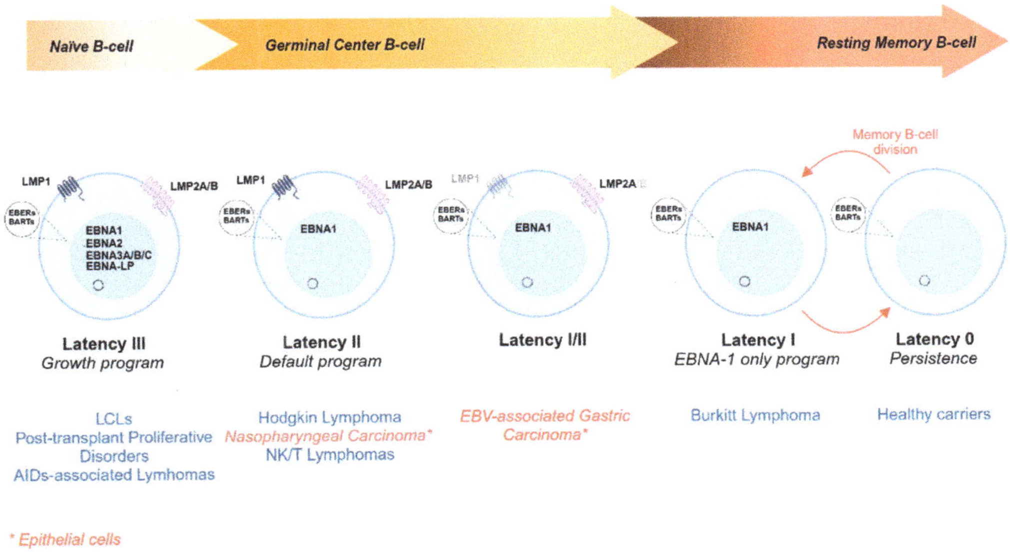

Figure 2. EBV latency programs and associated malignancies. After primary infection, the infected cell enters the Latency III program, also known as the growth program, expressing all latent genes, BARTs, and EBERs. This activates B cells promoting their transition toward a germinal center reaction and switching to Latency II or the default program, with a more restricted gene expression pattern. Finally, memory B cells exit the germinal center showing the Latency 0 program, also known as "Persistence", in which only ncRNAs are expressed. Eventually, the transition from Latency 0 to Latency I (or the EBNA-1 only program) may occur when memory B cells proliferate so that the EBV episome is properly segregated to the daughter cells. In blue, EBV-associated lymphomas with their corresponding latency program. In red, EBV-associated epithelial malignancies and their corresponding latency programs are included. Created in BioRender. Garcia, L. (2024) BioRender.com/y04j261.

3. EBV Receptors in Human Cells and Infection

Depending on the cell type, different viral surface glycoproteins are involved in the recognition and interaction with the host cell, leading to different EBV mechanisms of entrance in the cell. In the case of B cells, this process has been extensively studied and is very well known. However, the mechanism through which EBV interacts and enters epithelial cells is much more poorly understood, although several molecules and processes have already been reported. The EBV glycoproteins gB and the gH/gL complex are needed for the virus membrane fusion with both B cells and epithelial cells [38]. gB activates the viral membrane fusion with that of the host cell [39], and gH/gL regulates this fusion process [40]. While in B cells EBV entrance takes place through endocytosis followed by the fusion of the viral membrane with the endocytic vesicle membrane of the cell [41], it has been described that in epithelial cells, EBV can either enter the cell by direct fusion of the virus membrane with the cell membrane or via lipid rafts endocytosis or micropinocytosis [42].

The glycoprotein CR2 (Complement Receptor 2, also called CD21) is the EBV-cell receptor in B cells [43,44]. CR2 is expressed in mature B lymphocytes and B cell lines but not in early pre- and pro-B cells. It is also found in peripheral blood and thymic T cells and T cell lines [45]. CR1 (Complement Receptor 1, CD35) has also been reported to serve as an EBV receptor [46], but so far, the information on CR1 as an EBV receptor is limited.

CR2 recognizes specific C3d fragments and acts as a co-receptor for B cell receptors (BCR). It is found in a complex with CD19 or CD81 [47]. CR2 is also expressed in follicular dendritic cells [48]. The importance of the CD19-CR2 complex has been demonstrated in vivo as mice lacking either CD19 or CR2 showed reduced formation of germinal center and primary antibody responses [49–52]

CR2 plays an important role in enhancing the immune response promoting cell proliferation in preactivated B cells. In follicular dendritic cells, CR2 rescues antigen-activated B cells from apoptosis and promotes somatic hypermutations and class switch recombination [51]. CR2 has a very short intracellular domain, suggesting that it is very unlikely that it may act as a signal transducer. However, CR2 interaction with CD19 in close proximity with BCR enhances BCR activation and its downstream signaling [45].

The binding of EBV to B cells takes place through the EBV envelope glycoprotein gp350/220 interaction with CR2 [44,53,54], promoting the viral entrance into the host cell by endocytosis [55]. The viral envelope fuses with the cell membrane through a mechanism involving three other viral proteins: gp85 (gH), gp25 (gL), and gp42 [10,38,56]. Since gp42 also binds to the HLA class II, the virus uses this complex as a cofactor to infect B-lymphocytes [56–58] (Figure 3a). Although EBV is considered a B-lymphotropic virus, other cell types such as T lymphocytes and epithelial cells can be infected by EBV, since the virus has been detected in different T lymphomas, nasopharyngeal carcinomas, and gastric cancer, among others. Therefore, EBV can infect cells independently of CR2 [59], although with lower efficiency.

In the case of epithelial cells, the EBV can interact with the cell surface through gp350/220 when these cells express CR2 [60]. For epithelial cells that do not express CR2, different receptors have been described so far, including integrins, neuropilin-1 (NRP-1), non-muscle myosin heavy chain IIA (NMHC-IIA, also called MYH9), and EPHA2 (Figure 3b). Integrins $\alpha v\beta 5$, $\alpha v\beta 6$, and $\alpha v\beta 8$ interact with EBV through gH/gL [61,62]. Also, integrin $\beta 1$ is able to interact with BMRF-2, facilitating EBV attachment to epithelial cells. NRP-1 is a neuropilin able to interact with EBV gB in epithelial cells, promoting viral entrance into the host cell [42]. NMHC-IIA (MYH9) was described as another EBV receptor in nasopharyngeal carcinoma, which interacts with gH/gL of the EBV envelope [63]. Ephrin Receptor A2 (EPHA2) was the latest EBV receptor reported in epithelial cells. EPHA2 is a member of the RKT protein family and is known to serve as the entry molecule for KSHV (another human herpesvirus) and other pathogens, apart from EBV [64,65].

Whether EBV primary infection takes place through epithelial cells or B cells is a matter of debate. Different models have been reported, claiming that either B cells, epithelial cells, or both concomitantly are the first cells infected by EBV [66]. However, the most solid evidence suggests that the B cells in the submucosal secondary lymphoid tissues (such as tonsils) are the first contact with the EBV. This is based mainly on two proven facts: (1) polarized epithelial cells from the oral cavity are only infected through their apical surface by direct contact with EBV^+ lymphocytes [67]; (2) the presence of EBV is detected much earlier in the blood than in epithelial cells from the oral cavity [66]. In fact, the main reservoir of EBV-infected cells is the lymphatic tissue. Upon first contact with the virus after the eventual activation of EBV^+ lymphocytes, oropharynx cells undergo a lytic replication cycle. This lytic infection leads to the production of new viral particles which are spread through the throat. EBV particles infect resting B lymphocytes from the oropharynx lymphatic tissue, which undergo a growth-transforming (Latency III) infection, promoting the proliferation of B cells and activating the host immune response. B cells migrate toward the follicle, switching from latency III to Latency II, where they initiate the germinal center reaction. Latency 0 memory B cells exit the germinal center establishing the long persistent infection, having very restricted expression of viral genes to avoid the immune surveillance. Eventually, memory B cells can divide. In this case, the cells switch to Latency I, allowing the expression of EBNA1 for EBV replication and proper segregation to daughter cells. Occasionally, memory B cells undergo plasma cell differentiation in the tonsil, which leads to the activation of the lytic replication cycle, releasing the virus to the saliva for its spreading to new host individuals or the infection of new epithelial cells or B cells.

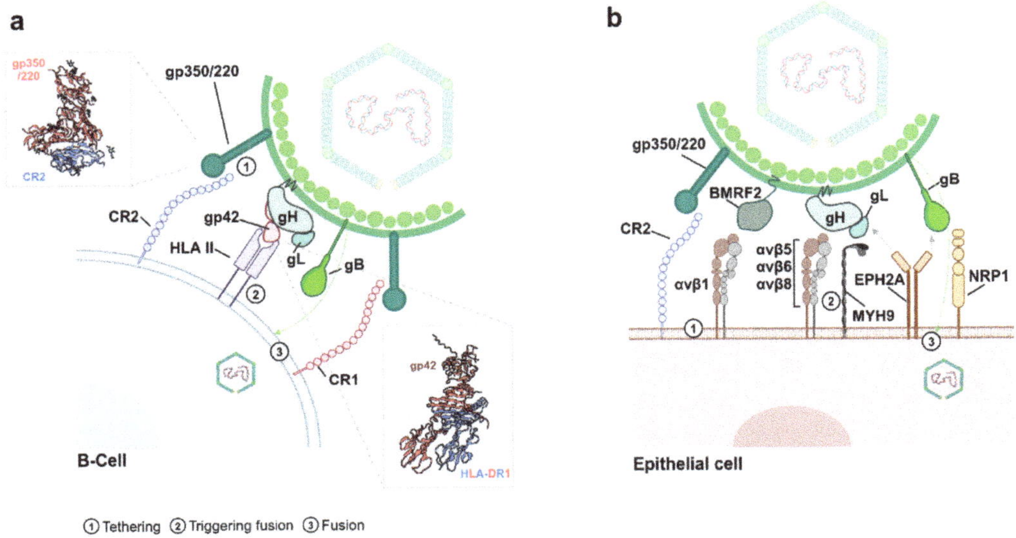

Figure 3. Molecules implicated in the interaction and entry of EBV with its host cell. (**a**) Scheme depicting EBV interaction with B cells. gp350/220 from the EBV envelope interacts with CR2 (or CR1), facilitating gp42 interaction with the HLA class II of the B cell surface. This triggers EBV fusion with the B cell membrane through gB to further deliver the virus genoma inside the cell. PDB: 8SM0 (structure of gp350/220 in complex with CR2). PDB: 1KG0 (Epstein–Barr virus gp42 bound to the MHC class II Receptor HLA-DR1). (**b**) Scheme depicting the EBV interaction with epithelial cells. gp350/220 or BMRF2 recognizing CR2 or β1-integrins, respectively. gH/gL interact with different β5, β6, and β8 integrins and MYH9 in the cell surface. gB, which drives the fusion with the cell membrane, interacts with NRP1. EPH2A binds both gH/gL and gB. Created in BioRender. Garcia, L. (2024) BioRender.com/y04j261.

4. Non-Malignant EBV-Associated Diseases

4.1. Infectious Mononucleosis

Primary infection with EBV during childhood is usually asymptomatic or associated with mild hepatitis. When this primary infection happens during adolescence it can be asymptomatic or associated with acute infectious mononucleosis (IM) [68]. Contact with the virus takes place through saliva. Infants under 2 years old infected with the EBV seroconvert asymptomatically, showing activated EBV-specific T cells and viral loads similar to those from adolescents that develop IM but lack lymphocytosis [69]. A racial disparity has been proposed, suggesting that black young children are more susceptible to EBV infection, most likely because of genetic reasons that are still unknown [70,71].

IM appears as a result of the immunological response triggered by EBV infection. Upon EBV-induced B cell proliferation, NK-cell and T-cell responses are activated [66,72,73]. NK-cell population expands in frequency and number showing differences among distinct NK subsets and their anatomical localizations. While there is little or no increase in CD4+ T cell response, the CD8+ T cell population expands concomitantly to the development of symptoms, leading to an increase in the number of mononucleated cells in the blood [66,74]. These are atypical T-cells that characterize IM, known as "Downey cells" [75]. IM is considered an immunopathological disease due to the high secretion of proinflammatory cytokines by the CD8+ T cells. This leads to the most common symptoms observed in IM patients, which include sore throat, fever, abdominal discomfort, and fatigue, due to swollen lymph glands and enlarged spleen and liver [76].

4.2. Multiple Sclerosis

Multiple sclerosis (MS) is a chronic inflammatory disease associated with neurodegeneration and central nervous system inflammation due to an altered T-cell response [77]. Among others, EBV is considered a major environmental risk factor for MS development [78–80]. Virtually all MS cases are positive for EBV infection and high titers of IgG anti-EBNA-1 before the onset of MS [81,82]. Because of this, EBV has been suggested as a prerequisite for MS. In fact, there is a direct correlation between IM and the risk of developing MS [81]. Also, the combination of EBV with other MS risk factors (such as the presence of the HLA allele *DRB15*) enhances the probability of developing MS [83–85]. Indeed, MS patients show elevated titers against EBV years before developing MS symptoms [86].

Whether EBV reaches the brain is unclear. While EBV DNA has been found postmortem in the brain of MS patients in a limited number of studies [87–89], other reports have not been able to reproduce those findings lacking convincing evidence of the presence of EBV DNA or infected B cells within the brain, central nervous system, or the cerebrospinal fluid [87,90]. Thus, those theories regarding the mechanism of action by which EBV causes MS that rely on the presence of the EBV within the central nervous system cannot be accepted until its presence within those tissues is demonstrated. One of the most accepted hypotheses by which EBV triggers MS is molecular mimicry. Molecular mimicry refers to the fact that pathogen antigens (EBV in this case) are very similar to specific proteins of the host, leading to an immune response that cannot distinguish between the pathogen and host. This is the case of myelin basic protein (MBP) which, due to its similarity with EBV antigens, is recognized and attacked by T cells from MS patients [91–94]. Antibodies against EBNA-1 and BFRF3 have been reported to cross-react with HNRNPL and Septin-9, respectively [95,96].

4.3. Rheumatoid Arthritis and Systemic Lupus Erythematosus

Besides MS, the infection with EBV is associated with two other autoimmune diseases: rheumatoid arthritis and systemic lupus erythematosus. The possible association of EBV with rheumatoid arthritis (RA) was originally suggested when high titers of anti-EBV antibodies were found in the sera and synovial liquid of RA patients (reviewed in [96,97]). It has been hypothesized that the mechanism can be related to molecular mimicry, as the EBV protein gp110 shares an amino acid sequence with a protein of the DRB1 genes, a Major Histocompatibility Complex (MHC) class II family, the *HLA-DRB1*04:01* allele. This epitope would be responsible for raising antibodies in EBV-infected patients [96,98].

The other autoimmune disease associated with EBV is systemic lupus erythematosus (SLR). These patients have autoantibodies against host nuclear antigens and suffer fatigue, systemic pain, skin rash, inflammation, and vascular lesions. Several viruses (e.g., cytomegalovirus, herpesvirus) are known to trigger an autoimmune response, an effect that may also be elicited by EBV [99]. EBV antigens exhibit structural molecular mimicry with common SLE antigens and functional molecular mimicry with critical immune-regulatory components. SLE patients, from a number of unique geographic regions, are shown to have higher rates of EBV seroconversion, especially against early EBV antigens, suggesting frequent viral reactivation. SLE patients also have an increased EBV viral load [96]. The similarity between EBNA1 and at least three SLE autoantigens (SmB/B′, SmD, and 60kdRo) may explain the association of EBV with SLE [100].

5. EBV-Associated Carcinogenesis

EBV is the first human oncogenic virus discovered and it has been classified as class I carcinogen by the WHO. It is estimated that EBV is associated with 1–2% of human cancers and 200,000 newly diagnosed cancers each year [101]. Among the different types of cancer associated with EBV, we can divide them into two main groups: non-lymphoma (epithelial cancers) and lymphomas. Non-lymphoma cancers include nasopharyngeal (NPC), lymphoepithelial-like (LELCs), and gastric (GC) carcinomas. Among the lymphomas associated with EBV, we highlight the following: T cell and NK cell lymphoma (TNKL), Hodgkin lymphoma (HL), diffuse large B cell lymphoma (DLBCL), immunosuppressed individual associated lymphomas, and BL. The latter will be discussed separately since it is the main focus of this review.

5.1. EBV-Associated Non-Lymphoid Cancers

5.1.1. Nasopharyngeal Carcinoma

Nasopharyngeal carcinoma (NPC) arises from the mucosal epithelium of the nasopharynx. NPCs are divided into three groups according to the latest WHO classification of Head and Neck Tumors: (I) keratinizing squamous cell carcinoma; (II) non-keratinizing differentiated carcinoma; and (III) non-keratinizing undifferentiated carcinoma [102]. Both types II and III (non-keratinizing carcinomas) are associated with high titers of EBV, while type I is not [103,104]. Similarly to BL, the association of NPC with EBV infection is more prevalent in specific geographical areas. In eastern and south-eastern Asia and some areas of the Middle East, where NPC incidence is very high, EBV is associated with >95% of the NPC cases, compared with areas where the NPC incidence is lower, in which EBV association is reduced to ~75% of the cases [104]. Higher incidence and worse prognosis of NPC have been reported for men compared to women [105].

NPC has been suggested to originate from the clonal expansion of a single EBV-infected epithelial cell in the nasopharynx [103]. Thus, the EBV infection occurs prior to the tumor development, which may be promoted by the EBV itself. Before the EBV infection, there is evidence of specific genetic events taking place, which may lead to predisposition to viral infection [106,107]. Healthy nasopharyngeal tissue studies and results from pre-malign tissue biopsies have revealed LOH in chromosomes 3p and 9p and hypermethylation of *RASSF1A* and *CDKN2A* tumor suppressors among other alterations, which may increase EBV infection predisposition (reviewed [108,109]). NPC express the Latency program II with the expression of EBNA-1, LMP-1, LMP-2A/B, EBERs, and BARTs. LMP-1 plays a key role in the regulation of transcription of genes that lead to NPC [110]. However, LMP1 protein expression has been detected only in 2/3 of the cases [111], while in in situ precancerous lesions, its expression has been always observed [103]. This suggests that LMP-1 may be essential during the early stages of NPC development. On the other hand, the intermittent reactivation of the lytic cycle in NPC is associated with tumor progression and immune cell infiltration [112].

Lymphoepithelial-like carcinomas (LELCs) are undifferentiated squamous cell carcinomas with lymphoplasmacytic infiltration of the stroma. LELCs are very similar to nonkeratinizing undifferentiated NPCs and arise at locations other than the nasopharynx. Their association with EBV is not uniform along the different LELCs subtypes, given that the association with EBV is more common in the LELC of the salivary glands, which is the most common LELC. EBV-associated salivary glands LELCs are found in endemic regions (southeast Asian and Arctic Inuit populations), with nearly 100% association, while in non-endemic areas, association with EBV is rarely found [113–115]. Sinonasal tract LELCs are also associated with EBV [116], while in other LELC types such as oropharyngeal and laryngeal, EBV association is rare [117–120].

5.1.2. Gastric Cancer

Gastric cancer (GC) is the fifth cancer most frequently diagnosed worldwide with more than one million new cases every year [121]. Different risk factors have been associated with GC development, including EBV infection [122]. In 1990, Burke and colleagues demonstrated, for the first time, the presence of EBV DNA in gastric carcinoma cells [123]. EBV-associated GC (EBVaGC) constitutes itself as a unique molecular subtype, completely different from EBV-non-associated GC (EBVnGC) [124], and comprises 10% of all GC [125,126]. GC has been classified into four major subtypes: EBV-associated, microsatellite instability, genomic stability, and chromosomal instability [127].

There are two theories regarding the primary contact of EBV with gastric epithelial cells: by swallowing saliva containing EBV from the lytic replication of oropharyngeal epithelial cells or by contact with EBV productive infected B cells circulating through the digestive system [128]. It is known that EBV infection of gastric epithelial cells takes place prior to cell transformation, and it is believed that EBVaGC arises from the clonal expansion of a single infected cell [129]. EBVaGC and EBVnGC are very different since each of them develops in anatomically different locations. EBVaGC shows much higher levels of tumor-infiltrating lymphocytes (TILs) (which makes it "immune hot" vs. "immune cold" EBVnGC), cystic gastritis is associated with EBVaGC, and EBVaGC has a better prognosis and overall survival [130]. The viral expression in EBVaGC consists of a Latency I/II intermediate state. EBNA-1, BART, and EBERs are constitutively expressed, while most of them have none or low LMP1 and LMP2B levels and 40% show LMP2A expression [131,132]. Among the genetic alterations and mutations of EBVaGC, *CDKN2A* is found silenced by promoter hypermethylation while MLHI is expressed (opposite to EBVnGC); amplifications of 9p24.1 have been reported leading to gains of *JAK2*, *PD-L1*, and *PD-L2* genes and mutations in PIK3CA and ARID1A have also been found (reviewed in [124,130,133]).

5.2. EBV-Associated Lymphomas

5.2.1. T Cell and NK Cell LYMPHOMA

Although EBV preferentially infects B-lymphocytes, the discovery of EBV DNA in malignant T cells of a rare disease within the nasal cavity broadened the spectrum EBV target cells [134]. EBV-2 is more effective than EBV-1 in infecting T-cells, opposite to what is described for B cells [135]. How EBV infects T lymphocytes is not fully understood yet but it has been described that EBV enters via CR2 [136].

Extranodal NK/T-cell lymphoma, nasal type (ENKTL-NT), is universally associated with EBV, which indicates that the virus may play an important role in the pathogenesis of this disease [134]. ENKTL-NT shows higher incidence in eastern Asia and Latin America compared to Western countries and it represents 1–2% of all NK/T-cell lymphomas [137,138]. ENKTL-NT shows EBV latency I/II, with the expression of EBNA1, LMP2, and variable LMP1. The exact mechanism of how EBV promotes ENKTL-NT is not known. EBV can integrate within specific genomic regions, including an intronic region of the NHEJ1 gene, which has been associated with genome instability [139]. Also, it is believed that the EBV latency program modulates cell signaling processes and prevents apoptosis [140]. Among the molecular signatures of ENKTL-NT, MYC, and NF-κB have been found activated, together with p53 alterations [141].

5.2.2. Hodgkin Lymphoma

Hodgkin lymphoma (HL) consists of a group of diseases arising from transformed B-lymphocytes that accumulate mainly within the lymph nodes, although extranodal disease may also occur. There are two major types of HL, which depend on the type of B cell that becomes transformed: classical HL (cHL), which consists of Hodgkin/Reed–Sternberg (HRS) cells, and nodular lymphocyte predominant HL (NLPHL), in which malignant cells are lymphocyte predominant (LP) [142].

EBV infection is associated with cHL but not with NLPHL, and it is detected in 20–50% of cHL [143]. Several risk factors contribute to EBV^+ cHL, including HIV infection, clinic history of IM, and the *HLA-A*01* allele [144]. Monoclonal EBV genomes have been detected in HRS cells, suggesting an important role of EBV in cHL onset [23,145]. EBV^+ HRS cells express the Latency II program: EBNA1, LMP1, LMP2A, EBERs, and BARTs (See Section 2; Figure 2). HRS cells show evidence of somatic hypermutation; therefore, they most likely derive from B cells, which have undergone germinal center reactions [146,147]. However, they lack BCR function, which would lead to apoptosis in these cells. The acquisition of anti-apoptotic alterations allows them to survive and contributes to tumor development [144]. EBV infection facilitates the survival of HRS cells. In fact, LMP2A mimics BCR, promoting B cell development, even in the absence of BCR function. In general, EBV$^+$ cHL presents a reduced number of chromosomal alterations and aneuploidies when compared to EBV$^-$ cHL, which suggests that the EBV latency genes promote the deregulation of the signaling pathways needed for malignant transformation (reviewed in [148]).

5.2.3. Diffuse Large B Cell Lymphoma

Diffuse large B cell lymphoma (DLBCL) is the non-Hodgkin lymphoma most frequently diagnosed (25–35% of all lymphoma cases) [142]. Between 5 to 15% of the DLBCL are EBV-infected and correlate with a poorer prognosis than EBV-negative DLBCL. While in Western countries, EBV$^+$ DLBCL is less frequent, the highest percentages of EBV-associated DLBCL are found in Asian and South American countries [149]. In general, older people (>50 years old) suffer an extranodal disease, while younger patients show a nodal disease.

EBV-infected DLBCL usually expresses either the latency II or the latency III programs [150]. This might reflect the potential dual theory supporting that EBV can infect both naïve B cells and memory B cells. EBV$^+$ DLBCL appears to have a lower dependency on cellular events since EBV genes likely contribute to carcinogenesis. EBV$^+$ is considered an independent DLBCL entity due to the extensive molecular differences with other subgroups of this disease [151].

5.2.4. Immunosuppressed Individual-Associated Lymphomas

Individuals who suffer immunodeficiencies are prone to develop EBV-related lymphoproliferative disorders (LPDs). Among these, the most common cases are those patients with AIDS or individuals who are receiving immunosuppressive therapies due to organ transplants. The latter, post-transplant lymphoproliferative disorders (PTLDs), are typical after both solid organ and hematopoietic stem cell transplantation [152–154]. In the majority of cases, PTLD is associated with active replication of EBV after either primary infection or reactivation during treatment with immunosuppressive drugs. EBV$^+$ PTLDs express the latency III program (see Section 2), in which the six viral latent genes are expressed, similar to LCLs.

6. Burkitt Lymphoma

Burkitt lymphoma (BL) is a very aggressive B cell non-Hodgkin lymphoma, characterized by its fast growth rate. BL was the first tumor associated with a virus. In 1964, Michael Anthony Epstein, Yvonne Barr, and Bert Achong identified viral particles in a lymphoma recurrently suffered by African children, the so-called Epstein–Barr virus (EBV) [1]. BL is one of the most prevalent cancers in children from the African equatorial belt, accounting for up to 74% of childhood malignancies [155,156].

The fifth edition of the WHO classification regarding hematolymphoid tumors currently distinguishes between EBV-positive BL (EBV$^+$ BL) and EBV-negative BL (EBV$^-$ BL), since they form discrete biological groups based on their molecular features and thus use the virology rather than the epidemiology as the criteria to classify the BL [142]. EBV infection confers BL cells mechanisms to evade apoptosis, an essential feature during the onset of this disease [157–161]. EBV$^-$ BL are considered "mutational driven" while EBV$^+$ BL are considered "virus driven". The main differences between both subtypes highlighted in the current WHO classification are the following: EBV$^+$ BL harbors higher somatic hypermutation levels, fewer driver mutations, and lower frequency of mutations in specific genes (see below) [162]. These differences between EBV$^+$ BL and EBV$^-$ BL are summarized in Table 1. However, it is worth mentioning that, for more than 50 years, BL has been classified into three subgroups depending on the epidemiological context and geographical location: endemic BL (eBL), sporadic BL (sBL), and immunodeficient-associated BL (iBL) [155,163]. The association of BL with EBV infection varies depending on geographical distribution and associated risk factors, accounting for 98% of BL in Africa (formerly known as eBL),

30–40% of the BL cases associated with immunodeficiencies (formerly known as iBL), and 5–10% of the rest of BL cases (formerly known as sBL). The eBL variant strikingly correlates with regions where malaria is endemic and with early EBV infection events, mainly Equatorial Africa and Papua New Guinea. This evidence indicates a strong association of malaria infection with EBV and BL pathogenesis. The peak of incidence is found at the age of 6 and it is more common in boys than in girls [164].

EBV$^-$ BL (sBL) is mainly found in North America, Europe, and some regions of Asia. Its incidence is low: two cases per million population <18 years old. In some regions of South America, southern Europe, North Africa, and the Middle East, there is an intermediate association of BL with EBV. Interestingly, overall EBV prevalence in patients with BL has decreased from 64% in 1969–82 to 54% in 2009–2021) [165]. The so-called iBL variant is mainly related to HIV infection, although this association is not as strong as the one with malaria. HIV-BL is more common in those patients who still show an intermediate CD4 T-cell response and its incidence is very variable [166,167]. BL is also found in organ transplant recipients.

The major common feature shared by all BLs (both EBV$^+$ and EBV$^-$ BL) consists of a translocation involving chromosome 8 [168], where the oncogenic transcription factor *MYC* locus is located [169], and most frequently chromosome 14 at the immunoglobulin heavy chain locus or, less commonly, chromosomes 2 or 22 at the immunoglobulin light chain loci (this will be discussed in the following sections). The molecular mechanisms leading to MYC translocation in BL remain largely unknown. Breakpoints in the 8;14 translocation affect the S (switch) regions of the heavy chain loci, suggesting that it originates from aberrant class switch recombination (reviewed in [170]). Also, it has been recently reported that EBV reactivation in lymphoblastoid cell lines induces *MYC* and *IGH* spatial proximity, which would facilitate the t(8;14) translocation [171]. The association between EBV and BL was reported after very early EBV's discovery. A study from 1978 that performed prospective serological investigation in a collection of blood samples of children from Uganda, where BL is endemic, showed that those who had developed BL displayed statistically significantly higher titers of antibodies against EBV viral capsid antigens compared to healthy children.

Interestingly, no differences were found in antibody titers against early antigens or nuclear antigens. These higher antibody titers were associated with a higher risk of developing BL [172,173]. The correlation between high antiviral-capsid antibody titers and BL was also reported for sporadic BL later on.

Together with EBV, there are other risk cofactors that contribute to the development of BL, some of which depend on the geographical location or genetic predispositions, and they vary depending on the BL subtype. Among them, chronic immune activation due to malaria or HIV infection is considered a high risk of BL development, especially malaria in eBL [174]. Repeated infections of *Plasmodium falciparum* (the parasite that causes malaria) during childhood in malaria-endemic regions together with the high incidence of EBV infection are directly associated with BL development. Chronic immune activation due to malaria weakens immune control over EBV and promotes B cell proliferation and EBV reactivation [175]. This environment of rapid B cell proliferation and increased germinal center formation increases the likelihood of oncogenic events, especially MYC translocations. The more the germinal center reactions, the higher the activation-induced cytidine deaminase (AID) activity. In the germinal center, AID is responsible for the somatic hypermutations (SHM) and class switch recombination (CSR) events needed for proper antibody diversification [176,177]. In BL, increased AID activity due to different factors, including chronic immune activation, leads to off-target breaks or errors, making *IG::MYC* translocations more likely [178]. Similar to malaria, the dysregulation of the immune system caused by HIV [179] increases the risk of developing BL (HIV$^+$ individuals are at 10–100 times higher risk of developing BL compared to HIV$^-$ individuals). HIV-BL is more similar to sBL in terms of clinical and pathological features [155,180]. The mycotoxin aflatoxin B1 is a food contaminant more common in the African regions with higher

EBV prevalence, and it has been shown to stimulate EBV-mediated transformation [181]. Recently, the secretion of the cytokine CCLC2 [182] and the downregulation of the TGB1 gene [183] have been shown as possible mechanisms to explain the aflatoxin B1 role in BL.

Table 1. Major differences between EBV$^+$ and EBV$^-$ BL. Modified from [155,156,180].

Current WHO Classification	EBV$^+$ BL	EBV$^-$ BL
BL subtype (traditional classification)	95% eBL 5–10% sBL 30–40% iBL	90–95% sBL 60–70% iBL
Geographical distribution	Equatorial belt of Africa and Papua New Guinea (eBL)	Worldwide (>North America, northern Europe and east Asia)
Age	Children Median age, 6 y/o (eBL)	Adults All ages
Presentation site	Jaw, eye, abdomen, kidneys and ovaries (eBL) Abdomen, lymph nodes and bone marrow (sBL)	Abdomen, lymph nodes and bone marrow
Annual incidence	40–50 per million children younger than 18 years. 50% of all childhood cancers and up to 90% of lymphoma Twice more frequent in boys than girls	2–3 per million 3–5 times more frequent in boys than girls
Cofactors	Malaria infection and aflatoxin B1 in eBL, HIV infection in iBL	
Precursor cell	Memory and germinal center B cell	Germinal center B cell
Prevalent Ig breakpoint	VDJ region, switch (s)μ in some cases in eBL Sμ, Sα or J region in sBL Sμ in HIV-associated BL	Sμ, Sα or J region in sBL Sμ in HIV-associated BL
Prevalent MYC breakpoint	>100 bp upstream of first exon in eBL Between exons 1 and 2 or in 5′ near the first exon in sBL and iBL	Between exons 1 and 2 or in 5′ near the first exon
Mechanism of BL pathogenesis	Virus-driven	Mutational
Mutational landscape	++ AID activity Higher levels SHM	More driver mutations (apoptotic pathway)

There are different hypotheses regarding the origin of BL. Whether it originates from germinal center B cells or mature memory B cells is unclear. Both sporadic and endemic BL present somatic mutations in the rearranged V region genes, meaning that they derive from either germinal center or germinal center descendants [184,185]. While the surface phenotype of BL resembles that of germinal center cells, the Ig features support the hypothesis that BL is a tumor arising from latently infected memory B cells. It has also been postulated that the cell of origin could be a memory B cell that re-enters the germinal center and that there may be differences depending on whether the BL is EBV$^+$ or EBV$^-$ [185]. In any case, the germinal center reaction is essential for the pathogenesis of this disease, since it is where key molecular events, such as *MYC* translocations, take place [177,178].

The interaction of EBV with B cells takes place through a gp350/220 EBV glycoprotein binding to CR2 in the surface of B cells [44,53]. It is worth mentioning that naïve B cells show the highest CR2 (and CR1) expression compared with germinal center and mature B cells [186], which leads to the idea that naïve B cells are the cells most likely to be infected by EBV, since EBV infection rates directly correlate with CR2 expression levels [187]. Of note, *CR2* has been reported as a direct MYC target gene, demonstrating that modulation of MYC expression directly affects the EBV loads found in the surface of B cells [188]. Thus, *MYC* translocations in BL occurring in germinal center cells would increase the density of CR2 in the plasma membrane of germinal center cells increasing the likelihood of EBV infection. Although this has not been proven, CR2 upregulation in germinal center cells by deregulated MYC is in line with the hypothesis in which *MYC* translocations favor EBV infection that will be discussed in the following sections.

Most BL cells display the Latency I program in which only EBNA1, EBERs, and BARTs are expressed (Figure 2) [156]. It is unclear whether BL originates from either B cells in Latency I or a specific clone of B cells in Latency III that switches to a Latency I program [189]. The latency I program of BL cells suggests that the B cells leaving the germinal center cannot completely differentiate into resting mature B cells since they keep proliferating (cannot transit to latency 0), most likely due to MYC deregulation [189,190]. *EBNA1* is mainly expressed from Qp, wherein Wp and Cp promoters are shut down. However, there are cases in which a different latency state of BL cells is found, where *EBNA1* is transcribed from Wp instead of Qp, together with *EBNA3A/B/C* and *EBNA-LP*. This is known as "Wp-restricted" BL [189]. These cells are infected with an EBNA2-deficient EBV in which neither EBNA2 nor LMP1 are expressed. Moreover, single-cell analysis of BL samples also revealed another form of latency in which EBNA2 but not LMP1 is expressed. This suggests that LMP1 or EBNA2-induced *LMP1* is not compatible with *MYC* translocation in BL. All three latency subtypes seem to be related with their ability to protect BL cells from the apoptosis induced by MYC, wherein the latency I is the most sensitive, Wp restriction is the most resistant, and EBNA2+/LMP1- shows an intermediate phenotype [191].

Several studies have focused their attention on deciphering the mutational landscape of BL, especially the differences between EBV$^+$ and EBV$^-$ BL [42,162,192,193]. While the two entities cannot be distinguished only on basis of the mutational profiling, mutations in specific genes have been reported to differentially occur in EBV$^+$ vs. EBV$^-$ BLs, especially in genes involved in BCR signaling, proliferation, TP53 pathway, chromatin remodeling, and sphingosine-1-phosphate signaling (reviewed in [163]). *MYC* is also frequently mutated (see Section 9). At the genomic level, EBV$^+$ BL shows higher mutational burden but fewer driving mutations than EBV$^-$ BL, especially in those genes involved in apoptosis [162]. One of the most striking differences is found in the TCF3 pathway, as EBV$^-$ BL has been reported to harbor higher mutational rates in the *TCF3* gene and its negative regulator *ID3* (activating and inactivating mutations, respectively), leading to increased TCF3 pathway activity [42,162,192]. Mutations in genes involved in epigenetic regulation processes were found significantly more frequently in EBV$^+$ BL compared with EBV$^-$ BL [162].

Transcriptomic studies have also revealed that EBV$^-$ BL shows higher levels of *MYC, DDX3X, CCND3,* and *TP53* mRNA compared to EBV$^+$ BL. Also, it has been recently reported that the expression of SOX11 (SRY-box transcription factor 11) is expressed only in EBV$^-$ BL, although its biological significance is still unknown [194]. On the other hand, AID activity has been recurrently reported to be much higher in EBV$^+$ BL, in line with the increased mutational rate of the AICDA gene compared with EBV$^-$ BL [162]. Interestingly, EBV$^+$ BL is characterized by increased PI3K activity via downregulation of PTEN, followed by mTORC1 activation [192].

LCLs generated by EBV infection show a global genome demethylation as compared to normal lymphocytes, but this has not been demonstrated in BL cells [195]. In contrast, in BL, NPC, and EBVaGC there is a tendency toward CpG island hypermethylation [196]. It has been shown that methylation changes in the promoters are responsible for the differential expression of genes between EBV$^+$ and EBV$^-$ BL. For example, ID3 promoter methylation is higher and its expression is lower in BL EBV$^+$ BL than in EBV$^-$ BL [197]. Several viral genes have functions linked to epigenetic regulation mediated by EBV, such as the recruitment of polycomb repressive complex 2 (PRC2) by EBNA3A, the recruitment of HDAC1/2 by EBNA3C, or the interaction of the SNF-SWI remodeler complex with EBNA2, but none of them are expressed in the genetic program latency I of BL (reviewed in [20,196]).

7. EBV Genes That Contribute to Lymphoma Progression

As already discussed above, latency I is the main latency program in BL, together with a minor percentage of cells showing the so-called "Wp-restricted" or the $LMP1^{-}/EBNA2^{+}$ expression profiles. We will focus on the oncogenic activities driven by the EBV genes expressed in BL. However, an effect of the rest of the latent proteins expressed only during the early process of the primary infection cannot be ruled out as an important event in the BL onset. This is the case of LMP1 and LMP2 membrane proteins, whose expression is restricted to latency III/II programs and thus not expressed in BL cells. LMP1 and LMP2 have the ability to mimic CD40 [198] and BCR [199] signaling, respectively, likely ensuring EBV-infected B cell survival during the germinal center reaction. The EBV genes more linked to B cell transformation are briefly reviewed below.

EBNA1: This gene encodes a transcription factor that binds both viral and host promoters and it is required for proper EBV episome segregation [200]. EBNA1 is the only gene expressed in all latency and lytic programs (except for latency 0) and thus the major latent gene expressed in BL. The potential oncogenic capacity of EBNA1 has been a matter of extensive study and debate throughout the last decades. In fact, EBV lacking EBNA1 is much less efficient in the immortalization of B cells than wild-type EBV [201]. Reports using transgenic mice claim that EBNA1 does not promote lymphomagenesis [202], although EBNA1 has been reported to induce B cell neoplasia in mice [203], in an MDM2-dependent manner [204]. EBNA1 inhibition increases cell death in BL cells; thus, it seems to have a pro-survival role in BL [205]. It has been reported that EBNA1 impairs p53 activity through different mechanisms [206] and that the EBNA1-VAV1 interaction impairs BIM expression [207], altogether conferring resistance to apoptosis in BL.

EBNA2: This encoded transcription factor, which does not directly bind to DNA, is essential for the transformation of B cells in vitro by EBV [208], while the rest of viral latent genes are dispensable [209]. EBNA2 has been reported to functionally replace intracellular Notch signaling [210], supporting cell survival and lymphomagenesis in BL. The MYC oncogene and LMP1 are directly activated by EBNA2 [211].

EBNA3: The EBNA3 family of transcription factors is composed of three isoforms: EBNA3A, EBNA3B, and EBNA3C. Among them, EBNA3A and EBNA3C are essential for B cell transformation, while EBNA3B is not [212]. In fact, EBNA3B has been proposed to have a tumor suppressive role, since its inactivation drives lymphomagenesis and immune evasion [19]. EBNA3A and EBNA3C confer survival advantage by blocking apoptosis through the inhibition of BIM [213]. This is interesting since MYC overexpression can induce apoptosis via BIM [214]. Finally, EBNA3A regulates MCL-1 mitochondrial organization through the upregulation of BLF-1 expression [215], impairing apoptosis. This would be an EBV-mediated survival mechanism that resembles the MCL-1 function during germinal center in mice [216].

BHRF1: This gene encodes for an early lytic gene thought to be only expressed during lytic replication from Wp. However, its expression has been detected in other latency programs and more importantly in "Wp-restricted" BL. As a BCL2 homologue, BHRF1 interferes with BIM and other proapoptotic proteins in BL cells, conferring resistance to apoptosis [101,158,217] and contributing to lymphomagenesis.

8. MYC Protein and Biology

MYC is the prototypical transcription factor of the superfamily of factors that contain the helix-loop-helix-leucine zipper domain (HLH-LZ). MYC (also called c-MYC) is one of the three members of the MYC family. The other two proteins of the family are MYCN (also called N-MYC) and MYCL1 (also called L-MYC). All paralogs share the HLH-LZ domain as well as the other four conserved regions of the molecule, called MYC boxes (Figure 4a). However, MYC is by far the member of this family more heavily involved in cancer and is also the most ubiquitously expressed in human tissues.

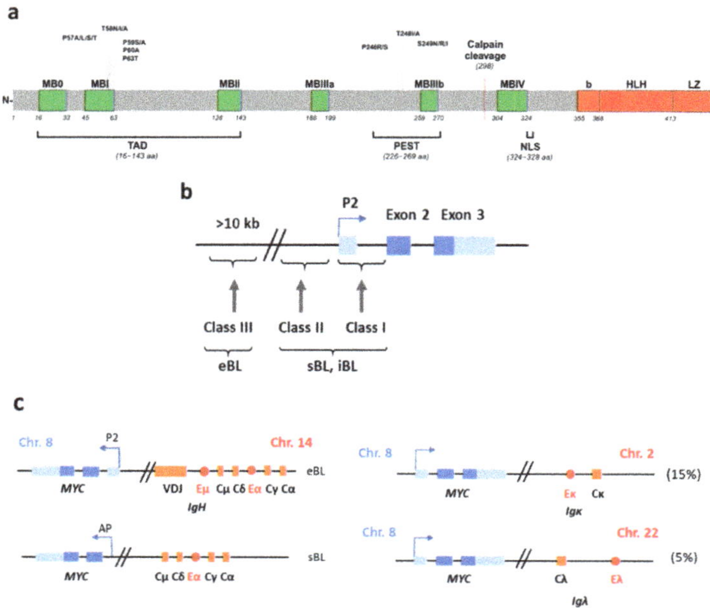

Figure 4. The oncogenic MYC transcription factor and its translocations within the different Ig loci found in Burkitt lymphoma. (**a**) Scheme of the MYC protein. The conserved boxes among MYC proteins are in green. The bHLHLZ domain is in red. The most prevalent mutations in BL are indicated. TAD, transactivation domain; PEST, degradation domain; NLS, nuclear localization signals. (**b**). Schematic representation (not to scale) of the breakpoints in the chromosome 8 in BL. The three exons of the MYC genes are the blue boxes, with the coding regions in darker blue. The major promoter, P2, is shown. The three classes of breakpoints with the chromosome 14 (immunoglobulin heavy chain genes) are indicated as class I, II, and III and the BL subtype where it is more prevalent. The breakpoint in the eBL typically occurs > 10 kb upstream of the first *MYC* exon and in the first noncoding exon or first intron in the sBL. (**c**) Schematic figures (not to scale) of prototypical translocations MYC::IGH to heavy chain genes (left schemes, 80% of the BL cases), MYC::IGK to light chain kappa genes (upper right scheme, 15%), and MYC::IGL light chain lambda genes (lower right scheme, 5%). In the t(8;14) translocation, the break points are the switch regions of the constant gene segments encoding the IgH isotypes and are upstream of *MYC*. In eBL, the breakpoint in the chromosome 14 is more frequently in the J segment. The predominant breakpoint in sBL occurs in the Cμ (M) regions of the H chain locus. In the t(2;8) and t(8;22) translocations, the kappa and lambda light chain lie far downstream of *MYC*. AP refers to an alternative promoter used in this type of translocations. The enhancers are denoted as red circles. Adapted from [163,170]. Partially Created in BioRender. Garcia, L. (2024) BioRender.com/y04j261.

All *MYC* genes operate as transcription factors forming heterodimers with MAX, also an HLH-LZ protein [218]. The MYC-MAX heterodimer is the active form, which binds to specific DNA sequences called E-boxes (canonical sequence CACGTG) in the regulatory regions of target genes. The MYC network includes other components of the HLH-LZ family such as the MXDs and MLX with different functions in gene expression regulation upon binding to E-boxes in the DNA [219,220].

Transcriptomic analyses revealed that MYC can regulate thousands of genes, when comparing cells with low vs. high MYC expression. However, the number of regulated genes depends on the MYC levels in a particular cell type. The number of MYC-binding sites revealed by genome-wide technologies ranks between 7000 and 15,000 in different models and regulates more than 1000 genes [221–225]. Indeed, MYC is bound at one or more sites of the regulatory regions of 10–15% of human genes [225]. In agreement with the large number of MYC target genes, the overexpression of MYC deregulates a series of biological functions such as cell cycle progression, nucleotide biosynthesis, energy metabolism, protein synthesis and ribosome genesis, genomic maintenance, immortalization, and differentiation block [226–228]. These series of acquired or enhanced functions in MYC-overexpressing cells would confer a competitive advantage triggering the MYC-mediated oncogenesis. Not surprisingly, *MYC* is dysregulated in more than 50% of human malignancies of all types [229,230], although it is particularly prevalent in lymphoma and leukemia [231,232].

9. MYC Deregulation in BL

MYC was the first oncogenic transcription factor identified, initially as the transforming genes of an avian retrovirus that induced several tumors, most prominently, myelocytomas [233–235]. Interestingly, the original virus did not induce lymphoma, although when later isolated, it induced T and B lymphomas [236]. In the 1970s, a recurrent translocation between the chromosomes 8 and 14 was described in BL [237]. In 1982, it was reported that the translocation involved the transforming sequences of the myelocytomatosis (MYC) retrovirus MC29, being the first consistent cancer-associated translocation described [169].

Virtually all BL carry a reciprocal translocation that places the *MYC* gene (in chromosome 8) in the vicinity of immunoglobulin regulatory regions. MYC plays a prominent role in germinal center formation [238,239]. In the germinal center, the activation-induced cytidine deaminase (AID) initiates somatic hypermutation and class switch recombination [240], but through a defective process, it can also induce Ig/MYC translocations [178,241].

The most common translocation in BL and other lymphomas is the t(8;14) where *MYC* is placed in the IgH region (80% of the BL cases) but it can also be translocated to the light chin Ig genes: the κ light chain in the t(2;8)(p11;q24) translocation (15% of the cases) and the λ light chain in the t(8;22)(q24;q11) translocation (5% of the cases) [242]. In the most common t(8;14) translocation, the breakpoint is upstream of the MYC gene and in the other two types, it is downstream (Figure 4b). Interestingly, there is a difference in the translocation point depending on the EBV association. In EBV^- sBL and HIV-BL, the breakpoint of the t(8;14) translocation in the *MYC* gene occurs usually between exons 1 and 2 of *MYC* (being exon 1 non-coding) or upstream in the proximity of the first exon. In the EBV^+ eBL, the breakpoint usually occurs far upstream of the *MYC* gene, in some cases more than 100 kb, and most often affects the VDJ region or the M (Cμ) region in some cases (reviewed in [170,242]) (Figure 4c). It is of notable that ~2% of BL carry *MYC* translocations not affecting the immunoglobulin genes [243].

The leukemic form of BL is the acute lymphoblastic leukemia type L3, which also carries the *MYC::IGH* translocation [244]. However, *MYC* translocation is also frequent in other B cell malignancies, more prominently in diffuse large B cell lymphoma (70%), mantle cell lymphoma (14–20%) [232,245,246], and multiple myeloma (15–50%) [247].

This translocation results in *MYC* expression under the control of immunoglobulin gene regulatory elements, including enhancers, which are very active in mature B cells. Thus, the translocation leads to the overexpression of *MYC* or to the loss of the physiological regulation of MYC. Interestingly, the normal *MYC* allele is transcriptionally silent in BL [248,249] and thus most of the MYC protein in most BL cells is derived from the translocated allele. Whereas *MYC* translocation is rare in solid tumors, *MYC* is amplified or transcriptionally deregulated in many tumors of virtually any type. Overall, the reported incidence of *MYC* deregulation through translocation, amplification, transcriptional deregulation, or mutation ranges from 50% to 70% [229,230].

Besides *MYC* translocations, 30–50% of BL carry mutations within the *MYC* gene [193,250–252]. T58 is one of the most prevalent mutated codons (Figure 4a). Phosphorylation of this threonine marks MYC for ubiquitylation and degradation [253], resulting in a more stable MYC protein and higher MYC levels in the cell. Other genes have been found to be recurrently mutated in BL, such as ID3, p53, GNA13, RET, PIK3R1, ARID1A, and SMARCA4 [193,252]. Interestingly, ID3 mutations are much more frequent in BL (30% of the cases) than in any other high-grade lymphoma such as DLBCL [252].

The first animal model generated for MYC-driven cancer was the *Eμ-Myc* transgenic mouse, in which MYC expression is targeted to the lymphoid compartment by the immunoglobulin heavy chain gene promoter and enhancer. These mice develop cell lymphomas [254,255] that do not faithfully reproduce the BL characteristics, as the tumor cells are at the pre-B developmental stage whereas BL is composed of more mature B cells. However, these transgenic mice demonstrated the relevance of *MYC* deregulation in B cell malignancies. Later on, additional transgenic mice lines have been generated that better reproduce human BL: (i) a mouse model carrying a single copy of the 240-kb IgH/c-*Myc* translocation region [256], (ii) a model that carries the murine *Myc* cDNA inserted in the *IgH* locus in a site that corresponds to the human t(8;14) translocation break [257], (iii) mice with MYC linked to the 3′ *IgH* locus control region (3′ LCR) [258], or (iv) mice with combined MYC overexpression that constitutive activation of the PI3K [259].

10. Roles of MYC in the Pathogenesis of BL

The role of EBV in BL pathogenesis has been a matter of debate. The canonical mechanism postulates that EBV infection triggers the malignant transformation of B cells in the germinal center of the lymph nodes. However, there is no doubt regarding the pivotal role of *MYC* translocations in BL being the hallmark of this disease and probably the main driver of BL. What happens first, whether it is *MYC* translocations or EBV infection during the onset of BL is not clear. The *"virus first"* is the most accepted hypothesis, at least for the endemic form affecting children in Africa and other malaria-exposed countries. The following data support this hypothesis:

(a) EBV antibodies in serum precede the onset of BL in African children. Although the original data are based on a limited number of cases, this is likely the most compelling argument. Indeed, the WHO declared EBV as the causing agent of BL [260];
(b) The analysis of clonality based on the number of terminal repeats of the EBV genome in BL cell lines suggests that the tumors arise clonally after infection by EBV [261];
(c) EBV transforms resting B cells to generate lymphoblastoid cell lines. These are immortalized B or semi-immortalized cell lines or LCLs. These cells are not fully immortal but can grow for a large number of population doublings retaining a diploid karyotype without becoming tumorigenic [256];
(d) EBV could induce *MYC* gene translocation via activation of AID. The viral EBNA3C protein induces AID expression, which could generate aberrant chromosomal translocations including those that involve *MYC* [178,241,262];

However, other evidence argues against this *"virus first"* model, such as the following:

(a) First, in most sporadic BL cases, EBV infection is absent while *MYC* is translocated in all cases. Moreover, although there are differences in the prevalence of some genetic alterations and gene expression patterns, there are no differential molecular markers in the mutational landscape between EBV^+ and EBV^- lymphomas, including mutations in the *MYC* gene [42,162,192,250,263];
(b) More than 90% of the human adult population has been infected by EBV, but oncogenic events related to EBV infection have a very low incidence, and in fact, the prevalence of EBV^+ BL is decreasing [165];
(c) Remarkably, *MYC* translocation is not found in other tumors associated with EBV infection, encompassing nasopharyngeal cancer (where EBV association is above 90%) [109], gastric cancer, and Hodgkin lymphoma [264–266];
(d) The EBV genes required for B cell transformation into lymphoblastoid cell lines are not expressed in BL. Indeed, only the latent EBNA1 protein is consistently expressed in BL (see Section 2). EBNA1 is required for viral DNA replication but has no transforming

effects, and it is unable to transform B cells. Moreover, MYC represses the EBV-transforming *LMP1* gene [267]. It has been reported that EBNA2 and EBNA3C induce *MYC* expression [211,268,269]. However, neither EBNA2 nor EBNA3C are expressed in most BL. Moreover, a relevant fraction of BLs (15%) carries a deletion of the *EBNA2* gene in the viral genome [189];

(e) EBV-generated lymphoblastoid cell lines express some viral oncogenes (EBNA2, LMP1) but do not carry *MYC* translocations [270];

(f) The receptor for EBV in B cells, *CR2*, is a direct MYC target gene. Thus, B cells expressing high MYC levels as a result of the translocation will also express higher CR2 levels and thus will be more efficiently infected by EBV [188].

The "MYC first" hypothesis has been proposed earlier [270,271] and the MYC-mediated up-regulation of the CR2 receptor gives further support to this hypothesis. According to this mechanism, *MYC* translocation (and thus *MYC* deregulation) will occur stochastically at one or a few cells that will express more CR2 in its membrane, which in turn will lead to a more efficient infection. The "MYC first" mechanism would be compatible with the epidemiological data and the viral clonality data, as the subsequent EBV infection will co-adjuvant to lymphoma development. Figure 5 schematizes both mechanisms to explain the EBV-BL association.

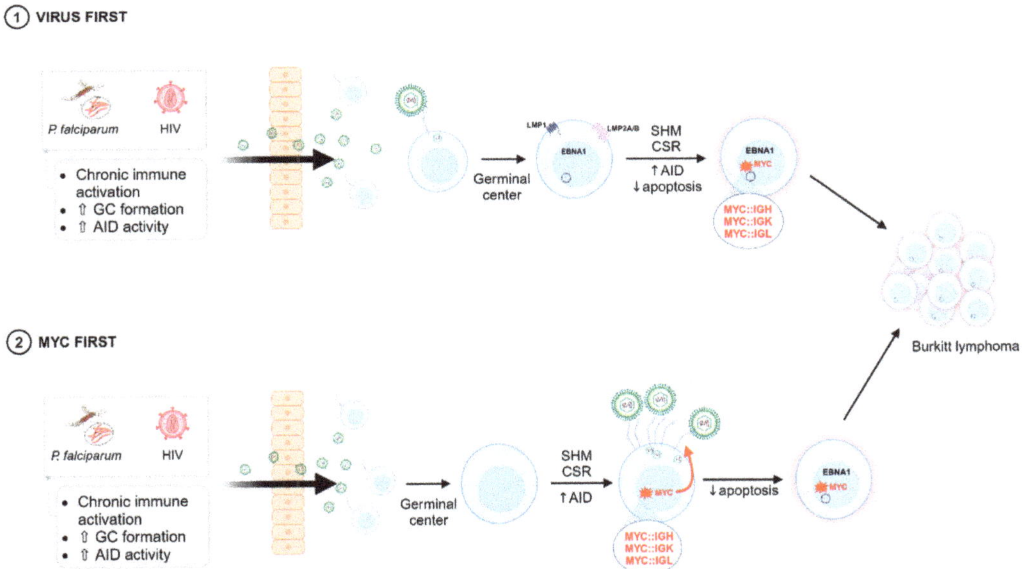

Figure 5. Two alternative models of EBV infection and *MYC* translocation in B cells prior to Burkitt lymphoma development. Chronic immune activation triggered by malaria or HIV leads to increased B cell proliferation, germinal center formation, and AID activity. In the *virus first* model, higher loads of EBV due to the weakened immune system favor EBV-mediated infection of B cells, leading to the increased susceptibility of MYC translocations during somatic hypermutation (SHM) and class switch recombination (CSR) in the germinal center. In the *MYC first* model, the chronic immune activation due to malaria or HIV makes the B cells prone to MYC translocations due to errors during SHM or CSR. Translocated *MYC* would lead to higher CR2 density on the surface of B cells, increasing the probability of EBV infection. In both models, EBV would confer the resistance mechanisms to apoptosis needed for the B cell to cope with the deregulated MYC activity, leading to Burkitt lymphoma development. Created in BioRender. Garcia, L. (2024) BioRender.com/y04j261.

It must be noted that the *virus fist* or *MYC first* mechanisms are not mutually exclusive. So, it is possible that one BL arises after the infection, and the virus would use the CR2 expressed at basal levels in the germinal center B cell. Although BL does not express EBV-transforming genes (e.g., EBNA2, LMP1), a "hit and run mechanism" by which these genes or the translocation-promoting EBNA3C is expressed after infection contributes to B cell proliferation and then becomes silenced in the growing lymphoma, cannot be ruled out. This may increase the pool of cells increasing the chance of *MYC* translocation. However, in other tumors, *MYC* translocation may come first, and the virus will find cells with a higher density of receptors in the membrane, which will be preferentially infected. Actually, at least one case of BL has been reported in which the EBV$^+$ cells and EBV$^-$ cells coexist in the same tumor, despite the fact that both cells show the same clonal origin [272].

Regardless of whether EBV is or is not the primary cause of BL, it is clearly a risk factor contributing to carcinogenesis even if the tumor is initiated by a *MYC* translocation event. A number of EBV-mediated mechanisms are described.

EBV confers resistance to apoptosis as described above (Section 6) and contributes to the metabolic reprogramming of B cells [273]. At least two viral genes have proven oncogenic properties (*EBNA2* and *LMP1*). Moreover, it has been recently shown that the latent EBV infection collaborates with Myc over-expression to induce BL-like lymphomas in mice [274].

EBV contributes to the scape from the immune surveillance of BL cells. This might also be the mechanism for the higher BL incidence in areas where malaria is endemic and for the HIV-associated BL.

Interestingly, lymphoblastoid cells (immortalized by EBV) are not tumorigenic in nude mice but when engineered to constitutively express a *MYC* gene, the cells become tumorigenic [275]. It is noteworthy that MYC represses the expression of EBV lytic genes by interacting with the EBV genome origin of lytic replication, preventing the expression of *BZLF1* and therefore the lytic cascade of gene expression [276]. Therefore, MYC would enhance viral infection on one hand but prevent the lytic cycle of infected cells on the other hand, thus contributing to the latency I program typical of BL cells.

11. Concluding Remarks

It is undisputed that the MYC oncogene, by chromosomal translocation to the immunoglobulin genes, is the hallmark and the cause for BL development. It is also established that a relevant fraction of BL is EBV$^+$. However, the molecular mechanisms that may explain the connection between EBV infection and *MYC* translocation are not yet clear. A role of MYC in BL pathogenesis is supported for the ample repertoire of biological effects elicited by MYC that impinge on cell proliferation. Indeed, MYC is deregulated in a majority of human cancers including lymphoma, and it is usually associated with a poorer outcome. However, in contrast to non-lymphoid tumors where translocations are rare, *MYC* is frequently translocated and deregulated in B cell lymphomas and in virtually all cases of Burkitt lymphoma. It is also clear that BL is associated to EBV infection in a relevant fraction of cases. This association reaches the 95% of the cases in the so-called endemic variant of the disease that occurs in geographical areas were malaria is prevalent, as equatorial Africa, whereas in other areas, the extent of this association is much less frequent. A debate regarding whether EBV is sufficient to trigger BL or if it is a risk factor exists. Some data support the most accepted mechanism, which is that viral infection of B cells in the lymph nodes is the triggering event that leads to *MYC* translocation followed by lymphomagenesis. However, other data argue about this "*virus first*" hypothesis and propose the "*MYC first*" hypothesis by which the *MYC* translocation is the first event and the antiapoptotic effect of EBV will critically favor lymphomagenesis. Moreover, the fact that MYC induces the expression of the virus receptor, CR2/CD21, will also favor the EBV infection in B cells already carrying the *MYC* translocation.

Author Contributions: Conceptualization, J.L. and L.G.-G.; writing—original draft preparation, S.S., J.L. and L.G.-G.; writing—review and editing, J.L. and L.G.-G.; visualization, J.L. and L.G.-G.; funding acquisition, J.L. All authors have read and agreed to the published version of the manuscript.

Funding: This research was funded by MICIU/AEI/10.13039/501100011033, grant number PID2020-115903GB-100 to J.L.

Acknowledgments: BioRender was used for the generation of most figures included in this manuscript (created in BioRender. Garcia, L. (2024) BioRender.com/y04j261).

Conflicts of Interest: The authors declare no conflicts of interest.

References

1. Epstein, M.A.; Achong, B.G.; Barr, Y.M. Virus particles in cultured lymphoblasts from Burkitt's lymphoma. *Lancet* **1964**, *1*, 702–703. [CrossRef] [PubMed]
2. Sample, J.; Young, L.; Martin, B.; Chatman, T.; Kieff, E.; Rickinson, A.; Kieff, E. Epstein-Barr virus types 1 and 2 differ in their EBNA-3A, EBNA-3B, and EBNA-3C genes. *J. Virol.* **1990**, *64*, 4084–4092. [CrossRef] [PubMed]
3. Gratama, J.W.; Ernberg, I. Molecular epidemiology of Epstein-Barr virus infection. *Adv. Cancer Res.* **1995**, *67*, 197–255. [PubMed]
4. Zimber, U.; Adldinger, H.K.; Lenoir, G.M.; Vuillaume, M.; Knebel-Doeberitz, M.V.; Laux, G.; Desgranges, C.; Wittmann, P.; Freese, U.K.; Schneider, U.; et al. Geographical prevalence of two types of Epstein-Barr virus. *Virology* **1986**, *154*, 56–66. [CrossRef]
5. de-Thé, G.; Day, N.E.; Geser, A.; Lavoué, M.F.; Ho, J.H.; Simons, M.J.; Sohier, R.; Tukei, P.; Vonka, V.; Zavadova, H. Sero-epidemiology of the Epstein-Barr virus: Preliminary analysis of an international study—A review. *IARC Sci. Publ.* **1975**, *11*, 3–16.
6. IARC Working Group on the Evaluation of Carcinogenic Risks to Humans. Epstein-Barr Virus and Kaposi's Sarcoma Herpesvirus/Human Herpesvirus 8. *IARC Monogr. Eval. Carcinog. Risks Hum.* **1997**, *70*, 1–492.
7. Young, L.S. Epstein–Barr Virus: General Features. In *Encylopedia of Virology*, 3rd ed.; Academic Press: Cambridge, MA, USA, 2008; pp. 148–157.
8. Dunmire, S.K.; Verghese, P.S.; Balfour, H.H., Jr. Primary Epstein-Barr virus infection. *J. Clin. Virol.* **2018**, *102*, 84–92. [CrossRef]
9. Liebowitz, D.K.E. Epstein-Barr virus. In *The HUman Herpesvirus*; Roizman, B., Whitley, R.J., Lopez, C., Eds.; Raven: New York, NY, USA, 1993; pp. 107–172.
10. Sathiyamoorthy, K.; Jiang, J.; Hu, Y.X.; Rowe, C.L.; Mohl, B.S.; Chen, J.; Jiang, W.; Mellins, E.D.; Longnecker, R.; Zhou, Z.H.; et al. Assembly and architecture of the EBV B cell entry triggering complex. *PLoS Pathog.* **2014**, *10*, e1004309. [CrossRef]
11. Lerner, M.R.; Andrews, N.C.; Miller, G.; Steitz, J.A. Two small RNAs encoded by Epstein-Barr virus and complexed with protein are precipitated by antibodies from patients with systemic lupus erythematosus. *Proc. Natl. Acad. Sci. USA* **1981**, *78*, 805–809. [CrossRef]
12. Pfeffer, S.; Zavolan, M.; Grasser, F.A.; Chien, M.; Russo, J.J.; Ju, J.; John, B.; Enright, A.J.; Marks, D.; Sander, C.; et al. Identification of virus-encoded microRNAs. *Science* **2004**, *304*, 734–736. [CrossRef]
13. Kintner, C.R.; Sugden, B. The structure of the termini of the DNA of Epstein-Barr virus. *Cell* **1979**, *17*, 661–671. [CrossRef] [PubMed]
14. Zimmermann, J.; Hammerschmidt, W. Structure and role of the terminal repeats of Epstein-Barr virus in processing and packaging of virion DNA. *J. Virol.* **1995**, *69*, 3147–3155. [CrossRef] [PubMed]
15. Jones, M.D.; Griffin, B.E. Clustered repeat sequences in the genome of Epstein Barr virus. *Nucleic Acids Res.* **1983**, *11*, 3919–3937. [CrossRef] [PubMed]
16. Cheung, A.; Kieff, E. Epstein-Barr virus DNA. X. Direct repeat within the internal direct repeat of Epstein-Barr virus DNA. *J. Virol.* **1981**, *40*, 501–507. [CrossRef]
17. Klein, G. Viral latency and transformation: The strategy of Epstein-Barr virus. *Cell* **1989**, *58*, 5–8. [CrossRef]
18. Munz, C. Latency and lytic replication in Epstein-Barr virus-associated oncogenesis. *Nat. Rev. Microbiol.* **2019**, *17*, 691–700. [CrossRef]
19. Kang, M.S.; Kieff, E. Epstein-Barr virus latent genes. *Exp. Mol. Med.* **2015**, *47*, e131. [CrossRef]
20. Torne, A.S.; Robertson, E.S. Epigenetic Mechanisms in Latent Epstein-Barr Virus Infection and Associated Cancers. *Cancers* **2024**, *16*, 991. [CrossRef]
21. Schaefer, B.C.; Strominger, J.L.; Speck, S.H. Redefining the Epstein-Barr virus-encoded nuclear antigen EBNA-1 gene promoter and transcription initiation site in group I Burkitt lymphoma cell lines. *Proc. Natl. Acad. Sci. USA* **1995**, *92*, 10565–10569. [CrossRef]
22. Schaefer, B.C.; Strominger, J.L.; Speck, S.H. A simple reverse transcriptase PCR assay to distinguish EBNA1 gene transcripts associated with type I and II latency from those arising during induction of the viral lytic cycle. *J. Virol.* **1996**, *70*, 8204–8208. [CrossRef]
23. Coates, P.J.; Slavin, G.; D'Ardenne, A.J. Persistence of Epstein-Barr virus in Reed-Sternberg cells throughout the course of Hodgkin's disease. *J. Pathol.* **1991**, *164*, 291–297. [CrossRef] [PubMed]
24. Schlager, S.; Speck, S.H.; Woisetschlager, M. Transcription of the Epstein-Barr virus nuclear antigen 1 (EBNA1) gene occurs before induction of the BCR2 (Cp) EBNA gene promoter during the initial stages of infection in B cells. *J. Virol.* **1996**, *70*, 3561–3570. [CrossRef] [PubMed]
25. Elliott, J.; Goodhew, E.B.; Krug, L.T.; Shakhnovsky, N.; Yoo, L.; Speck, S.H. Variable methylation of the Epstein-Barr virus Wp EBNA gene promoter in B-lymphoblastoid cell lines. *J. Virol.* **2004**, *78*, 14062–14065. [CrossRef]
26. Young, L.S.; Arrand, J.R.; Murray, P.G. *EBV Gene Expression and Regulation*; Cambridge University Press: Cambridge, UK, 2007.

27. Laichalk, L.L.; Thorley-Lawson, D.A. Terminal differentiation into plasma cells initiates the replicative cycle of Epstein-Barr virus in vivo. *J. Virol.* **2005**, *79*, 1296–1307. [CrossRef]
28. Thorley-Lawson, D.A. EBV Persistence--Introducing the Virus. *Curr. Top. Microbiol. Immunol.* **2015**, *390*, 151–209. [CrossRef]
29. Tempera, I.; Lieberman, P.M. Epigenetic regulation of EBV persistence and oncogenesis. *Semin. Cancer Biol.* **2014**, *26*, 22–29. [CrossRef]
30. Greenspan, J.S.; Greenspan, D.; Lennette, E.T.; Abrams, D.I.; Conant, M.A.; Petersen, V.; Freese, U.K. Replication of Epstein-Barr virus within the epithelial cells of oral "hairy" leukoplakia, an AIDS-associated lesion. *N. Engl. J. Med.* **1985**, *313*, 1564–1571. [CrossRef]
31. Rooney, C.M.; Rowe, D.T.; Ragot, T.; Farrell, P.J. The spliced BZLF1 gene of Epstein-Barr virus (EBV) transactivates an early EBV promoter and induces the virus productive cycle. *J. Virol.* **1989**, *63*, 3109–3116. [CrossRef]
32. Countryman, J.; Miller, G. Activation of expression of latent Epstein-Barr herpesvirus after gene transfer with a small cloned subfragment of heterogeneous viral DNA. *Proc. Natl. Acad. Sci. USA* **1985**, *82*, 4085–4089. [CrossRef]
33. Biggin, M.; Bodescot, M.; Perricaudet, M.; Farrell, P. Epstein-Barr virus gene expression in P3HR1-superinfected Raji cells. *J. Virol.* **1987**, *61*, 3120–3132. [CrossRef]
34. Sinclair, A.J.; Brimmell, M.; Shanahan, F.; Farrell, P.J. Pathways of activation of the Epstein-Barr virus productive cycle. *J. Virol.* **1991**, *65*, 2237–2244. [CrossRef] [PubMed]
35. Aubry, V.; Mure, F.; Mariamé, B.; Deschamps, T.; Wyrwicz, L.S.; Manet, E.; Gruffat, H. Epstein-Barr virus late gene transcription depends on the assembly of a virus-specific preinitiation complex. *J. Virol.* **2014**, *88*, 12825–12838. [CrossRef] [PubMed]
36. Fixman, E.D.; Hayward, G.S.; Hayward, S.D. trans-acting requirements for replication of Epstein-Barr virus ori-Lyt. *J. Virol.* **1992**, *66*, 5030–5039. [CrossRef]
37. Yap, L.F.; Wong, A.K.C.; Paterson, I.C.; Young, L.S. Functional Implications of Epstein-Barr Virus Lytic Genes in Carcinogenesis. *Cancers* **2022**, *14*, 5780. [CrossRef]
38. Kirschner, A.N.; Lowrey, A.S.; Longnecker, R.; Jardetzky, T.S. Binding-site interactions between Epstein-Barr virus fusion proteins gp42 and gH/gL reveal a peptide that inhibits both epithelial and B-cell membrane fusion. *J. Virol.* **2007**, *81*, 9216–9229. [CrossRef]
39. Backovic, M.; Jardetzky, T.S.; Longnecker, R. Hydrophobic residues that form putative fusion loops of Epstein-Barr virus glycoprotein B are critical for fusion activity. *J. Virol.* **2007**, *81*, 9596–9600. [CrossRef]
40. Omerovic, J.; Lev, L.; Longnecker, R. The amino terminus of Epstein-Barr virus glycoprotein gH is important for fusion with epithelial and B cells. *J. Virol.* **2005**, *79*, 12408–12415. [CrossRef]
41. Miller, N.; Hutt-Fletcher, L.M. Epstein-Barr virus enters B cells and epithelial cells by different routes. *J. Virol.* **1992**, *66*, 3409–3414. [CrossRef]
42. Abate, F.; Ambrosio, M.R.; Mundo, L.; Laginestra, M.A.; Fuligni, F.; Rossi, M.; Zairis, S.; Gazaneo, S.; De Falco, G.; Lazzi, S.; et al. Distinct Viral and Mutational Spectrum of Endemic Burkitt Lymphoma. *PLoS Pathog.* **2015**, *11*, e1005158. [CrossRef]
43. Nemerow, G.R.; Wolfert, R.; McNaughton, M.E.; Cooper, N.R. Identification and characterization of the Epstein-Barr virus receptor on human B lymphocytes and its relationship to the C3d complement receptor (CR2). *J. Virol.* **1985**, *55*, 347–351. [CrossRef]
44. Fingeroth, J.D.; Weis, J.J.; Tedder, T.F.; Strominger, J.L.; Biro, P.A.; Fearon, D.T. Epstein-Barr virus receptor of human B lymphocytes is the C3d receptor CR2. *Proc. Natl. Acad. Sci. USA* **1984**, *81*, 4510–4514. [CrossRef] [PubMed]
45. Rozaleen Dash, N.D. Complement receptor 2 (CR2/CD21). *Int. J. Res. Med. Sci.* **2017**, *5*, 4. [CrossRef]
46. Ogembo, J.G.; Kannan, L.; Ghiran, I.; Nicholson-Weller, A.; Finberg, R.W.; Tsokos, G.C.; Fingeroth, J.D. Human complement receptor type 1/CD35 is an Epstein-Barr Virus receptor. *Cell Rep.* **2013**, *3*, 371–385. [CrossRef]
47. Fearon, D.T.; Carter, R.H. The CD19/CR2/TAPA-1 complex of B lymphocytes: Linking natural to acquired immunity. *Annu. Rev. Immunol.* **1995**, *13*, 127–149. [CrossRef]
48. Lottin-Divoux, S.; Jean, D.; Le Romancer, M.; Frade, R. Activation of Epstein-Barr virus/C3d receptor (gp140, CR2, CD21) on human B lymphoma cell surface triggers Cbl tyrosine phosphorylation, its association with p85 subunit, Crk-L and Syk and its dissociation with Vav. *Cell Signal* **2006**, *18*, 1219–1225. [CrossRef]
49. Rickert, R.C.; Rajewsky, K.; Roes, J. Impairment of T-cell-dependent B-cell responses and B-1 cell development in CD19-deficient mice. *Nature* **1995**, *376*, 352–355. [CrossRef]
50. Engel, P.; Zhou, L.J.; Ord, D.C.; Sato, S.; Koller, B.; Tedder, T.F. Abnormal B lymphocyte development, activation, and differentiation in mice that lack or overexpress the CD19 signal transduction molecule. *Immunity* **1995**, *3*, 39–50. [CrossRef]
51. Ahearn, J.M.; Fischer, M.B.; Croix, D.; Goerg, S.; Ma, M.; Xia, J.; Zhou, X.; Howard, R.G.; Rothstein, T.L.; Carroll, M.C. Disruption of the Cr2 locus results in a reduction in B-1a cells and in an impaired B cell response to T-dependent antigen. *Immunity* **1996**, *4*, 251–262. [CrossRef]
52. Cherukuri, A.; Cheng, P.C.; Pierce, S.K. The role of the CD19/CD21 complex in B cell processing and presentation of complement-tagged antigens. *J. Immunol.* **2001**, *167*, 163–172. [CrossRef]
53. Nemerow, G.R.; Mold, C.; Schwend, V.K.; Tollefson, V.; Cooper, N.R. Identification of gp350 as the viral glycoprotein mediating attachment of Epstein-Barr virus (EBV) to the EBV/C3d receptor of B cells: Sequence homology of gp350 and C3 complement fragment C3d. *J. Virol.* **1987**, *61*, 1416–1420. [CrossRef]
54. Nemerow, G.R.; Houghten, R.A.; Moore, M.D.; Cooper, N.R. Identification of an epitope in the major envelope protein of Epstein-Barr virus that mediates viral binding to the B lymphocyte EBV receptor (CR2). *Cell* **1989**, *56*, 369–377. [CrossRef] [PubMed]

55. Tanner, J.; Weis, J.; Fearon, D.; Whang, Y.; Kieff, E. Epstein-Barr virus gp350/220 binding to the B lymphocyte C3d receptor mediates adsorption, capping, and endocytosis. *Cell* **1987**, *50*, 203–213. [CrossRef] [PubMed]
56. Kirschner, A.N.; Sorem, J.; Longnecker, R.; Jardetzky, T.S. Structure of Epstein-Barr virus glycoprotein 42 suggests a mechanism for triggering receptor-activated virus entry. *Structure* **2009**, *17*, 223–233. [CrossRef] [PubMed]
57. Mullen, M.M.; Haan, K.M.; Longnecker, R.; Jardetzky, T.S. Structure of the Epstein-Barr virus gp42 protein bound to the MHC class II receptor HLA-DR1. *Mol. Cell* **2002**, *9*, 375–385. [CrossRef]
58. Adamson, A.L.; Kenney, S.C. Rescue of the Epstein-Barr virus BZLF1 mutant, Z(S186A), early gene activation defect by the BRLF1 gene product. *Virology* **1998**, *251*, 187–197. [CrossRef]
59. Janz, A.; Oezel, M.; Kurzeder, C.; Mautner, J.; Pich, D.; Kost, M.; Hammerschmidt, W.; Delecluse, H.J. Infectious Epstein-Barr virus lacking major glycoprotein BLLF1 (gp350/220) demonstrates the existence of additional viral ligands. *J. Virol.* **2000**, *74*, 10142–10152. [CrossRef]
60. Birkenbach, M.; Tong, X.; Bradbury, L.E.; Tedder, T.F.; Kieff, E. Characterization of an Epstein-Barr virus receptor on human epithelial cells. *J. Exp. Med.* **1992**, *176*, 1405–1414. [CrossRef]
61. Chesnokova, L.S.; Nishimura, S.L.; Hutt-Fletcher, L.M. Fusion of epithelial cells by Epstein-Barr virus proteins is triggered by binding of viral glycoproteins gHgL to integrins alphavbeta6 or alphavbeta8. *Proc. Natl. Acad. Sci. USA* **2009**, *106*, 20464–20469. [CrossRef]
62. Chesnokova, L.S.; Hutt-Fletcher, L.M. Fusion of Epstein-Barr virus with epithelial cells can be triggered by alphavbeta5 in addition to alphavbeta6 and alphavbeta8, and integrin binding triggers a conformational change in glycoproteins gHgL. *J. Virol.* **2011**, *85*, 13214–13223. [CrossRef]
63. Wang, H.B.; Zhang, H.; Zhang, J.P.; Li, Y.; Zhao, B.; Feng, G.K.; Du, Y.; Xiong, D.; Zhong, Q.; Liu, W.L.; et al. Neuropilin 1 is an entry factor that promotes EBV infection of nasopharyngeal epithelial cells. *Nat. Commun.* **2015**, *6*, 6240. [CrossRef]
64. Coller, H.A.; Grandori, C.; Tamayo, P.; Colbert, T.; Lander, E.S.; Eisenman, R.N.; Golub, T.R. Expression analysis with oligonucleotide microarrays reveals that MYC regulates genes involved in growth, cell cycle, signaling, and adhesion. *Proc. Natl. Acad. Sci. USA* **2000**, *97*, 3260–3265. [CrossRef] [PubMed]
65. Chen, J.; Sathiyamoorthy, K.; Zhang, X.; Schaller, S.; Perez White, B.E.; Jardetzky, T.S.; Longnecker, R. Ephrin receptor A2 is a functional entry receptor for Epstein-Barr virus. *Nat. Microbiol.* **2018**, *3*, 172–180. [CrossRef] [PubMed]
66. Dunmire, S.K.; Grimm, J.M.; Schmeling, D.O.; Balfour, H.H., Jr.; Hogquist, K.A. The Incubation Period of Primary Epstein-Barr Virus Infection: Viral Dynamics and Immunologic Events. *PLoS Pathog.* **2015**, *11*, e1005286. [CrossRef] [PubMed]
67. Tugizov, S.M.; Berline, J.W.; Palefsky, J.M. Epstein-Barr virus infection of polarized tongue and nasopharyngeal epithelial cells. *Nat. Med.* **2003**, *9*, 307–314. [CrossRef]
68. Krabbe, S.; Hesse, J.; Uldall, P. Primary Epstein-Barr virus infection in early childhood. *Arch. Dis. Child.* **1981**, *56*, 49–52. [CrossRef]
69. Jayasooriya, S.; de Silva, T.I.; Njie-jobe, J.; Sanyang, C.; Leese, A.M.; Bell, A.I.; McAulay, K.A.; Yanchun, P.; Long, H.M.; Dong, T.; et al. Early virological and immunological events in asymptomatic Epstein-Barr virus infection in African children. *PLoS Pathog.* **2015**, *11*, e1004746. [CrossRef]
70. Condon, L.M.; Cederberg, L.E.; Rabinovitch, M.D.; Liebo, R.V.; Go, J.C.; Delaney, A.S.; Schmeling, D.O.; Thomas, W.; Balfour, H.H., Jr. Age-specific prevalence of Epstein-Barr virus infection among Minnesota children: Effects of race/ethnicity and family environment. *Clin. Infect. Dis.* **2014**, *59*, 501–508. [CrossRef]
71. Balfour, H.H., Jr.; Sifakis, F.; Sliman, J.A.; Knight, J.A.; Schmeling, D.O.; Thomas, W. Age-specific prevalence of Epstein-Barr virus infection among individuals aged 6–19 years in the United States and factors affecting its acquisition. *J. Infect. Dis.* **2013**, *208*, 1286–1293. [CrossRef]
72. Hislop, A.D.; Annels, N.E.; Gudgeon, N.H.; Leese, A.M.; Rickinson, A.B. Epitope-specific evolution of human $CD8^+$ T cell responses from primary to persistent phases of Epstein-Barr virus infection. *J. Exp. Med.* **2002**, *195*, 893–905. [CrossRef]
73. Azzi, T.; Lunemann, A.; Murer, A.; Ueda, S.; Beziat, V.; Malmberg, K.J.; Staubli, G.; Gysin, C.; Berger, C.; Munz, C.; et al. Role for early-differentiated natural killer cells in infectious mononucleosis. *Blood* **2014**, *124*, 2533–2543. [CrossRef]
74. Taylor, G.S.; Long, H.M.; Brooks, J.M.; Rickinson, A.B.; Hislop, A.D. The immunology of Epstein-Barr virus-induced disease. *Annu. Rev. Immunol.* **2015**, *33*, 787–821. [CrossRef] [PubMed]
75. Downey, H.; Mckinlay, C.A. Acute lymphadenosis compared with acute lymphatic leukemia. *JAMA Intern. Med.* **1923**, *32*, 82–112. [CrossRef]
76. Leung, A.K.C.; Lam, J.M.; Barankin, B. Infectious Mononucleosis: An Updated Review. *Curr. Pediatr. Rev.* **2024**, *20*, 305–322. [CrossRef]
77. Carbajal, K.S.; Mironova, Y.; Ulrich-Lewis, J.T.; Kulkarni, D.; Grifka-Walk, H.M.; Huber, A.K.; Shrager, P.; Giger, R.J.; Segal, B.M. Th Cell Diversity in Experimental Autoimmune Encephalomyelitis and Multiple Sclerosis. *J. Immunol.* **2015**, *195*, 2552–2559. [CrossRef]
78. Ascherio, A.; Munger, K.L.; Lünemann, J.D. The initiation and prevention of multiple sclerosis. *Nat. Rev. Neurol.* **2012**, *8*, 602–612. [CrossRef]
79. Burnard, S.; Lechner-Scott, J.; Scott, R.J. EBV and MS: Major cause, minor contribution or red-herring? *Mult. Scler. Relat. Disord.* **2017**, *16*, 24–30. [CrossRef]

80. Banwell, B.; Krupp, L.; Kennedy, J.; Tellier, R.; Tenembaum, S.; Ness, J.; Belman, A.; Boiko, A.; Bykova, O.; Waubant, E.; et al. Clinical features and viral serologies in children with multiple sclerosis: A multinational observational study. *Lancet Neurol.* **2007**, *6*, 773–781. [CrossRef]
81. Mouhieddine, T.H.; Darwish, H.; Fawaz, L.; Yamout, B.; Tamim, H.; Khoury, S.J. Risk factors for multiple sclerosis and associations with anti-EBV antibody titers. *Clin. Immunol.* **2015**, *158*, 59–66. [CrossRef]
82. Pakpoor, J.; Disanto, G.; Gerber, J.E.; Dobson, R.; Meier, U.C.; Giovannoni, G.; Ramagopalan, S.V. The risk of developing multiple sclerosis in individuals seronegative for Epstein-Barr virus: A meta-analysis. *Mult. Scler. J.* **2013**, *19*, 162–166. [CrossRef]
83. De Jager, P.L.; Simon, K.C.; Munger, K.L.; Rioux, J.D.; Hafler, D.A.; Ascherio, A. Integrating risk factors: HLA-DRB1*1501 and Epstein-Barr virus in multiple sclerosis. *Neurology* **2008**, *70*, 1113–1118. [CrossRef]
84. Sundström, P.; Nyström, M.; Ruuth, K.; Lundgren, E. Antibodies to specific EBNA-1 domains and HLA DRB1*1501 interact as risk factors for multiple sclerosis. *J. Neuroimmunol.* **2009**, *215*, 102–107. [CrossRef] [PubMed]
85. Pender, M.P. The essential role of Epstein-Barr virus in the pathogenesis of multiple sclerosis. *Neuroscientist* **2011**, *17*, 351–367. [CrossRef] [PubMed]
86. Bagert, B.A. Epstein-Barr virus in multiple sclerosis. *Curr. Neurol. Neurosci. Rep.* **2009**, *9*, 405–410. [CrossRef] [PubMed]
87. Peferoen, L.A.N.; Lamers, F.; Lodder, L.N.R.; Gerritsen, W.H.; Huitinga, I.; Melief, J.; Giovannoni, G.; Meier, U.; Hintzen, R.Q.; Verjans, G.M.G.M.; et al. Epstein Barr virus is not a characteristic feature in the central nervous system in established multiple sclerosis. *Brain* **2009**, *133*, e137. [CrossRef]
88. Serafini, B.; Severa, M.; Columba-Cabezas, S.; Rosicarelli, B.; Veroni, C.; Chiappetta, G.; Magliozzi, R.; Reynolds, R.; Coccia, E.M.; Aloisi, F. Epstein-Barr virus latent infection and BAFF expression in B cells in the multiple sclerosis brain: Implications for viral persistence and intrathecal B-cell activation. *J. Neuropathol. Exp. Neurol.* **2010**, *69*, 677–693. [CrossRef]
89. Tzartos, J.S.; Khan, G.; Vossenkamper, A.; Cruz-Sadaba, M.; Lonardi, S.; Sefia, E.; Meager, A.; Elia, A.; Middeldorp, J.M.; Clemens, M.; et al. Association of innate immune activation with latent Epstein-Barr virus in active MS lesions. *Neurology* **2012**, *78*, 15–23. [CrossRef]
90. Sargsyan, S.A.; Shearer, A.J.; Ritchie, A.M.; Burgoon, M.P.; Anderson, S.; Hemmer, B.; Stadelmann, C.; Gattenlöhner, S.; Owens, G.P.; Gilden, D.; et al. Absence of Epstein-Barr virus in the brain and CSF of patients with multiple sclerosis. *Neurology* **2010**, *74*, 1127–1135. [CrossRef]
91. Lomakin, Y.; Arapidi, G.P.; Chernov, A.; Ziganshin, R.; Tcyganov, E.; Lyadova, I.; Butenko, I.O.; Osetrova, M.; Ponomarenko, N.; Telegin, G.; et al. Exposure to the Epstein-Barr Viral Antigen Latent Membrane Protein 1 Induces Myelin-Reactive Antibodies In Vivo. *Front. Immunol.* **2017**, *8*, 777. [CrossRef]
92. Jog, N.R.; McClain, M.T.; Heinlen, L.D.; Gross, T.; Towner, R.; Guthridge, J.M.; Axtell, R.C.; Pardo, G.; Harley, J.B.; James, J.A. Epstein Barr virus nuclear antigen 1 (EBNA-1) peptides recognized by adult multiple sclerosis patient sera induce neurologic symptoms in a murine model. *J. Autoimmun.* **2020**, *106*, 102332. [CrossRef]
93. Wucherpfennig, K.W.; Strominger, J.L. Molecular mimicry in T cell-mediated autoimmunity: Viral peptides activate human T cell clones specific for myelin basic protein. *Cell* **1995**, *80*, 695–705. [CrossRef]
94. Lünemann, J.D.; Jelčić, I.; Roberts, S.; Lutterotti, A.; Tackenberg, B.r.; Martin, R.; Münz, C. EBNA1-specific T cells from patients with multiple sclerosis cross react with myelin antigens and co-produce IFN-γ and IL-2. *J. Exp. Med.* **2008**, *205*, 1763–1773. [CrossRef] [PubMed]
95. Lindsey, J.W. Antibodies to the Epstein-Barr virus proteins BFRF3 and BRRF2 cross-react with human proteins. *J. Neuroimmunol.* **2017**, *310*, 131–134. [CrossRef] [PubMed]
96. Zhao, Y.; Zhang, Q.; Zhang, B.; Dai, Y.; Gao, Y.; Li, C.; Yu, Y.; Li, C. Epstein-Barr Viruses: Their Immune Evasion Strategies and Implications for Autoimmune Diseases. *Int. J. Mol. Sci.* **2024**, *25*, 8160. [CrossRef] [PubMed]
97. Toussirot, E.; Roudier, J. Epstein-Barr virus in autoimmune diseases. *Best. Pract. Res. Clin. Rheumatol.* **2008**, *22*, 883–896. [CrossRef]
98. Roudier, J.; Petersen, J.; Rhodes, G.H.; Luka, J.; Carson, D.A. Susceptibility to rheumatoid arthritis maps to a T-cell epitope shared by the HLA-Dw4 DR beta-1 chain and the Epstein-Barr virus glycoprotein gp110. *Proc. Natl. Acad. Sci. USA* **1989**, *86*, 5104–5108. [CrossRef]
99. Draborg, A.H.; Duus, K.; Houen, G. Epstein-Barr virus and systemic lupus erythematosus. *Clin. Dev. Immunol.* **2012**, *2012*, 370516. [CrossRef]
100. James, J.A.; Harley, J.B.; Scofield, R.H. Epstein-Barr virus and systemic lupus erythematosus. *Curr. Opin. Rheumatol.* **2006**, *18*, 462–467. [CrossRef]
101. Fitzsimmons, L.; Cartlidge, R.; Chang, C.; Sejic, N.; Galbraith, L.C.A.; Suraweera, C.D.; Croom-Carter, D.; Dewson, G.; Tierney, R.J.; Bell, A.I.; et al. EBV BCL-2 homologue BHRF1 drives chemoresistance and lymphomagenesis by inhibiting multiple cellular pro-apoptotic proteins. *Cell Death Differ.* **2020**, *27*, 1554–1568. [CrossRef]
102. Rueda Domínguez, A.; Cirauqui, B.; García Castaño, A.; Alvarez Cabellos, R.; Carral Maseda, A.; Castelo Fernández, B.; Iglesias Rey, L.; Rubió-Casadevall, J.; Arrazubi, V.; Mesía, R. SEOM-TTCC clinical guideline in nasopharynx cancer (2021). *Clin. Transl. Oncol.* **2022**, *24*, 670–680. [CrossRef]
103. Pathmanathan, R.; Prasad, U.; Chandrika, G.; Sadler, R.; Flynn, K.; Raab-Traub, N. Undifferentiated, nonkeratinizing, and squamous cell carcinoma of the nasopharynx. Variants of Epstein-Barr virus-infected neoplasia. *Am. J. Pathol.* **1995**, *146*, 1355–1367.
104. Chen, Y.-P.; Chan, A.T.C.; Le, Q.-T.; Blanchard, P.; Sun, Y.; Ma, J. Nasopharyngeal carcinoma. *Lancet* **2019**, *394*, 64–80. [CrossRef] [PubMed]

105. Bray, F.; Laversanne, M.; Sung, H.; Ferlay, J.; Siegel, R.L.; Soerjomataram, I.; Jemal, A. Global cancer statistics 2022: GLOBOCAN estimates of incidence and mortality worldwide for 36 cancers in 185 countries. *CA Cancer J. Clin.* **2024**, *74*, 229–263. [CrossRef] [PubMed]
106. Chan, A.S.C.; To, K.F.; Lo, K.W.; Ding, M.; Li, X.; Johnson, P.; Huang, D.P. Frequent chromosome 9p losses in histologically normal nasopharyngeal epithelia from southern Chinese. *Int. J. Cancer* **2002**, *102*, 300–303. [CrossRef] [PubMed]
107. Chan, A.S.C.; To, K.F.; Lo, K.W.; Mak, K.F.; Pak, W.; Chiu, B.; Tse, G.M.K.; Ding, M.; Li, X.; Lee, J.C.K.; et al. High Frequency of Chromosome 3p Deletion in Histologically Normal Nasopharyngeal Epithelia from Southern Chinese1. *Cancer Res.* **2000**, *60*, 5365–5370.
108. Lo, K.-W.; Chung, G.T.-Y.; To, K.-F. Deciphering the molecular genetic basis of NPC through molecular, cytogenetic, and epigenetic approaches. *Semin. Cancer Biol.* **2012**, *22*, 79–86. [CrossRef]
109. Su, Z.Y.; Siak, P.Y.; Leong, C.O.; Cheah, S.C. The role of Epstein-Barr virus in nasopharyngeal carcinoma. *Front. Microbiol.* **2023**, *14*, 1116143. [CrossRef]
110. Thornburg, N.J.; Raab-Traub, N. Induction of Epidermal Growth Factor Receptor Expression by Epstein-Barr Virus Latent Membrane Protein 1 C-Terminal-Activating Region 1 Is Mediated by NF-κB p50 Homodimer/Bcl-3 Complexes. *J. Virol.* **2007**, *81*, 12954–12961. [CrossRef]
111. Hu, L.-F.; Chen, F.; Zhen, Q.-F.; Zhang, Y.-W.; Luo, Y.; Zheng, X.; Winberg, G.; Ernberg, I.; Klein, G. Differences in the growth pattern and clinical course of EBV-LMP1 expressing and non-expressing nasopharyngeal carcinomas. *Eur. J. Cancer* **1995**, *31*, 658–660. [CrossRef]
112. Guo, X.; Li, T.; Li, F.; Xu, Y.; Wang, H.; Cheng, W.; Tang, J.; Zhou, G.; Chen, H.; Ng, M.; et al. Intermittent abortive reactivation of Epstein-Barr virus during the progression of nasopharyngeal cancer as indicated by elevated antibody levels. *Oral. Oncol.* **2019**, *93*, 85–90. [CrossRef]
113. Chan, J.K.; Yip, T.T.; Tsang, W.Y.; Poon, Y.F.; Wong, C.S.; Ma, V.W. Specific association of Epstein-Barr virus with lymphoepithelial carcinoma among tumors and tumorlike lesions of the salivary gland. *Arch. Pathol. Lab. Med.* **1994**, *118*, 994–997.
114. Leung, S.Y.; Chung, L.P.; Yuen, S.T.; Ho, C.M.; Wong, M.P.; Chan, S.Y. Lymphoepithelial carcinoma of the salivary gland: In situ detection of Epstein-Barr virus. *J. Clin. Pathol.* **1995**, *48*, 1022–1027. [CrossRef] [PubMed]
115. Gallo, O.; Santucci, M.; Calzolari, A.; Storchi, O.F. Epstein-Barr virus (EBV) infection and undifferentiated carcinoma of the parotid gland in Caucasian patients. *Acta Otolaryngol.* **1994**, *114*, 572–575. [CrossRef] [PubMed]
116. Zong, Y.; Liu, K.; Zhong, B.; Chen, G.; Wu, W. Epstein-Barr virus infection of sinonasal lymphoepithelial carcinoma in Guangzhou. *Chin. Med. J.* **2001**, *114*, 132–136.
117. MacMillan, C.; Kapadia, S.B.; Finkelstein, S.D.; Nalesnik, M.A.; Barnes, L. Lymphoepithelial carcinoma of the larynx and hypopharynx: Study of eight cases with relationship to Epstein-Barr virus and p53 gene alterations, and review of the literature. *Hum. Pathol.* **1996**, *27*, 1172–1179. [CrossRef]
118. Kermani, W.; Belcadhi, M.; Sriha, B.; Abdelkéfi, M. Epstein-Barr virus-associated lymphoepithelial carcinoma of the larynx. *Eur. Ann. Otorhinolaryngol. Head. Neck Dis.* **2015**, *132*, 231–233. [CrossRef]
119. Coskun, B.U.; Cinar, U.; Sener, B.M.; Dadas, B. Lymphoepithelial carcinoma of the larynx. *Auris Nasus Larynx* **2005**, *32*, 189–193. [CrossRef]
120. Carpenter, D.H.; El-Mofty, S.K.; Lewis, J.S., Jr. Undifferentiated carcinoma of the oropharynx: A human papillomavirus-associated tumor with a favorable prognosis. *Mod. Pathol.* **2011**, *24*, 1306–1312. [CrossRef]
121. Smyth, E.C.; Nilsson, M.; Grabsch, H.I.; van Grieken, N.C.T.; Lordick, F. Gastric cancer. *Lancet* **2020**, *396*, 635–648. [CrossRef]
122. Yusefi, A.R.; Bagheri Lankarani, K.; Bastani, P.; Radinmanesh, M.; Kavosi, Z. Risk Factors for Gastric Cancer: A Systematic Review. *Asian Pac. J. Cancer Prev.* **2018**, *19*, 591–603. [CrossRef]
123. Burke, A.P.; Yen, T.S.; Shekitka, K.M.; Sobin, L.H. Lymphoepithelial carcinoma of the stomach with Epstein-Barr virus demonstrated by polymerase chain reaction. *Mod. Pathol.* **1990**, *3*, 377–380.
124. Yang, J.; Liu, Z.; Zeng, B.; Hu, G.; Gan, R. Epstein–Barr virus-associated gastric cancer: A distinct subtype. *Cancer Lett.* **2020**, *495*, 191–199. [CrossRef] [PubMed]
125. Shibata, D.; Weiss, L.M. Epstein-Barr virus-associated gastric adenocarcinoma. *Am. J. Pathol.* **1992**, *140*, 769–774. [PubMed]
126. Takada, K. Epstein-Barr virus and gastric carcinoma. *Mol. Pathol.* **2000**, *53*, 255–261. [CrossRef] [PubMed]
127. Bass, A.J.; Thorsson, V.; Shmulevich, I.; Reynolds, S.M.; Miller, M.; Bernard, B.; Hinoue, T.; Laird, P.W.; Curtis, C.; Shen, H.; et al. Comprehensive molecular characterization of gastric adenocarcinoma. *Nature* **2014**, *513*, 202–209. [CrossRef]
128. Imai, S.; Nishikawa, J.; Takada, K. Cell-to-Cell Contact as an Efficient Mode of Epstein-Barr Virus Infection of Diverse Human Epithelial Cells. *J. Virol.* **1998**, *72*, 4371–4378. [CrossRef] [PubMed]
129. Imai, S.; Koizumi, S.; Sugiura, M.; Tokunaga, M.; Uemura, Y.; Yamamoto, N.; Tanaka, S.; Sato, E.; Osato, T. Gastric carcinoma: Monoclonal epithelial malignant cells expressing Epstein-Barr virus latent infection protein. *Proc. Natl. Acad. Sci. USA* **1994**, *91*, 9131–9135. [CrossRef] [PubMed]
130. Salnikov, M.Y.; MacNeil, K.M.; Mymryk, J.S. The viral etiology of EBV-associated gastric cancers contributes to their unique pathology, clinical outcomes, treatment responses and immune landscape. *Front. Immunol.* **2024**, *15*, 1358511. [CrossRef] [PubMed]

131. Sugiura, M.; Imai, S.; Tokunaga, M.; Koizumi, S.; Uchizawa, M.; Okamoto, K.; Osato, T. Transcriptional analysis of Epstein-Barr virus gene expression in EBV-positive gastric carcinoma: Unique viral latency in the tumour cells. *Br. J. Cancer* **1996**, *74*, 625–631. [CrossRef]
132. Luo, B.; Wang, Y.; Wang, X.F.; Liang, H.; Yan, L.P.; Huang, B.H.; Zhao, P. Expression of Epstein-Barr virus genes in EBV-associated gastric carcinomas. *World J. Gastroenterol.* **2005**, *11*, 629–633. [CrossRef]
133. Sun, K.; Jia, K.; Lv, H.; Wang, S.Q.; Wu, Y.; Lei, H.; Chen, X. EBV-Positive Gastric Cancer: Current Knowledge and Future Perspectives. *Front. Oncol.* **2020**, *10*, 583463. [CrossRef]
134. Harabuchi, Y.; Yamanaka, N.; Kataura, A.; Imai, S.; Kinoshita, T.; Mizuno, F.; Osato, T. Epstein-Barr virus in nasal T-cell lymphomas in patients with lethal midline granuloma. *Lancet* **1990**, *335*, 128–130. [CrossRef] [PubMed]
135. Coleman, C.B.; Wohlford, E.M.; Smith, N.A.; King, C.A.; Ritchie, J.A.; Baresel, P.C.; Kimura, H.; Rochford, R. Epstein-Barr virus type 2 latently infects T cells, inducing an atypical activation characterized by expression of lymphotactic cytokines. *J. Virol.* **2015**, *89*, 2301–2312. [CrossRef] [PubMed]
136. Smith, N.A.; Coleman, C.B.; Gewurz, B.E.; Rochford, R. CD21 (Complement Receptor 2) is the receptor for Epstein-Barr virus entry into T cells. *J. Virol.* **2020**, *94*, 1110–1128. [CrossRef] [PubMed]
137. Fox, C.P.; Civallero, M.; Ko, Y.-H.; Manni, M.; Skrypets, T.; Pileri, S.; Kim, S.J.; Cabrera, M.E.; Shustov, A.R.; Chiattone, C.S.; et al. Survival outcomes of patients with extranodal natural-killer T-cell lymphoma: A prospective cohort study from the international T-cell Project. *Lancet Haematol.* **2020**, *7*, e284–e294. [CrossRef]
138. Takahara, M.; Kumai, T.; Kishibe, K.; Nagato, T.; Harabuchi, Y. Extranodal NK/T-Cell Lymphoma, Nasal Type: Genetic, Biologic, and Clinical Aspects with a Central Focus on Epstein–Barr Virus Relation. *Microorganisms* **2021**, *9*, 1381. [CrossRef]
139. Peng, R.J.; Han, B.W.; Cai, Q.Q.; Zuo, X.Y.; Xia, T.; Chen, J.R.; Feng, L.N.; Lim, J.Q.; Chen, S.W.; Zeng, M.S.; et al. Genomic and transcriptomic landscapes of Epstein-Barr virus in extranodal natural killer T-cell lymphoma. *Leukemia* **2019**, *33*, 1451–1462. [CrossRef]
140. Gru, A.A.; Haverkos, B.H.; Freud, A.G.; Hastings, J.; Nowacki, N.B.; Barrionuevo, C.; Vigil, C.E.; Rochford, R.; Natkunam, Y.; Baiocchi, R.A.; et al. The Epstein-Barr Virus (EBV) in T Cell and NK Cell Lymphomas: Time for a Reassessment. *Curr. Hematol. Malig. Rep.* **2015**, *10*, 456–467. [CrossRef]
141. Ng, S.B.; Selvarajan, V.; Huang, G.; Zhou, J.; Feldman, A.L.; Law, M.; Kwong, Y.L.; Shimizu, N.; Kagami, Y.; Aozasa, K.; et al. Activated oncogenic pathways and therapeutic targets in extranodal nasal-type NK/T cell lymphoma revealed by gene expression profiling. *J. Pathol.* **2011**, *223*, 496–510. [CrossRef]
142. Alaggio, R.; Amador, C.; Anagnostopoulos, I.; Attygalle, A.D.; Araujo, I.B.d.O.; Berti, E.; Bhagat, G.; Borges, A.M.; Boyer, D.; Calaminici, M.; et al. The 5th edition of the World Health Organization Classification of Haematolymphoid Tumours: Lymphoid Neoplasms. *Leukemia* **2022**, *36*, 1720–1748. [CrossRef]
143. Nohtani, M.; Vrzalikova, K.; Ibrahim, M.; Powell, J.E.; Fennell, É.; Morgan, S.; Grundy, R.; McCarthy, K.; Dewberry, S.; Bouchal, J.; et al. Impact of Tumour Epstein-Barr Virus Status on Clinical Outcome in Patients with Classical Hodgkin Lymphoma (cHL): A Review of the Literature and Analysis of a Clinical Trial Cohort of Children with cHL. *Cancers* **2022**, *14*, 4297. [CrossRef]
144. Vrzalikova, K.; Pugh, M.; Mundo, L.; Murray, P. The contribution of ebv to the pathogenesis of classical hodgkin lymphoma. *Ann. Lymphoma* **2021**, *5*, 1–13. [CrossRef]
145. Anagnostopoulos, I.; Herbst, H.; Niedobitek, G.; Stein, H. Demonstration of monoclonal EBV genomes in Hodgkin's disease and Ki-1-positive anaplastic large cell lymphoma by combined Southern blot and in situ hybridization. *Blood* **1989**, *74*, 810–816. [CrossRef]
146. Marafioti, T.; Hummel, M.; Foss, H.-D.; Laumen, H.; Korbjuhn, P.; Anagnostopoulos, I.; Lammert, H.; Demel, G.; Theil, J.; Wirth, T.; et al. Hodgkin and Reed-Sternberg cells represent an expansion of a single clone originating from a germinal center B-cell with functional immunoglobulin gene rearrangements but defective immunoglobulin transcription. *Blood* **2000**, *95*, 1443–1450. [CrossRef]
147. Kanzler, H.; Küppers, R.; Hansmann, M.L.; Rajewsky, K. Hodgkin and Reed-Sternberg cells in Hodgkin's disease represent the outgrowth of a dominant tumor clone derived from (crippled) germinal center B cells. *J. Exp. Med.* **1996**, *184*, 1495–1505. [CrossRef]
148. Murray, P.G.; Young, L.S. An etiological role for the Epstein-Barr virus in the pathogenesis of classical Hodgkin lymphoma. *Blood* **2019**, *134*, 591–596. [CrossRef]
149. Chabay, P. Advances in the Pathogenesis of EBV-Associated Diffuse Large B Cell Lymphoma. *Cancers* **2021**, *13*, 2717. [CrossRef]
150. Cohen, M.; De Matteo, E.; Narbaitz, M.; Carreño, F.A.; Preciado, M.V.; Chabay, P.A. Epstein-Barr virus presence in pediatric diffuse large B-cell lymphoma reveals a particular association and latency patterns: Analysis of viral role in tumor microenvironment. *Int. J. Cancer* **2013**, *132*, 1572–1580. [CrossRef]
151. Gebauer, N.; Künstner, A.; Ketzer, J.; Witte, H.M.; Rausch, T.; Benes, V.; Zimmermann, J.; Gebauer, J.; Merz, H.; Bernard, V.; et al. Genomic insights into the pathogenesis of Epstein-Barr virus-associated diffuse large B-cell lymphoma by whole-genome and targeted amplicon sequencing. *Blood Cancer J.* **2021**, *11*, 102. [CrossRef]
152. Nijland, M.L.; Kersten, M.J.; Pals, S.T.; Bemelman, F.J.; Ten Berge, I.J. Epstein-Barr Virus-Positive Posttransplant Lymphoproliferative Disease After Solid Organ Transplantation: Pathogenesis, Clinical Manifestations, Diagnosis, and Management. *Transplant. Direct* **2016**, *2*, e48. [CrossRef]

153. Al Hamed, R.; Bazarbachi, A.H.; Mohty, M. Epstein-Barr virus-related post-transplant lymphoproliferative disease (EBV-PTLD) in the setting of allogeneic stem cell transplantation: A comprehensive review from pathogenesis to forthcoming treatment modalities. *Bone Marrow Transplant.* **2020**, *55*, 25–39. [CrossRef]
154. Allen, U.; Preiksaitis, J. Epstein-Barr Virus and Posttransplant Lymphoproliferative Disorder in Solid Organ Transplant Recipients. *Am. J. Transplant.* **2009**, *9*, S87–S96. [CrossRef]
155. Brady, G.; MacArthur, G.J.; Farrell, P.J. Epstein-Barr virus and Burkitt lymphoma. *J. Clin. Pathol.* **2007**, *60*, 1397–1402. [CrossRef]
156. Thorley-Lawson, D.A.; Allday, M.J. The curious case of the tumour virus: 50 years of Burkitt's lymphoma. *Nat. Rev. Microbiol.* **2008**, *6*, 913–924. [CrossRef]
157. Mancao, C.; Hammerschmidt, W. Epstein-Barr virus latent membrane protein 2A is a B-cell receptor mimic and essential for B-cell survival. *Blood* **2007**, *110*, 3715–3721. [CrossRef]
158. Kelly, G.L.; Long, H.M.; Stylianou, J.; Thomas, W.A.; Leese, A.; Bell, A.I.; Bornkamm, G.W.; Mautner, J.; Rickinson, A.B.; Rowe, M. An Epstein-Barr virus anti-apoptotic protein constitutively expressed in transformed cells and implicated in burkitt lymphomagenesis: The Wp/BHRF1 link. *PLoS Pathog.* **2009**, *5*, e1000341. [CrossRef]
159. Kelly, G.L.; Milner, A.E.; Baldwin, G.S.; Bell, A.I.; Rickinson, A.B. Three restricted forms of Epstein-Barr virus latency counteracting apoptosis in c-myc-expressing Burkitt lymphoma cells. *Proc. Natl. Acad. Sci. USA* **2006**, *103*, 14935–14940. [CrossRef]
160. Ruf, I.K.; Rhyne, P.W.; Yang, H.; Borza, C.M.; Hutt-Fletcher, L.M.; Cleveland, J.L.; Sample, J.T. EBV regulates c-MYC, apoptosis, and tumorigenicity in Burkitt's lymphoma. *Curr. Top. Microbiol. Immunol.* **2001**, *258*, 153–160. [CrossRef]
161. Fitzsimmons, L.; Boyce, A.J.; Wei, W.; Chang, C.; Croom-Carter, D.; Tierney, R.J.; Herold, M.J.; Bell, A.I.; Strasser, A.; Kelly, G.L.; et al. Coordinated repression of BIM and PUMA by Epstein-Barr virus latent genes maintains the survival of Burkitt lymphoma cells. *Cell Death Differ.* **2018**, *25*, 241–254. [CrossRef]
162. Grande, B.M.; Gerhard, D.S.; Jiang, A.; Griner, N.B.; Abramson, J.S.; Alexander, T.B.; Allen, H.; Ayers, L.W.; Bethony, J.M.; Bhatia, K.; et al. Genome-wide discovery of somatic coding and noncoding mutations in pediatric endemic and sporadic Burkitt lymphoma. *Blood* **2019**, *133*, 1313–1324. [CrossRef]
163. López, C.; Burkhardt, B.; Chan, J.K.C.; Leoncini, L.; Mbulaiteye, S.M.; Ogwang, M.D.; Orem, J.; Rochford, R.; Roschewski, M.; Siebert, R. Burkitt lymphoma. *Nat. Rev. Dis. Primers* **2022**, *8*, 78. [CrossRef]
164. Hämmerl, L.; Colombet, M.; Rochford, R.; Ogwang, D.M.; Parkin, D.M. The burden of Burkitt lymphoma in Africa. *Infect. Agent. Cancer* **2019**, *14*, 17. [CrossRef]
165. Al-Khreisat, M.J.; Ismail, N.H.; Tabnjh, A.; Hussain, F.A.; Mohamed Yusoff, A.A.; Johan, M.F.; Islam, M.A. Worldwide Prevalence of Epstein-Barr Virus in Patients with Burkitt Lymphoma: A Systematic Review and Meta-Analysis. *Diagnostics* **2023**, *13*, 2068. [CrossRef]
166. Guech-Ongey, M.; Simard, E.P.; Anderson, W.F.; Engels, E.A.; Bhatia, K.; Devesa, S.S.; Mbulaiteye, S.M. AIDS-related Burkitt lymphoma in the United States: What do age and CD4 lymphocyte patterns tell us about etiology and/or biology? *Blood* **2010**, *116*, 5600–5604. [CrossRef]
167. Hernández-Ramírez, R.U.; Qin, L.; Lin, H.; Leyden, W.; Neugebauer, R.S.; Althoff, K.N.; Achenbach, C.J.; Hessol, N.A.; D'Souza, G.; Gebo, K.A.; et al. Association of immunosuppression and HIV viraemia with non-Hodgkin lymphoma risk overall and by subtype in people living with HIV in Canada and the USA: A multicentre cohort study. *Lancet HIV* **2019**, *6*, e240–e249. [CrossRef]
168. Manolov, G.; Manolova, Y. Marker band in one chromosome 14 from Burkitt lymphomas. *Nature* **1972**, *237*, 33–34. [CrossRef]
169. Dalla-Favera, R.; Bregni, M.; Erikson, J.; Patterson, D.; Gallo, R.C.; Croce, C.M. Human c-myc onc gene is located on the region of chromosome 8 that is translocated in Burkitt lymphoma cells. *Proc. Natl. Acad. Sci. USA* **1982**, *79*, 7824–7827. [CrossRef]
170. Guikema, J.E.; Schuuring, E.; Kluin, P.M. Structure and consequences of IGH switch breakpoints in Burkitt lymphoma. *J. Natl. Cancer Inst. Monogr.* **2008**, *2008*, 32–36. [CrossRef]
171. Sall, F.B.; Shmakova, A.; Karpukhina, A.; Tsfasman, T.; Lomov, N.; Canoy, R.J.; Boutboul, D.; Oksenhendler, E.; Toure, A.O.; Lipinski, M.; et al. Epstein-Barr virus reactivation induces MYC-IGH spatial proximity and t(8;14) in B cells. *J. Med. Virol.* **2023**, *95*, e28633. [CrossRef]
172. de-Thé, G.; Geser, A.; Day, N.E.; Tukei, P.M.; Williams, E.H.; Beri, D.P.; Smith, P.G.; Dean, A.G.; Bronkamm, G.W.; Feorino, P.; et al. Epidemiological evidence for causal relationship between Epstein-Barr virus and Burkitt's lymphoma from Ugandan prospective study. *Nature* **1978**, *274*, 756–761. [CrossRef]
173. Geser, A.; de The, G.; Lenoir, G.; Day, N.E.; Williams, E.H. Final case reporting from the Ugandan prospective study of the relationship between EBV and Burkitt's lymphoma. *Int. J. Cancer* **1982**, *29*, 397–400. [CrossRef]
174. Chene, A.; Donati, D.; Orem, J.; Björkman, A.; Mbidde, E.R.; Kironde, F.; Wahlgren, M.; Bejarano, M.T. Endemic Burkitt's lymphoma as a polymicrobial disease: New insights on the interaction between Plasmodium falciparum and Epstein–Barr virus. *Semin. Cancer Biol.* **2009**, *19*, 411–420. [CrossRef] [PubMed]
175. Rochford, R.; Cannon, M.J.; Moormann, A.M. Endemic Burkitt's lymphoma: A polymicrobial disease? *Nat. Rev. Microbiol.* **2005**, *3*, 182–187. [CrossRef] [PubMed]
176. Muramatsu, M.; Kinoshita, K.; Fagarasan, S.; Yamada, S.; Shinkai, Y.; Honjo, T. Class Switch Recombination and Hypermutation Require Activation-Induced Cytidine Deaminase (AID), a Potential RNA Editing Enzyme. *Cell* **2000**, *102*, 553–563. [CrossRef]
177. Pasqualucci, L.; Bhagat, G.; Jankovic, M.; Compagno, M.; Smith, P.; Muramatsu, M.; Honjo, T.; Morse, H.C.; Nussenzweig, M.C.; Dalla-Favera, R. AID is required for germinal center–derived lymphomagenesis. *Nat. Genet.* **2008**, *40*, 108–112. [CrossRef]

178. Ramiro, A.R.; Jankovic, M.; Eisenreich, T.; Difilippantonio, S.; Chen-Kiang, S.; Muramatsu, M.; Honjo, T.; Nussenzweig, A.; Nussenzweig, M.C. AID is required for c-myc/IgH chromosome translocations in vivo. *Cell* **2004**, *118*, 431–438. [CrossRef]
179. Shirai, A.; Cosentino, M.; Leitman-Klinman, S.F.; Klinman, D.M. Human immunodeficiency virus infection induces both polyclonal and virus-specific B cell activation. *J. Clin. Investig.* **1992**, *89*, 561–566. [CrossRef]
180. Molyneux, E.M.; Rochford, R.; Griffin, B.; Newton, R.; Jackson, G.; Menon, G.; Harrison, C.J.; Israels, T.; Bailey, S. Burkitt's lymphoma. *Lancet* **2012**, *379*, 1234–1244. [CrossRef]
181. Accardi, R.; Gruffat, H.; Sirand, C.; Fusil, F.; Gheit, T.; Hernandez-Vargas, H.; Le Calvez-Kelm, F.; Traverse-Glehen, A.; Cosset, F.L.; Manet, E.; et al. The mycotoxin aflatoxin B1 stimulates Epstein-Barr virus-induced B-cell transformation in in vitro and in vivo experimental models. *Carcinogenesis* **2015**, *36*, 1440–1451. [CrossRef]
182. Maroui, M.A.; Odongo, G.A.; Mundo, L.; Manara, F.; Mure, F.; Fusil, F.; Jay, A.; Gheit, T.; Michailidis, T.M.; Ferrara, D.; et al. Aflatoxin B1 and Epstein-Barr virus-induced CCL22 expression stimulates B cell infection. *Proc. Natl. Acad. Sci. USA* **2024**, *121*, e2314426121. [CrossRef]
183. Manara, F.; Jay, A.; Odongo, G.A.; Mure, F.; Maroui, M.A.; Diederichs, A.; Sirand, C.; Cuenin, C.; Granai, M.; Mundo, L.; et al. Epigenetic Alteration of the Cancer-Related Gene TGFBI in B Cells Infected with Epstein-Barr Virus and Exposed to Aflatoxin B1: Potential Role in Burkitt Lymphoma Development. *Cancers* **2022**, *14*, 1284. [CrossRef]
184. Chapman, C.J.; Wright, D.; Stevenson, F.K. Insight into Burkitt's lymphoma from immunoglobulin variable region gene analysis. *Leuk. Lymphoma* **1998**, *30*, 257–267. [CrossRef] [PubMed]
185. Bellan, C.; Lazzi, S.; Hummel, M.; Palummo, N.; de Santi, M.; Amato, T.; Nyagol, J.; Sabattini, E.; Lazure, T.; Pileri, S.A.; et al. Immunoglobulin gene analysis reveals 2 distinct cells of origin for EBV-positive and EBV-negative Burkitt lymphomas. *Blood* **2005**, *106*, 1031–1036. [CrossRef] [PubMed]
186. Cumpelik, A.; Heja, D.; Hu, Y.; Varano, G.; Ordikhani, F.; Roberto, M.P.; He, Z.; Homann, D.; Lira, S.A.; Dominguez-Sola, D.; et al. Dynamic regulation of B cell complement signaling is integral to germinal center responses. *Nat. Immunol.* **2021**, *22*, 757–768. [CrossRef] [PubMed]
187. Jiang, R.; Gu, X.; Moore-Medlin, T.N.; Nathan, C.-A.; Hutt-Fletcher, L.M. Oral dysplasia and squamous cell carcinoma: Correlation between increased expression of CD21, Epstein-Barr virus and CK19. *Oral. Oncol.* **2012**, *48*, 836–841. [CrossRef]
188. Molina, E.; García-Gutiérrez, L.; Junco, V.; Perez-Olivares, M.; de Yébenes, V.G.; Blanco, R.; Quevedo, L.; Acosta, J.C.; Marín, A.V.; Ulgiati, D.; et al. MYC directly transactivates CR2/CD21, the receptor of the Epstein-Barr virus, enhancing the viral infection of Burkitt lymphoma cells. *Oncogene* **2023**, *42*, 3358–3370. [CrossRef]
189. Kelly, G.; Bell, A.; Rickinson, A. Epstein-Barr virus-associated Burkitt lymphomagenesis selects for downregulation of the nuclear antigen EBNA2. *Nat. Med.* **2002**, *8*, 1098–1104. [CrossRef]
190. Thorley-Lawson, D.A.; Gross, A. Persistence of the Epstein-Barr virus and the origins of associated lymphomas. *N. Engl. J. Med.* **2004**, *350*, 1328–1337. [CrossRef]
191. Kelly, G.L.; Stylianou, J.; Bell, A.I.; Wei, W.; Rowe, M.; Rickinson, A.B. Three Restricted Forms of Epstein-Barr Virus Latency Counteracting Apoptosis in c-Myc Expressing Burkitt Lymphoma Cells. *Blood* **2007**, *110*, 1572. [CrossRef]
192. Kaymaz, Y.; Oduor, C.I.; Yu, H.; Otieno, J.A.; Ong'echa, J.M.; Moormann, A.M.; Bailey, J.A. Comprehensive Transcriptome and Mutational Profiling of Endemic Burkitt Lymphoma Reveals EBV Type-Specific Differences. *Mol. Cancer Res.* **2017**, *15*, 563–576. [CrossRef]
193. Panea, R.I.; Love, C.L.; Shingleton, J.R.; Reddy, A.; Bailey, J.A.; Moormann, A.M.; Otieno, J.A.; Ong'echa, J.M.; Oduor, C.I.; Schroeder, K.M.S.; et al. The whole-genome landscape of Burkitt lymphoma subtypes. *Blood* **2019**, *134*, 1598–1607. [CrossRef]
194. Sureda-Gómez, M.; Iaccarino, I.; De Bolòs, A.; Meyer, M.; Balsas, P.; Richter, J.; Rodríguez, M.L.; López, C.; Carreras-Caballé, M.; Glaser, S.; et al. SOX11 expression is restricted to EBV-negative Burkitt lymphoma and is associated with molecular genetic features. *Blood* **2024**, *144*, 187–200. [CrossRef] [PubMed]
195. Hansen, K.D.; Sabunciyan, S.; Langmead, B.; Nagy, N.; Curley, R.; Klein, G.; Klein, E.; Salamon, D.; Feinberg, A.P. Large-scale hypomethylated blocks associated with Epstein-Barr virus-induced B-cell immortalization. *Genome Res.* **2014**, *24*, 177–184. [CrossRef] [PubMed]
196. Niller, H.H.; Banati, F.; Salamon, D.; Minarovits, J. Epigenetic Alterations in Epstein-Barr Virus-Associated Diseases. *Adv. Exp. Med. Biol.* **2016**, *879*, 39–69. [CrossRef] [PubMed]
197. Hernandez-Vargas, H.; Gruffat, H.; Cros, M.P.; Diederichs, A.; Sirand, C.; Vargas-Ayala, R.C.; Jay, A.; Durand, G.; Le Calvez-Kelm, F.; Herceg, Z.; et al. Viral driven epigenetic events alter the expression of cancer-related genes in Epstein-Barr-virus naturally infected Burkitt lymphoma cell lines. *Sci. Rep.* **2017**, *7*, 5852. [CrossRef]
198. Gires, O.; Zimber-Strobl, U.; Gonnella, R.; Ueffing, M.; Marschall, G.; Zeidler, R.; Pich, D.; Hammerschmidt, W. Latent membrane protein 1 of Epstein-Barr virus mimics a constitutively active receptor molecule. *Embo J.* **1997**, *16*, 6131–6140. [CrossRef]
199. Caldwell, R.G.; Wilson, J.B.; Anderson, S.J.; Longnecker, R. Epstein-Barr virus LMP2A drives B cell development and survival in the absence of normal B cell receptor signals. *Immunity* **1998**, *9*, 405–411. [CrossRef]
200. Yates, J.L.; Warren, N.; Sugden, B. Stable replication of plasmids derived from Epstein–Barr virus in various mammalian cells. *Nature* **1985**, *313*, 812–815. [CrossRef]
201. Humme, S.; Reisbach, G.; Feederle, R.; Delecluse, H.J.; Bousset, K.; Hammerschmidt, W.; Schepers, A. The EBV nuclear antigen 1 (EBNA1) enhances B cell immortalization several thousandfold. *Proc. Natl. Acad. Sci. USA* **2003**, *100*, 10989–10994. [CrossRef]

202. Kang, M.S.; Lu, H.; Yasui, T.; Sharpe, A.; Warren, H.; Cahir-McFarland, E.; Bronson, R.; Hung, S.C.; Kieff, E. Epstein-Barr virus nuclear antigen 1 does not induce lymphoma in transgenic FVB mice. *Proc. Natl. Acad. Sci. USA* **2005**, *102*, 820–825. [CrossRef]
203. Wilson, J.B.; Bell, J.L.; Levine, A.J. Expression of Epstein-Barr virus nuclear antigen-1 induces B cell neoplasia in transgenic mice. *Embo J.* **1996**, *15*, 3117–3126. [CrossRef]
204. AlQarni, S.; Al-Sheikh, Y.; Campbell, D.; Drotar, M.; Hannigan, A.; Boyle, S.; Herzyk, P.; Kossenkov, A.; Armfield, K.; Jamieson, L.; et al. Lymphomas driven by Epstein-Barr virus nuclear antigen-1 (EBNA1) are dependant upon Mdm2. *Oncogene* **2018**, *37*, 3998–4012. [CrossRef] [PubMed]
205. Kennedy, G.; Komano, J.; Sugden, B. Epstein-Barr virus provides a survival factor to Burkitt's lymphomas. *Proc. Natl. Acad. Sci. USA* **2003**, *100*, 14269–14274. [CrossRef] [PubMed]
206. Saridakis, V.; Sheng, Y.; Sarkari, F.; Holowaty, M.N.; Shire, K.; Nguyen, T.; Zhang, R.G.; Liao, J.; Lee, W.; Edwards, A.M.; et al. Structure of the p53 binding domain of HAUSP/USP7 bound to Epstein-Barr nuclear antigen 1 implications for EBV-mediated immortalization. *Mol. Cell* **2005**, *18*, 25–36. [CrossRef] [PubMed]
207. Wang, R.; Wang, J.; Zhang, N.; Wan, Y.; Liu, Y.; Zhang, L.; Pan, S.; Zhang, C.; Zhang, H.; Cao, Y. The interaction between Vav1 and EBNA1 promotes survival of Burkitt's lymphoma cells by down-regulating the expression of Bim. *Biochem. Biophys. Res. Commun.* **2019**, *511*, 787–793. [CrossRef]
208. Hammerschmidt, W.; Sugden, B. Genetic analysis of immortalizing functions of Epstein-Barr virus in human B lymphocytes. *Nature* **1989**, *340*, 393–397. [CrossRef]
209. Pich, D.; Mrozek-Gorska, P.; Bouvet, M.; Sugimoto, A.; Akidil, E.; Grundhoff, A.; Hamperl, S.; Ling, P.D.; Hammerschmidt, W. First Days in the Life of Naive Human B Lymphocytes Infected with Epstein-Barr Virus. *mBio* **2019**, *10*, 1110–1128. [CrossRef]
210. Höfelmayr, H.; Strobl, L.J.; Marschall, G.; Bornkamm, G.W.; Zimber-Strobl, U. Activated Notch1 can transiently substitute for EBNA2 in the maintenance of proliferation of LMP1-expressing immortalized B cells. *J. Virol.* **2001**, *75*, 2033–2040. [CrossRef]
211. Kaiser, C.; Laux, G.; Eick, D.; Jochner, N.; Bornkamm, G.W.; Kempkes, B. The proto-oncogene c-myc is a direct target gene of Epstein-Barr virus nuclear antigen 2. *J. Virol.* **1999**, *73*, 4481–4484. [CrossRef]
212. Tomkinson, B.; Kieff, E. Use of second-site homologous recombination to demonstrate that Epstein-Barr virus nuclear protein 3B is not important for lymphocyte infection or growth transformation in vitro. *J. Virol.* **1992**, *66*, 2893–2903. [CrossRef]
213. Anderton, E.; Yee, J.; Smith, P.; Crook, T.; White, R.E.; Allday, M.J. Two Epstein-Barr virus (EBV) oncoproteins cooperate to repress expression of the proapoptotic tumour-suppressor Bim: Clues to the pathogenesis of Burkitt's lymphoma. *Oncogene* **2008**, *27*, 421–433. [CrossRef]
214. Muthalagu, N.; Junttila, M.R.; Wiese, K.E.; Wolf, E.; Morton, J.; Bauer, B.; Evan, G.I.; Eilers, M.; Murphy, D.J. BIM is the primary mediator of MYC-induced apoptosis in multiple solid tissues. *Cell Rep.* **2014**, *8*, 1347–1353. [CrossRef] [PubMed]
215. Price, A.M.; Dai, J.; Bazot, Q.; Patel, L.; Nikitin, P.A.; Djavadian, R.; Winter, P.S.; Salinas, C.A.; Barry, A.P.; Wood, K.C.; et al. Epstein-Barr virus ensures B cell survival by uniquely modulating apoptosis at early and late times after infection. *Elife* **2017**, *6*, e22509. [CrossRef] [PubMed]
216. Vikstrom, I.; Carotta, S.; Lüthje, K.; Peperzak, V.; Jost, P.J.; Glaser, S.; Busslinger, M.; Bouillet, P.; Strasser, A.; Nutt, S.L.; et al. Mcl-1 is essential for germinal center formation and B cell memory. *Science* **2010**, *330*, 1095–1099. [CrossRef]
217. Desbien, A.L.; Kappler, J.W.; Marrack, P. The Epstein-Barr virus Bcl-2 homolog, BHRF1, blocks apoptosis by binding to a limited amount of Bim. *Proc. Natl. Acad. Sci. USA* **2009**, *106*, 5663–5668. [CrossRef]
218. Blackwood, E.M.; Eisenman, R.N. Max: A helix-loop-helix zipper protein that forms a sequence-specific DNA-binding complex with Myc. *Science* **1991**, *251*, 1211–1217. [CrossRef]
219. Diolaiti, D.; McFerrin, L.; Carroll, P.A.; Eisenman, R.N. Functional interactions among members of the MAX and MLX transcriptional network during oncogenesis. *Biochim. Biophys. Acta* **2015**, *1849*, 484–500. [CrossRef]
220. Carroll, P.A.; Diolaiti, D.; McFerrin, L.; Gu, H.; Djukovic, D.; Du, J.; Cheng, P.F.; Anderson, S.; Ulrich, M.; Hurley, J.B.; et al. Deregulated Myc requires MondoA/Mlx for metabolic reprogramming and tumorigenesis. *Cancer Cell* **2015**, *27*, 271–285. [CrossRef]
221. Zeller, K.I.; Zhao, X.; Lee, C.W.; Chiu, K.P.; Yao, F.; Yustein, J.T.; Ooi, H.S.; Orlov, Y.L.; Shahab, A.; Yong, H.C.; et al. Global mapping of c-Myc binding sites and target gene networks in human B cells. *Proc. Natl. Acad. Sci. USA* **2006**, *103*, 17834–17839. [CrossRef]
222. Sabo, A.; Kress, T.R.; Pelizzola, M.; de Pretis, S.; Gorski, M.M.; Tesi, A.; Morelli, M.J.; Bora, P.; Doni, M.; Verrecchia, A.; et al. Selective transcriptional regulation by Myc in cellular growth control and lymphomagenesis. *Nature* **2014**, *511*, 488–492. [CrossRef]
223. Nie, Z.; Hu, G.; Wei, G.; Cui, K.; Yamane, A.; Resch, W.; Wang, R.; Green, D.R.; Tessarollo, L.; Casellas, R.; et al. c-Myc is a universal amplifier of expressed genes in lymphocytes and embryonic stem cells. *Cell* **2012**, *151*, 68–79. [CrossRef]
224. Lin, C.Y.; Loven, J.; Rahl, P.B.; Paranal, R.M.; Burge, C.B.; Bradner, J.E.; Lee, T.I.; Young, R.A. Transcriptional amplification in tumor cells with elevated c-Myc. *Cell* **2012**, *151*, 56–67. [CrossRef] [PubMed]
225. Kress, T.R.; Sabò, A.; Amati, B. MYC: Connecting selective transcriptional control to global RNA production. *Nat. Rev. Cancer* **2015**, *15*, 593–607. [CrossRef] [PubMed]
226. Dang, C.V. MYC on the path to cancer. *Cell* **2012**, *149*, 22–35. [CrossRef]
227. Conacci-Sorrell, M.; McFerrin, L.; Eisenman, R.N. An Overview of MYC and Its Interactome. *Cold Spring Harb. Perspect. Med.* **2014**, *4*, 1–24. [CrossRef]
228. Bretones, G.; Delgado, M.D.; Leon, J. Myc and cell cycle control. *Biochim. Biophys. Acta* **2015**, *1849*, 506–516. [CrossRef]

229. Kalkat, M.; De Melo, J.; Hickman, K.A.; Lourenco, C.; Redel, C.; Resetca, D.; Tamachi, A.; Tu, W.B.; Penn, L.Z. MYC Deregulation in Primary Human Cancers. *Genes* **2017**, *8*, 151. [CrossRef]
230. Schaub, F.X.; Dhankani, V.; Berger, A.C.; Trivedi, M.; Richardson, A.B.; Shaw, R.; Zhao, W.; Zhang, X.; Ventura, A.; Liu, Y.; et al. Pan-cancer Alterations of the MYC Oncogene and Its Proximal Network across the Cancer Genome Atlas. *Cell Syst.* **2018**, *6*, 282–300.e2. [CrossRef]
231. Delgado, M.D.; Leon, J. Myc roles in hematopoiesis and leukemia. *Genes. Cancer* **2010**, *1*, 605–616. [CrossRef]
232. Ott, G.; Rosenwald, A.; Campo, E. Understanding MYC-driven aggressive B-cell lymphomas: Pathogenesis and classification. *Blood* **2013**, *122*, 3884–3891. [CrossRef]
233. Sheiness, D.; Bishop, J.M. DNA and RNA from uninfected vertebrate cells contain nucleotide sequences related to the putative transforming gene of avian myelocytomatosis virus. *J. Virol.* **1979**, *31*, 514–521. [CrossRef]
234. Sheiness, D.K.; Hughes, S.H.; Varmus, H.E.; Stubblefield, E.; Bishop, J.M. The vertebrate homolog of the putative transforming gene of avian myelocytomatosis virus: Characteristics of the DNA locus and its RNA transcript. *Virology* **1980**, *105*, 415–424. [CrossRef] [PubMed]
235. Duesberg, P.H.; Vogt, P.K. Avian acute leukemia viruses MC29 and MH2 share specific RNA sequences: Evidence for a second class of transforming genes. *Proc. Natl. Acad. Sci. USA* **1979**, *76*, 1633–1637. [CrossRef] [PubMed]
236. Enrietto, P.J.; Payne, L.N.; Hayman, M.J. A recovered avian myelocytomatosis virus that induces lymphomas in chickens: Pathogenic properties and their molecular basis. *Cell* **1983**, *35*, 369–379. [CrossRef] [PubMed]
237. Zech, L.; Haglund, U.; Nilsson, K.; Klein, G. Characteristic chromosomal abnormalities in biopsies and lymphoid-cell lines from patients with Burkitt and non-Burkitt lymphomas. *Int. J. Cancer* **1976**, *17*, 47–56. [CrossRef]
238. Dominguez-Sola, D.; Victora, G.D.; Ying, C.Y.; Phan, R.T.; Saito, M.; Nussenzweig, M.C.; Dalla-Favera, R. The proto-oncogene MYC is required for selection in the germinal center and cyclic reentry. *Nat. Immunol.* **2012**, *13*, 1083–1091. [CrossRef]
239. Calado, D.P.; Sasaki, Y.; Godinho, S.A.; Pellerin, A.; Köchert, K.; Sleckman, B.P.; de Alborán, I.M.; Janz, M.; Rodig, S.; Rajewsky, K. The cell-cycle regulator c-Myc is essential for the formation and maintenance of germinal centers. *Nat. Immunol.* **2012**, *13*, 1092–1100. [CrossRef]
240. Methot, S.P.; Di Noia, J.M. Molecular Mechanisms of Somatic Hypermutation and Class Switch Recombination. *Adv. Immunol.* **2017**, *133*, 37–87. [CrossRef]
241. Robbiani, D.F.; Bothmer, A.; Callen, E.; Reina-San-Martin, B.; Dorsett, Y.; Difilippantonio, S.; Bolland, D.J.; Chen, H.T.; Corcoran, A.E.; Nussenzweig, A.; et al. AID is required for the chromosomal breaks in c-myc that lead to c-myc/IgH translocations. *Cell* **2008**, *135*, 1028–1038. [CrossRef]
242. Boxer, L.M.; Dang, C.V. Translocations involving c-myc and c-myc function. *Oncogene* **2001**, *20*, 5595–5610. [CrossRef]
243. Boerma, E.G.; Siebert, R.; Kluin, P.M.; Baudis, M. Translocations involving 8q24 in Burkitt lymphoma and other malignant lymphomas: A historical review of cytogenetics in the light of todays knowledge. *Leukemia* **2009**, *23*, 225–234. [CrossRef]
244. Terwilliger, T.; Abdul-Hay, M. Acute lymphoblastic leukemia: A comprehensive review and 2017 update. *Blood Cancer J.* **2017**, *7*, e577. [CrossRef] [PubMed]
245. Dunleavy, K. Double-hit lymphomas: Current paradigms and novel treatment approaches. *Hematol. Am. Soc. Hematol. Educ. Program.* **2014**, *2014*, 107–112. [CrossRef] [PubMed]
246. Karube, K.; Campo, E. MYC Alterations in Diffuse Large B-Cell Lymphomas. *Semin. Hematol.* **2015**, *52*, 97–106. [CrossRef] [PubMed]
247. Rodrigues, J.M.; Hollander, P.; Schmidt, L.; Gkika, E.; Razmara, M.; Kumar, D.; Geisler, C.; Grønbæk, K.; Eskelund, C.W.; Räty, R.; et al. MYC protein is a high-risk factor in mantle cell lymphoma and identifies cases beyond morphology, proliferation and TP53/p53—A Nordic Lymphoma Group study. *Haematologica* **2024**, *109*, 1171–1183. [CrossRef]
248. Nishikura, K.; Erikson, J.; ar-Rushdi, A.; Huebner, K.; Croce, C.M. The translocated c-myc oncogene of Raji Burkitt lymphoma cells is not expressed in human lymphoblastoid cells. *Proc. Natl. Acad. Sci. USA* **1985**, *82*, 2900–2904. [CrossRef]
249. Hayday, A.C.; Gillies, S.D.; Saito, H.; Wood, C.; Wiman, K.; Hayward, W.S.; Tonegawa, S. Activation of a translocated human c-myc gene by an enhancer in the immunoglobulin heavy-chain locus. *Nature* **1984**, *307*, 334–340. [CrossRef]
250. Lopez, C.; Kleinheinz, K.; Aukema, S.M.; Rohde, M.; Bernhart, S.H.; Hubschmann, D.; Wagener, R.; Toprak, U.H.; Raimondi, F.; Kreuz, M.; et al. Genomic and transcriptomic changes complement each other in the pathogenesis of sporadic Burkitt lymphoma. *Nat. Commun.* **2019**, *10*, 1459. [CrossRef]
251. Bhatia, K.; Huppi, K.; Spangler, G.; Siwarski, D.; Iyer, R.; Magrath, I. Point mutations in the c-Myc transactivation domain are common in Burkitt's lymphoma and mouse plasmacytomas. *Nat. Genet.* **1993**, *5*, 56–61. [CrossRef]
252. Love, C.; Sun, Z.; Jima, D.; Li, G.; Zhang, J.; Miles, R.; Richards, K.L.; Dunphy, C.H.; Choi, W.W.; Srivastava, G.; et al. The genetic landscape of mutations in Burkitt lymphoma. *Nat. Genet.* **2012**, *44*, 1321–1325. [CrossRef]
253. Farrell, A.S.; Sears, R.C. MYC degradation. *Cold Spring Harb. Perspect. Med.* **2014**, *4*, a014365. [CrossRef]
254. Adams, J.M.; Harris, A.W.; Pinkert, C.A.; Corcoran, L.M.; Alexander, W.S.; Cory, S.; Palmiter, R.D.; Brinster, R.L. The c-myc oncogene driven by immunoglobulin enhancers induces lymphoid malignancy in transgenic mice. *Nature* **1985**, *318*, 533–538. [CrossRef] [PubMed]
255. Schmidt, E.V.; Pattengale, P.K.; Weir, L.; Leder, P. Transgenic mice bearing the human c-myc gene activated by an immunoglobulin enhancer: A pre-B-cell lymphoma model. *Proc. Natl. Acad. Sci. USA* **1988**, *85*, 6047–6051. [CrossRef] [PubMed]

256. Sugimoto, M.; Tahara, H.; Ide, T.; Furuichi, Y. Steps involved in immortalization and tumorigenesis in human B-lymphoblastoid cell lines transformed by Epstein-Barr virus. *Cancer Res.* **2004**, *64*, 3361–3364. [CrossRef] [PubMed]
257. Park, S.S.; Kim, J.S.; Tessarollo, L.; Owens, J.D.; Peng, L.; Han, S.S.; Tae Chung, S.; Torrey, T.A.; Cheung, W.C.; Polakiewicz, R.D.; et al. Insertion of c-Myc into Igh induces B-cell and plasma-cell neoplasms in mice. *Cancer Res.* **2005**, *65*, 1306–1315. [CrossRef]
258. Truffinet, V.; Pinaud, E.; Cogné, N.; Petit, B.; Guglielmi, L.; Cogné, M.; Denizot, Y. The 3′ IgH locus control region is sufficient to deregulate a c-myc transgene and promote mature B cell malignancies with a predominant Burkitt-like phenotype. *J. Immunol.* **2007**, *179*, 6033–6042. [CrossRef]
259. Sander, S.; Calado, D.P.; Srinivasan, L.; Kochert, K.; Zhang, B.; Rosolowski, M.; Rodig, S.J.; Holzmann, K.; Stilgenbauer, S.; Siebert, R.; et al. Synergy between PI3K signaling and MYC in Burkitt lymphomagenesis. *Cancer Cell* **2012**, *22*, 167–179. [CrossRef]
260. IARC. Biological Agentes. A Review of Human Carcinogenesis. Epstein-Barr Virus. *IARC Monogr.* **2012**, *100B*, 49–92.
261. Raab-Traub, N.; Flynn, K. The structure of the termini of the Epstein-Barr virus as a marker of clonal cellular proliferation. *Cell* **1986**, *47*, 883–889. [CrossRef]
262. Kalchschmidt, J.S.; Bashford-Rogers, R.; Paschos, K.; Gillman, A.C.; Styles, C.T.; Kellam, P.; Allday, M.J. Epstein-Barr virus nuclear protein EBNA3C directly induces expression of AID and somatic mutations in B cells. *J. Exp. Med.* **2016**, *213*, 921–928. [CrossRef]
263. Schmitz, R.; Young, R.M.; Ceribelli, M.; Jhavar, S.; Xiao, W.; Zhang, M.; Wright, G.; Shaffer, A.L.; Hodson, D.J.; Buras, E.; et al. Burkitt lymphoma pathogenesis and therapeutic targets from structural and functional genomics. *Nature* **2012**, *490*, 116–120. [CrossRef]
264. Lin, D.C.; Meng, X.; Hazawa, M.; Nagata, Y.; Varela, A.M.; Xu, L.; Sato, Y.; Liu, L.Z.; Ding, L.W.; Sharma, A.; et al. The genomic landscape of nasopharyngeal carcinoma. *Nat. Genet.* **2014**, *46*, 866–871. [CrossRef] [PubMed]
265. Wang, K.; Yuen, S.T.; Xu, J.; Lee, S.P.; Yan, H.H.; Shi, S.T.; Siu, H.C.; Deng, S.; Chu, K.M.; Law, S.; et al. Whole-genome sequencing and comprehensive molecular profiling identify new driver mutations in gastric cancer. *Nat. Genet.* **2014**, *46*, 573–582. [CrossRef] [PubMed]
266. Piris, M.A.; Medeiros, L.J.; Chang, K.C. Hodgkin lymphoma: A review of pathological features and recent advances in pathogenesis. *Pathology* **2019**, *52*, 154–165. [CrossRef]
267. Price, A.M.; Messinger, J.E.; Luftig, M.A. c-Myc Represses Transcription of the Epstein-Barr Virus Latent Membrane Protein 1 Early After Primary B Cell Infection. *J. Virol.* **2017**, *92*, 1110–1128. [CrossRef]
268. Jochner, N.; Eick, D.; Zimber-Strobl, U.; Pawlita, M.; Bornkamm, G.W.; Kempkes, B. Epstein-Barr virus nuclear antigen 2 is a transcriptional suppressor of the immunoglobulin mu gene: Implications for the expression of the translocated c-myc gene in Burkitt's lymphoma cells. *Embo J.* **1996**, *15*, 375–382. [CrossRef]
269. Bajaj, R.; Xu, F.; Xiang, B.; Wilcox, K.; Diadamo, A.J.; Kumar, R.; Pietraszkiewicz, A.; Halene, S.; Li, P. Evidence-based genomic diagnosis characterized chromosomal and cryptic imbalances in 30 elderly patients with myelodysplastic syndrome and acute myeloid leukemia. *Mol. Cytogenet.* **2011**, *4*, 3. [CrossRef]
270. Allday, M.J. How does Epstein-Barr virus (EBV) complement the activation of Myc in the pathogenesis of Burkitt's lymphoma? *Semin. Cancer Biol.* **2009**, *19*, 366–376. [CrossRef]
271. Bornkamm, G.W. The role of Epstein-Barr virus in human B cell malignancies. *Haematologica* **1987**, *72*, 67–70.
272. Trivedi, P.; Zhang, Q.J.; Chen, F.; Minarovits, J.; Ekman, M.; Biberfeld, P.; Klein, G.; Winberg, G. Parallel existence of Epstein-Barr virus (EBV) positive and negative cells in a sporadic case of Burkitt lymphoma. *Oncogene* **1995**, *11*, 505–510.
273. Wang, L.W.; Shen, H.; Nobre, L.; Ersing, I.; Paulo, J.A.; Trudeau, S.; Wang, Z.; Smith, N.A.; Ma, Y.; Reinstadler, B.; et al. Epstein-Barr-Virus-Induced One-Carbon Metabolism Drives B Cell Transformation. *Cell Metab.* **2019**, *30*, 539–555. [CrossRef]
274. Bristol, J.A.; Nelson, S.E.; Ohashi, M.; Casco, A.; Hayes, M.; Ranheim, E.A.; Pawelski, A.S.; Singh, D.R.; Hodson, D.J.; Johannsen, E.C.; et al. Latent Epstein-Barr virus infection collaborates with Myc over-expression in normal human B cells to induce Burkitt-like Lymphomas in mice. *PLoS Pathog.* **2024**, *20*, e1012132. [CrossRef] [PubMed]
275. Lombardi, L.; Newcomb, E.W.; Dalla-Favera, R. Pathogenesis of Burkitt lymphoma: Expression of an activated c-myc oncogene causes the tumorigenic conversion of EBV-infected human B lymphoblasts. *Cell* **1987**, *49*, 161–170. [CrossRef] [PubMed]
276. Guo, R.; Jiang, C.; Zhang, Y.; Govande, A.; Trudeau, S.J.; Chen, F.; Fry, C.J.; Puri, R.; Wolinsky, E.; Schineller, M.; et al. MYC Controls the Epstein-Barr Virus Lytic Switch. *Mol. Cell* **2020**, *78*, 653–669.e658. [CrossRef] [PubMed]

cancers

MDPI

Review

Molecular Mechanisms of Kaposi Sarcoma-Associated Herpesvirus (HHV8)-Related Lymphomagenesis

Caroline J. Yu and Blossom Damania *

Department of Microbiology & Immunology, Lineberger Comprehensive Cancer Center, University of North Carolina, Chapel Hill, NC 27599, USA; cjyu@unc.edu
* Correspondence: damania@med.unc.edu

Simple Summary: Kaposi sarcoma-associated herpesvirus (KSHV) is associated with lymphoproliferative disorders, including primary effusion lymphoma and multicentric Castleman disease. KSHV expresses viral proteins that aid in the evasion of antiviral immune responses and manipulate host factors and signaling to promote cell survival. By altering the cell environment, KSHV contributes to lymphomagenesis.

Abstract: Approximately 15–20% of cancers are caused by viruses. Kaposi sarcoma-associated herpesvirus (KSHV), also known as human herpesvirus 8 (HHV8), is an oncogenic virus that is the etiologic agent of not only Kaposi sarcoma but also the lymphoproliferative disorders, primary effusion lymphoma (PEL) and multicentric Castleman disease (MCD). KSHV can infect a broad tropism of cells, including B lymphocytes, wherein KSHV encodes specific viral proteins that can transform the cell. KSHV infection precedes the progression of PEL and MCD. KSHV establishes lifelong infection and has two phases of its lifecycle: latent and lytic. During the latent phase, viral genomes are maintained episomally with limited gene expression. Upon sporadic reactivation, the virus enters its replicative lytic phase to produce infectious virions. KSHV relies on its viral products to modulate host factors to evade immune detection or to co-opt their function for KSHV persistence. These manipulations dysregulate normal cell pathways to ensure cell survival and inhibit antiviral immune responses, which in turn, contribute to KSHV-associated malignancies. Here, we highlight the known molecular mechanisms of KSHV that promote lymphomagenesis and how these findings identify potential therapeutic targets for KSHV-associated lymphomas.

Keywords: Kaposi sarcoma-associated herpesvirus; KSHV; HHV8; lymphoma; lymphomagenesis; viral oncogenesis

1. Introduction

Kaposi sarcoma-associated herpesvirus (KSHV) is a member of the gamma-herpesvirus family, a subgroup of herpesviruses (including Epstein–Barr virus) with oncogenic potential. KSHV is a large double-stranded DNA virus with a 165 kb genome encoding over 100 genes. KSHV is associated with three malignancies, Kaposi sarcoma (KS) and two lymphoproliferative disorders, including primary effusion lymphoma (PEL) and multicentric Castleman disease (MCD). PEL and MCD are B cell lymphoproliferative diseases.

KSHV seroprevalence is dependent on the geographic region of the world. Seroprevalence is greater than 30% in sub-Saharan Africa, 15–20% in the Mediterranean region, and 2–5% in other regions around the world [1]. KSHV was first detected in a KS lesion from an acquired immunodeficiency syndrome (AIDS) patient [2]. Subsequently, KSHV was also detected in tissues from PEL and MCD patients [3,4].

KSHV establishes lifelong infection. Like all herpesviruses, KSHV has a biphasic life cycle with latent and lytic phases. During the latent phase, viral genomes are maintained episomally with limited gene expression. Latent genes help maintain viral DNA and prevent apoptosis, ensuring KSHV persistence. Upon sporadic reactivation, the virus enters its replicative lytic phase with elevated viral gene expression. Reactivation can be triggered by environmental stimuli such as stress and UV radiation, to name a few. Lytic replication allows for the temporally orchestrated expression of immediate early (IE), early (E), and late (L) viral genes for structural proteins to produce infectious virions (reviewed in [5]).

KSHV modulates host factors to evade immune detection or to co-opt their function for KSHV persistence. These manipulations dysregulate typical cell processes to ensure cell survival and inhibit antiviral immune responses, which, in turn, contribute to KSHV-associated malignancies. Both latent and lytic genes contribute to oncogenesis. Latent genes are continuously expressed in the tumor cell, and a small percentage of cells undergo lytic reactivation, where lytic proteins are thought to condition the microenvironment to promote KSHV-associated tumorigenesis (reviewed in [6]).

KSHV can infect a broad tropism of cells (reviewed in [7]), including B lymphocytes. B cells are the cell type of origin for both PEL and MCD. Here, we provide an overview of KSHV lymphoproliferative diseases and highlight the known molecular mechanisms of KSHV viral products that promote lymphomagenesis, and we explore how these findings identify potential therapeutic targets for KSHV-associated lymphomas.

2. Non-Hodgkin Lymphoma

Lymphomas are divided into Hodgkin or nonHodgkin lymphomas. Non-Hodgkin lymphoma (NHL) is the seventh most prevalent cancer and sixth most common cause of cancer-related death in the United States [8].

Non-Hodgkin lymphomas occur when lymphocytes grow abnormally and accumulate to form tumors. NHLs are a heterogeneous group of lymphoproliferative malignancies with over 60 subtypes, and the subtypes are further categorized by disease progression, either indolent (slow) or aggressive (fast). NHL subtypes mostly comprise diffuse large B cell lymphomas (~30%) and follicular lymphomas (~20%), while all other subtypes are <10% [9]. A total of 85–90% of NHL cases originate in B cells [10]. The most recent classification developed by the World Health Organization classifies types of NHL by type of lymphocyte (B, T, or NK cell), by the location the lymphoma originates from, and by the features of the cells, such as their morphology, specific surface protein markers, and genetic features of chromosomes [11].

The biggest risk factors for NHL are immunosuppression, UV radiation, viral infection, autoimmune and chronic inflammatory disorders, and occupational exposures [12]. Infections that weaken the immune system or cause chronic immune stimulation increase the risk. Chronic antigenic stimulation can also increase B cell proliferation, thus increasing the likelihood of mutations, especially immunoglobulin rearrangements. The cause of lymphomagenesis can be related to mutations in oncogenes, tumor suppressor genes, DNA repair genes, or viral infection.

Primary Effusion Lymphoma

Primary effusion lymphoma is a highly aggressive NHL with a poor prognosis, a 6-month median survival post diagnosis, and very limited efficacious therapeutics. PEL is rare, and it is primarily seen in HIV-positive individuals; however, KSHV is the causative agent [13]. PEL cells are predominantly latently infected with KSHV. PELs are B-cell-derived lymphomas that primarily develop in body cavities like the pleural, peritoneal, and pericardial spaces [14]. PEL patients show symptoms from fluid accumulation, such as difficulty with breathing, abdominal distension, and chest pain. PEL cells are large and anaplastic with irregular nuclei and prominent nucleoli. Based on VDJ arrangements, PEL cells are clonal in nature and express some B cell markers, including CD45, CD30, CD138, and other activation markers that resemble partially differentiated post-germinal center B cells, but they are CD20-negative [13]. Pleural and peritoneal effusions from patients with PEL generally contain high concentrations of viral IL-6 (vIL-6), as well as human IL-6 (hIL-6) and IL-10 [15]. These cytokines promote the autocrine growth of PEL cells.

KSHV is the etiologic agent of PEL, where 100% of PEL cells contain KSHV genomes. However, approximately 80% of PELs are co-infected with the related gamma-herpesvirus, Epstein–Barr virus (EBV) (reviewed in [16]). Both viruses typically persist in cells in their latent, quiescent state, with their viral genomes tethered to host chromatin as episomes, implying there may be a fitness advantage for the cell to retain these viruses. A selection of KSHV genes are known to be critical to the survival or proliferation of PEL cells, so KSHV is considered a genetic driver of PEL [17,18]. KSHV viral genes have been shown to be important for PEL survival, and the depletion of viral genes, e.g., KSHV *vFLIP*, *vCyclin*, and *vIRF3*, causes apoptosis of PEL and inhibits PEL growth in immunodeficient mice [17,18].

When EBV infection is also present in PEL, EBV latency I pattern is observed, where EBNA-1, EBER RNA, and low levels of LMP2a are expressed [19,20]. A study showed that co-infection with EBV allows for KSHV persistence in humanized mice, resulting in PEL-like lymphomagenesis [21]. EBV supports KSHV persistence after primary B cell infection and improves KSHV DNA maintenance after infection of EBV-negative PEL in vitro [22]. KSHV and EBV influence each other, where KSHV has been shown to enhance EBV lytic replication. Increased EBV lytic infection contributes to more frequent lymphomagenesis, which has been observed in both doubly infected B cells of humanized mice and co-infected human PEL cells [23,24]. Another study defined conditions to stably infect human B cells with KSHV and showed that optimal infection required coinfection by EBV [25]. Dually infected B cells were stably transformed, with the maintenance of both EBV and KSHV.

3. Multicentric Castleman Disease

Half of multicentric Castleman disease (MCD) cases are associated with KSHV, while the other half are idiopathic [26]. MCD is a lymphoproliferative disorder and is characterized by lymphadenopathy with polyclonal hypergammaglobulinemia and high serum levels of IL-6. KSHV-associated MCD is distinct from other forms of MCD, with a more aggressive clinical presentation compared to idiopathic MCD [27,28]. In 1995, KSHV was discovered as the etiologic agent of a plasmablastic variant of MCD occurring most commonly in HIV-infected patients (90%). MCD in HIV-infected patients usually develops 3 years after HIV diagnosis and is commonly identified alongside KS [29]. In HIV-negative patients, KSHV-MCD is largely seen in populations with a high prevalence of KSHV infection, such as men who have sex with men and populations from sub-Saharan Africa, where KSHV is endemic [30,31]. A minority of KSHV-infected individuals develop MCD years later, indicating the relevance of cofactors in the development of disease. Additionally, patients with MCD are more susceptible to other NHLs and organ failure [26].

Plasmablasts in MCD have an abnormal proliferation of immunoglobulin-restricted naïve B cells in the B cell mantle zone and interfollicular regions [13]. In KSHV-MCD, both viral and human IL-6 contribute to the plasmacytic differentiation of these cells, and high levels of vIL-6 are detected in the serum of MCD patients [32,33]. At times, both KS and KSHV-MCD can be observed in the same lymph node; however, plasmablasts are rarely coinfected with EBV [34]. While the expansion of clusters of plasmablasts can occur, these lesions are not clonal and do not always progress to lymphoma [26].

IL-6 is regarded as the driver of the proinflammatory state of MCD. It has been proposed that both nodal and systemic manifestations are reactive changes to elevated levels of IL-6 and other circulating cytokines and chemokines. Additionally, KSHV encodes a viral homolog of IL-6, vIL-6, with optimized IL-6 functions. vIL-6 has been validated as a driver in the development of MCD-like disease in mouse models [35,36]. Patients with MCD have prominent inflammatory manifestations that are attributed to vIL-6, hIL-6, and IL-10 cytokine levels. In MCD, the affected lymph nodes contain vIL-6-positive cells, and flares of the disease correlate with vIL-6 levels [37].

4. Mechanisms of KSHV-Associated Lymphomagenesis

4.1. KSHV Manipulates Host Machinery in B Cells

The molecular pathogenesis of B cell lymphomas includes aberrant intracellular signaling, disordered transcriptional and epigenetic regulation, and immune evasion [38]. B cells receive signals through receptors to induce transcriptional activity; thus, transcription factors are critical in B cell responses and are exploited by KSHV to reprogram the cell for survival. Commonly affected pathways include the NF-κB, PI3K/Akt/mTOR, and JAK/STAT pathways. KSHV is known to hijack these pathways [39–41]. Constitutive activation of NF-κB signaling often occurs, promoting survival, where in the context of lymphomas, negative regulators of NF-κB are often rendered inoperative [39].

Despite multiple cell cycle checkpoints, KSHV promotes the proliferation of cancer cells. Tumors form by overcoming replicative senescence. Throughout cell divisions, the telomere length of primary cells shortens. Shortened telomeres eventually activate the DNA damage response and tumor suppressor proteins p53 and retinoblastoma protein (Rb). Tumor suppressor proteins induce senescence, such that cells halt cell division but are metabolically active. For primary human cells to transform, both p53 and Rb must be inactivated. KSHV uses viral gene products and manipulates host machinery to inactivate these tumor suppressor proteins [42,43]. To sustain uncontrolled proliferation, cancer cells reprogram metabolic pathways. KSHV reprograms the cell to support its persistence and proliferation, including glucose metabolism, where KSHV-infected B and PEL cells have increased glucose uptake, increased lactate secretion, and decreased oxygen consumption. KSHV also upregulates glycolysis, promotes fatty acid synthesis, and alters metabolite sensing [44]. KSHV targets tumor suppressors and metabolic programs to funnel resources and support the continued growth of KSHV-infected cells.

KSHV infection is known to activate innate immune responses through both RNA and DNA sensors like toll-like receptors (TLRs), RIG-I-like receptors (RLRs), AIM2-like receptors (ALRs), and cyclic GMP-AMP synthase (cGAS) [45–51]. Once a foreign pathogen like KSHV is detected, the host innate immune response typically induces the production of type I interferons, priming cells to an antiviral state. However, KSHV evades immune detection by subverting these immune responses. The primary method of KSHV persistence is by defaulting to its latent state. Due to the low level of latent viral gene expression, the immune system may not be able to detect KSHV. KSHV also encodes viral products and co-opts cellular proteins as proviral factors to evade detection. For example, KSHV encodes a viral deubiquitinase, ORF64, which inhibits RLR-induced interferon production [52]. KSHV also uses cellular adenosine deaminase acting on RNA 1 (ADAR1) to further dampen RLR-dependent innate immune responses [53]. KSHV encodes a virion tegument protein, ORF52, LANA, and viral interform regulatory factor (vIRF1) to prevent cGAS-STING sensing [50,51,54]. Furthermore, KSHV uses cellular barrier-to-autointegration factor 1 (BAF) to compete with cGAS to prevent the cGAS-STING-mediated type I interferon response [55]. By counteracting antiviral immune responses, KSHV can persist and continue to disrupt cell function, contributing to the development of tumorigenesis.

Thus, KSHV manipulates canonical cell survival and proliferation pathways, reprograms metabolic activity, evades immune detection, and contributes to lymphomagenesis.

4.2. KSHV Latency-Associated Nuclear Antigen (LANA) Supports Lymphomagenesis

Viral oncogenes work together to promote cell survival and proliferation, creating a cell environment conducive to both KSHV persistence and lymphomagenesis. KSHV LANA is expressed in all latently infected cells and is a universal marker of KSHV infection, as it is present in KS spindle cells and MCD and PEL B cells. LANA is considered the master regulator of KSHV latency and relies on recruiting and manipulating host machinery. LANA has many functions, including maintenance of the viral genome, inhibition of tumor suppressors, activation of cellular signaling pathways to promote cell survival and proliferation, modulation of host immune responses, epigenetic regulation, and interaction with cellular proteins (Figure 1) [56].

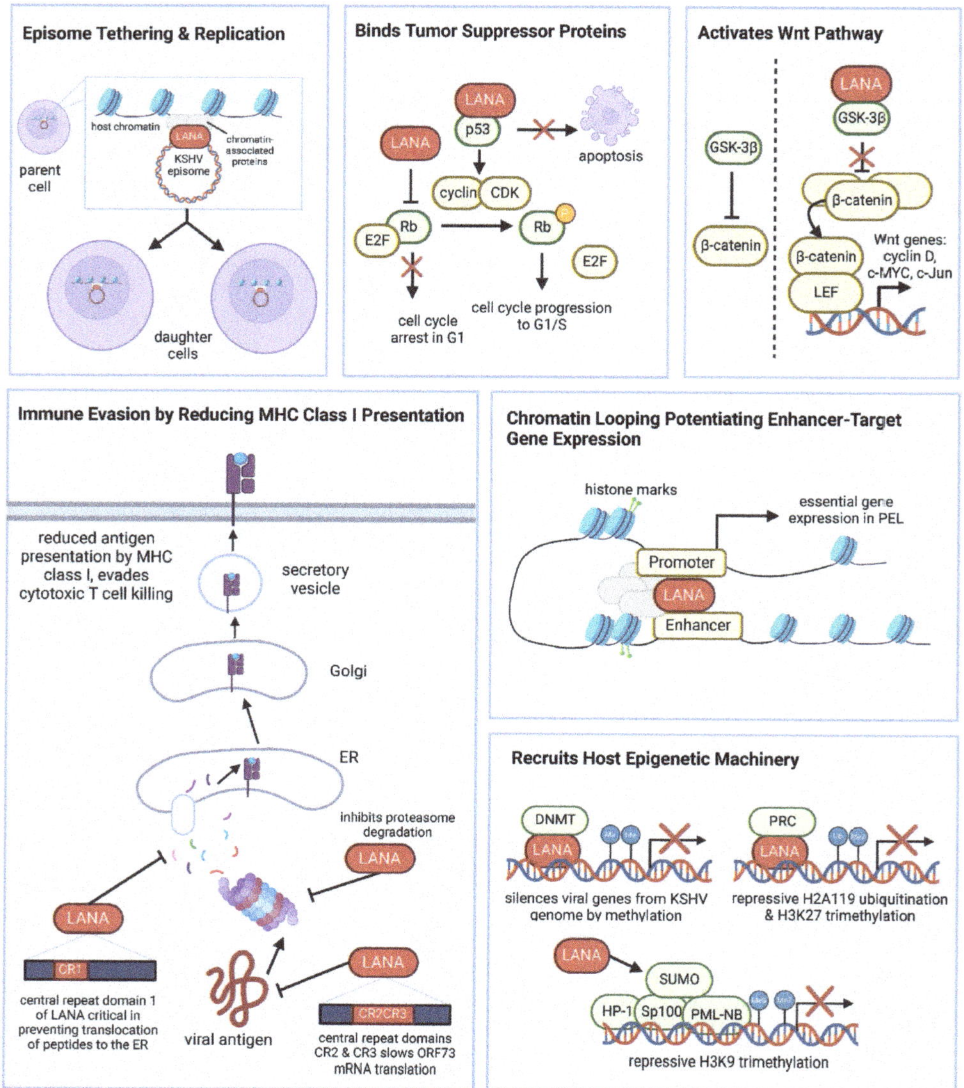

Figure 1. LANA manipulates host machinery for viral persistence. LANA is the master regulator of latency. LANA interacts with host proteins and has numerous functions, including, but not limited to, episome tethering for viral genome maintenance, inhibiting tumor suppressor proteins, activating pathways like Wnt, immune evasion, and controlling gene expression. The various roles of LANA contribute to KSHV-associated lymphomagenesis.

LANA localizes to the nucleus and tethers the KSHV episome to host cell chromosomes [57], ensuring that the viral genome is replicated in tandem with host DNA during normal cell division. LANA both binds to the sequences in the terminal repeats of the KSHV genome and the mitotic chromosomes. Additionally, LANA interacts with host chromatin and replication machinery to ensure that the genome is replicated and segregated appropriately. Daughter cells stably inherit the KSHV genome, which is essential for maintaining latency. Beyond tethering, a recent analysis profiling the genome-wide occupancy of LANA and epigenomic landscape of PEL cells found that LANA participates in chromatin looping and potentiates enhancer target gene expression [58]. The study observed increased levels of enhancer RNA transcription and activated chromatin marks at LANA-bound enhancers. The study examined LANA-occupied enhancers target genes and found that the identified targets are genes known to be critical for latency and tumorigenesis.

LANA also binds tumor suppressor proteins p53 and Rb [42,43]. When LANA binds p53, it inhibits p53's transcriptional activity and prevents p53-mediated apoptosis [42,59]. Infected cells can then evade cell cycle arrest and apoptosis, leading to uncontrolled cell proliferation. LANA interacts with Rb, which dysregulates Rb-mediated control of the cell cycle, leading to unrestrained cell division [43]. The full contribution of LANA binding Rb in the context of PEL is not known because while a proportion of Rb complexes to LANA, PEL cell lines show a lack of cyclin D-CDK inhibitor, p16INK4a, and BC3 PEL cells lack Rb altogether [60]. PEL cells remain sensitive to growth inhibition when p16INK4a expression is restored, implying that either Rb function in PEL cells is not completely inactive or that KSHV-infected cells are resistant to p16INK4a. Thus, LANA may only partly inhibit Rb, offering a predisposition to tumorigenesis, but other factors are likely needed for lymphomagenesis.

LANA is also known to activate the Wnt and Notch signaling pathways. It interacts with glycogen synthetase kinase 3β (GSK-3β), which typically phosphorylates and inactivates β-catenin through ubiquitin-mediated proteolysis [61]. However, upon LANA binding of GSK-3β, GSK-3β is relocated from the cytosol to the nucleus, which allows for the accumulation of β-catenin. The dysregulated β-catenin can then oligomerize the transcription factor LEF for the targeted transcription of Wnt genes, such as cyclin D, c-Myc, and c-Jun, which promote cell proliferation and survival. PEL cells show an upregulation of β-catenin [62]. The LANA-mediated alteration in the Wnt/β-catenin pathway further implicates LANA as a viral oncogene. LANA modulates Notch signaling, and the dysregulation of Notch signaling is implicated in its contribution to oncogenesis because Notch is important in the transcriptional activation of genes related to cell cycle, differentiation, and proliferation. The stability of Notch is negatively regulated by the E3 ligase Sel10. LANA has been shown to bind Sel10, suppressing the degradation of Notch, thus enhancing KSHV-infected cell proliferation [63]. Another function of LANA is upregulating telomerase activity by its interaction with transcription factor Sp1, contributing to cellular immortalization by maintaining telomere length [64].

LANA evades the host immune response by downregulating the expression of major histocompatibility complex (MHC) class I molecules [65], allowing infected cells to evade recognition and destruction from cytotoxic T cells. Epigenetically, LANA manipulates the cell through histone modifications and altering DNA methylation patterns. By interacting with histone-modifying enzymes, LANA alters chromatin structure, critical to regulating the expression of viral and host genes. During de novo infection, LANA recruits DNA methyltransferase 3A (DNMT3A) to broadly silence gene activity of the KSHV genome, allowing it to limit viral gene expression [66]. In summary, LANA interacts with a variety of cellular proteins to modulate their function and create a cell environment favoring viral latency and oncogenesis.

4.3. Viral Homologs Mimic Cellular Counterparts, Promoting Lymphomagenesis

Similar to other herpesviruses, KSHV has co-evolved with its human host and adapted a strategy of molecular mimicry by encoding a number of viral genes, such as *vCyclin*, *vFLIP*, *vGPCR*, and *vIRFs*, with homology to human genes (Table 1) [13,67].

In this review, we chose to focus on vIL-6, vCyclin, vFLIP, and vIRF3 because they are viral mimics that are well studied in the context of KSHV-associated lymphomagenesis. We also describe the viral microRNAs.

Table 1. KSHV viral genes with human gene homologs.

Human Gene	Viral Gene	KSHV ORF	Viral Function
Bcl-2	*vBcl-2*	ORF16	Pro-apoptotic factor signaling through caspases
CyclinD	*vCyclin*	ORF72	Complexes with CDK6 to drive G1 to S phase cell cycle transition
DUB	*ORF64*	ORF64	Viral deubiquitinase
FLIP	*vFLIP*	ORF71	Activates NF-κB signaling
GPCR	*vGPCR*	ORF74	Constitutively activates G-protein-coupled receptor for angiogenesis
IL-6	*vIL-6*	ORF K2	Cytokine that activates the JAK/STAT pathway
IRF4/IRF8	*vIRF3*	ORF K10.5 and K10.6	
	vIRF1	ORF K9	The vIRFs are transcription factors that bind cellular IRFs and inhibit
	vIRF2	ORF K11 and K11.1	interferon signaling, p53, and apoptosis
	vIRF4	ORF K10	Ubiquitin E3 ligases that target MHC class I molecules for degradation
MARCH ligases	*MIR1*	ORF K3	Binds CCR8 for Ca^{2+}-ion-dependent signaling
	MIR2	ORF K5	Agonist of CCR3 and CCR8 and antagonist of chemokine receptors
MIP1α	*vMIP-I*	ORF K6	Binds CCR4 as chemoattractant of Th2 T cells
	vMIP-II	ORF K4	Modulates inflammatory responses
	vMIP-III	ORF K4.1	
OX2	*vOX2*	ORF K14	

4.3.1. Viral Interleukin-6 (vIL-6)

KSHV encodes vIL-6, a viral homolog of human IL-6 (hIL-6), which has optimized IL-6 functions (reviewed in [37]). vIL-6 shares 25% sequence identity and 63% amino acid similarity with hIL-6; however, it can stimulate all known hIL-6 pathways [68]. vIL-6 differs from hIL-6 in several ways. vIL-6 can signal by engaging merely the gp130 subunit of the IL-6 receptor, whereas hIL-6 requires the full-length IL-6 receptor with both gp130 and gp80. vIL-6 also has broader capabilities in a wider variety of cell types compared to hIL-6 due to its ability to engage the ubiquitously expressed gp130 alone. Relative to hIL-6, vIL-6 secretion is considered inefficient because most of the protein is retained within the endoplasmic reticulum (ER). vIL-6 is expressed during both latency and lytic replication but to a much higher degree during the lytic phase. vIL-6 is detectable in sera and/or tissue samples from all three KSHV-associated malignancies [69].

Within the ER, the ER transmembrane protein vitamin K epoxide reductase complex subunit 1 variant 2 (VKORC1v2) is a binding partner of vIL-6, and the interaction contributes to growth and anti-apoptosis of PEL cells [70]. Additionally, vIL-6 has been shown to promote productive replication and latent PEL cell viability by upregulating the mannose-6-phosphate- and peptide hormone-interacting receptor IGF2R by preventing IGF2R binding with VKORC1cv2, rescuing IGF2R from degradation [71]. The rescued IGF2R, in turn, promotes PEL cell viability.

The role of vIL-6 in tumor development has been evaluated in several studies. vIL-6 has been shown to transform murine fibroblast NIH 3T3 cells, and when mice were inoculated with vIL-6-producing clones, the injected vIL-6-positive cells formed larger tumors relative to control cells [72]. Another group generated a vIL-6 transgenic mouse model using the BALB/c (C) genetic background because of its hyper-susceptibility to malignant plasma cell tumors. The C.vIL6 mice were prone to severe and sometimes fatal MCD-like disease [35]. The C.vIL6iMyc mice with transgenes for vIL6 and a chromosomal translocation to deregulate *Myc* in B cells developed aggressive plasmablastic neoplasms with both clinical and histopathological similarities to PEL, where immunodeficient $RAG2^{-/-}$ mice were transplanted with the BJAB B cell line containing either WT KSHV or a mutant strain without vIL-6 [36]. While the vIL-6 gene did not affect tumor mass, when vIL-6 was present, significantly more tumors were detected [73]. This indicates the critical role of vIL-6 in KSHV tumor formation in B cells in vivo.

vIL-6 contributes to lymphomagenesis through autocrine and paracrine signaling, activation of the JAK/STAT pathway, angiogenesis, inflammation, immune modulation, synergistic effects with other viral proteins, and direct effects on B cells. vIL-6 acts in both an autocrine and paracrine manner, where it can act on the cell producing vIL-6 as well as neighboring cells (whether infected or not) to create a pro-inflammatory and tumorigenic microenvironment by promoting the proliferation and survival of cells. Importantly, vIL-6 activates STAT3 through the JAK/STAT pathway [37]. Activated STAT3 translocates to the nucleus and promotes the transcription of genes involved in cell proliferation, survival, and immune evasion, including Bcl-2, Bcl-xL, cyclin D1, and c-Myc. vIL-6 signaling both phosphorylates and acetylates STAT3. Phosphorylated STAT3 transcriptionally activates DNA methyltransferase 1 (DNMT1) expression. Acetylated STAT3 forms a transcription factor complex with DNMT1 to bind and methylate the *CAV1* promoter for *CAV1* silencing. Downregulated CAV1 leads to activation of Akt signaling for increased cell invasion and growth transformation induced by KSHV [74]. Additionally, vIL-6 induces the production of a key angiogenesis factor, vascular endothelial growth factor (VEGF) [72]. Enhanced angiogenesis supplies the growing tumor with oxygen and nutrients, contributing to tumor growth and metastasis.

Recently, vIL-6 has been shown to promote the SIRT3-induced deacetylation of SERBP1 [75]. This prevents SERBP1 binding to and the protection of lipoyltransferase 2 mRNA from degradation, thus preventing ferroptosis. Ferroptosis is a form of cell death characterized by iron accumulation and lipid peroxidation that is known to restrict pathogenic infections and act as an innate tumor suppressor mechanism. Through SIRT3 and SERBP1, vIL-6 inhibits ferroptosis, thus allowing for KSHV-induced cell transformation.

Given that KSHV-associated PEL and MCD are often observed in the context of HIV patients, it has been proposed that HIV-1 and KSHV may synergize to promote angiogenesis and tumorigenesis. Studies show that both HIV-1 trans-activator of transcription (Tat) and negative factor (Nef) synergize with KSHV vIL-6 to promote vascular tube formation and cell proliferation, enhancing oncogenesis by activating the AKT pathway [76,77].

4.3.2. Viral Cyclin (vCyclin)

vCyclin mimics cellular cyclin D and E and activates cyclin-dependent kinases (CDK), namely CDK6, to drive cell cycle progression from the G1 to S phase (Figure 2) [78,79]. Human cyclin D1 is an important regulator of the cell cycle, where in normal cells, cyclin D1 expression levels are strictly regulated. However, when cyclin D1 expression is high, cyclin D1 drives unchecked cell proliferation, promoting tumor growth, and it is central to the pathogenesis of cancer. Like the cyclin D1-CDK6 complex, the vCyclin-CDK6 complex can phosphorylate and inactivate Rb, a critical regulator of the G1/S checkpoint [79]. Rb inactivation allows the cell to override cell cycle checkpoints and enter the S phase, leading to uncontrolled cell proliferation. Distinct from its cellular counterpart, the vCyclin-CDK6 complex is resistant to inhibition by cyclin-dependent kinase inhibitors like p21 and p27 [80]. Insensitivity allows for continued cell cycle progression, even though these inhibitory signals would typically halt cell division. Furthermore, while cyclin D expression is cell-cycle-dependent, vCyclin levels remain stable regardless of the cell cycle phase. The extended expression of vCyclin could contribute to the constitutive activation of vCyclin-CDK complexes in KSHV-infected cells [81,82].

vCyclin synergizes with other viral oncogenes, LANA and vFLIP. vCyclin and vFLIP are co-expressed from a bicistronic mRNA and are functionally interdependent on one another. vFLIP expression results in p21-/p27-mediated G1 arrest due to NF-kB hyper-activation; however, vCyclin prevents G1 arrest, promoting cell cycle entry [83]. Thus, vCyclin and vFLIP cooperate with one another. A study in lymphatic endothelial cells demonstrated that vCyclin is essential for bypassing senescence and thus may play a role in KS tumor development [84]; however, further studies are needed to show if this applies to B cells for PEL and MCD tumorigenesis.

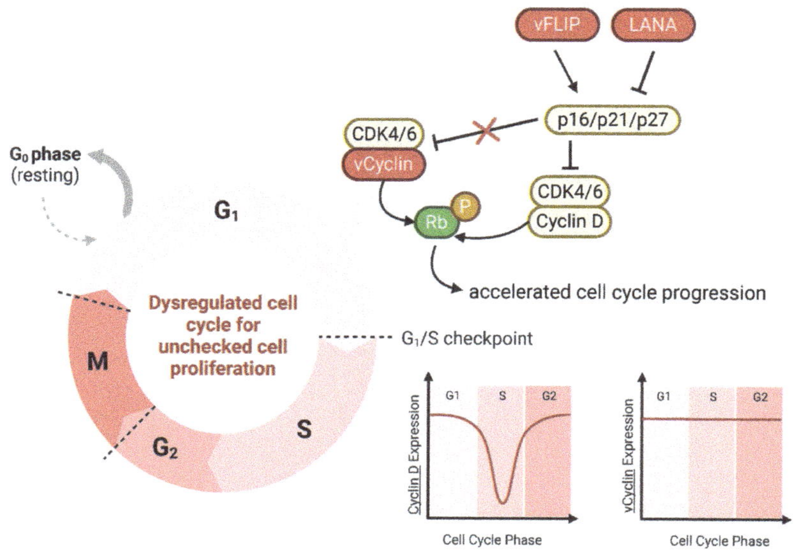

Figure 2. vCyclin dysregulates cell cycle for uncontrolled cell proliferation. vCyclin is the viral homolog of cellular cyclin D. vCyclin complexes with CDK6 and phosphorylates tumor suppressor protein Rb, bypassing the G1/S checkpoint. Cellular tumor suppressor proteins p16, p21, and p27 are cyclin-dependent kinase inhibitors that block cyclin D. However, unlike cyclin D, vCyclin is resistant to these CDK inhibitors, which contributes to cell cycle dysregulation. Additionally, other KSHV viral proteins like vFLIP and LANA regulate these cellular CDK inhibitors. Cyclin D expression is cell-cycle-phase-dependent; however, vCyclin expression remains stabilized throughout, further allowing for accelerated cell cycle progression and unchecked cell proliferation.

vCyclin can also alter cell metabolism. Insufficient oxygen levels in tissues, or hypoxia, induces pathways that promote tumor growth. Several studies show that hypoxia plays an important role in KSHV-induced tumors, where KS and PEL preferentially develop in locations with hypoxia [85]. Hypoxia is mediated by hypoxia-inducible factors (HIFs). Typically, the cell response to hypoxia with high HIF-1α levels induces metabolic reprogramming that restricts DNA replication and arrests cell division; however, KSHV-infected cells can maintain replication in hypoxia. A study showed that vCyclin can modulate HIF-1α signaling by binding HIF-1α for eventual lysosomal degradation through cytosolic translocation to restrict HIF-1α transcriptional activity [86]. Thus, vCyclin overcomes HIF-1α-mediated replication arrest and induces an oncogenic phenotype. A recent study found another way that KSHV alters metabolic pathways by hijacking nucleotide biosynthesis and deamidation-mediated glycolysis [87]. vCyclin binds CDK6 to phosphorylate carbamoyl-phosphate synthetase 2, aspartate transcarbamoylase, and dihydroorotase (CAD), a key enzyme of pyrimidine synthesis that deamidates the NF-kB subunit RelA to promote aerobic glycolysis and cell proliferation. KSHV-mediated metabolic reprogramming through vCyclin contributes to PEL tumorigenesis, shown in vitro and in vivo.

4.3.3. Viral FLICE Inhibitory Protein (vFLIP)

vFLIP is encoded from the same major latency locus as LANA and vCyclin and is the viral homolog of cellular FLICE inhibitory protein (FLIP). vFLIP is expressed during latency and undergoes RTA-induced degradation during lytic reactivation. Cellular FLIPs bind adaptor proteins in the Fas pathway through their death effector domains [88]. This binding impairs the recruitment and activation of caspase 8, blocking the caspase activation cascade, resulting in programmed cell death. Thus, vFLIP inhibits caspase-8, a key initiator of the extrinsic apoptosis pathway, allowing for survival and proliferation of infected cells (reviewed in [89]).

vFLIP also blocks cell death by activating NF-kB signaling [88,90,91]. vFLIP complexes with IKKγ, activating IKK, which leads to the phosphorylation of IkB and the release of active NF-kB. PEL cells display high levels of NF-kB activity [39]. Additionally, vFLIP promotes NF-kB signaling by upregulating IKKε and CADM1 and via the inhibition of the SAP18/HDAC1 complex, leading to NF-kB activation via the acetylation of p65 [92]. Thus, vFLIP is important in PEL cells to prevent cell death and upregulate NF-kB. NF-kB signaling leads to the transcription of cell proliferation and survival genes. Chronic activation of NF-kB augments the inflammatory tumor microenvironment that contributes to the progression of KSHV-associated lymphoproliferation. In a transgenic model of vFLIP, there was persistent NF-kB activation and an increased incidence of lymphomas as well as B cell abnormalities, consistent with those seen in MCD [90]. Another study showed that vFLIP increases LINE-1 retrotransposition, promoting genomic instability, a hallmark of cancer [93].

Hyperinflammation is a hallmark of KS-associated malignancies. A screen for inhibitors of KSHV-induced oncogenesis identified several anti-inflammatory candidates, including dexamethasone, which activates glucocorticoid receptor signaling. The group found that in KSHV-transformed cells, particularly, interleukin-1 alpha (IL-1α) was most induced, while IL-1 receptor antagonist (IL-1Ra) was highly suppressed [94]. KSHV micro RNAs and vFLIP were found to suppress IL-1Ra. Furthermore, vFLIP and vCyclin were found to inhibit glucocorticoid signaling in KSHV-transformed cells.

4.3.4. Viral Interferon Regulatory Factor 3 (vIRF3)

KSHV encodes multiple viral interferon regulatory factors (IRF) that interfere with the host interferon response. The KSHV genome has at least four genes encoding IRFs [95]. Cellular IRFs are transcription factors that promote the activation of type I interferons. However, vIRFs inhibit type I interferon signaling, which allows KSHV to evade a host's antiviral immune response. Thus, vIRFs promote cell survival and proliferation by modulating immune and inflammatory pathways.

While most of the vIRFs are lytic proteins, vIRF3, also known as LANA-2, encodes a nuclear protein expressed in latency [18]. However, latent vIRF3 gene expression differs by cell type. vIRF3 is expressed in latently infected B cells, including both PEL cells and cells from MCD but not KS spindle cells [96]. This implicates vIRF3 in the progression of KSHV-associated lymphomagenesis. vIRF3 is known to interact with p53 and reduce its stability, preventing its DNA binding activity [97]. Thus, the pro-apoptotic function of p53 is blocked, prolonging the lifespan of KSHV-infected B cells. Additionally, vIRF3 stimulates transcriptional activity of the oncogene Myc. vIRF3 competes with c-Myc to bind Myc modulator 1 (MM-1α), which reverses the MM-1α suppression of c-Myc. Furthermore, vIRF3 binds to the F-box of Skp2, part of a ubiquitin ligase complex, to recruit it to c-Myc-regulated promotors for the activation of c-Myc dependent transcription [98].

4.3.5. Viral microRNAs (miRNA)

miRNAs are non-coding single-stranded RNAs that are approximately 20–24 nucleotides long. They primarily function post-transcriptionally to inhibit target gene expression by binding to the complementary sequence of the target gene to compromise its expression by mRNA degradation or inhibition of translation. KSHV encodes 12 distinct pre-miRNAs that are later processed into 25 mature miRNAs in its latency-associated region of the viral genome [99–101]. KSHV-infected cells can release exosomes carrying viral miRNAs, affecting neighboring B cells in a paracrine manner, contributing to KSHV pathogenesis in PEL and MCD [102]. KSHV miRNAs are expressed at different levels, and the miRNA expression level also differs among PEL cell lines, where some PEL lines express low levels (BC-1, BCBL-1) while others express high levels (BC-3) [103]. The expression level of miRNAs correlates to the impact of their function [104]. A study using PAR-CLIP allowed for the transcriptome-wide identification of over 2000 cellular targets of

KSHV miRNAs in PEL cells (BC-1, BCBL1, and BC-3), implicating KSHV miRNAs in virus infection latency and lymphomagenesis [103].

Several KSHV miRNAs are expressed in latency and support latency maintenance. miR-K9-5p targets the KSHV replication and transcription activator (RTA), which is critical for initiating KSHV lytic reactivation [105]. Both miR-K3 and miR-K1 stabilize KSHV latency by regulating the RTA promoter, where miR-K3 targets nuclear factor I/B and miR-K1 regulates NF-κB signaling by targeting IκBα [106,107]. miR-K11 also stabilizes KSHV latency by attenuating the interferon response induced by pattern recognition receptors [108]. A study using CRISPR/Cas9 to ablate KSHV miRNA expression in PEL cell lines, BCBL-1 and BCP-1, led to the upregulation of KSHV lytic genes, further demonstrating the significance of KSHV miRNAs in maintaining latency [109].

There are KSHV miRNAs that share seed homology to cellular miRNAs, suggesting that these viral miRNAs mimic their human counterpart. miR-K11 has sequence similarity to human miR-155, which plays a role in B cell development and oncogenesis [110,111]. miR-K11 has been shown to target *BACH1* mRNA for increased viability under oxidative stress and the enhanced proliferation of $CD19^+$ B cells in the spleen [110–112]. Thus, it is thought that miR-K11 has a role in KSHV-induced lymphomagenesis. miR-K10a and cellular miR-142-3p are also analogs and share some seed homology. miR-142-3p is known to have high abundance in PEL cells. miR-K10a has been shown to overlap in function with miR-142-3p, and it represses cell cycle inhibitor p27 in murine fibroblasts and PEL cells; thus, it has transforming activity [24]. Additionally, miR-K10 and its variants inhibit TGF-β-regulated cell growth and apoptosis in PEL cells, contributing to the dysregulation of normal cell growth control [113]. Other KSHV miRNAs that inhibit cell death include miR-K1, K3, and K4. These miRNAs target *CASP3* to block caspase-induced apoptosis [114]. miR-K1 also targets p21, an inhibitory regulator of the cell cycle; thus, it blocks cell cycle arrest [115]. KSHV miRNAs maintain latency and dysregulate cell growth and death, promoting lymphomagenesis.

5. Therapeutics

Based on studies of KSHV, PEL, and MCD, there has been progress in the development of therapeutics. PEL is highly aggressive and frequently resistant to standard chemotherapy regimens like CHOP (cyclophosphamide, doxorubicin, vincristine, and prednisone). Thus, therapeutics have been developed that target cellular products (Figure 3).

Hsp90 is a chaperone protein that has been shown to be essential in the survival of KSHV-associated malignancies. vFLIP forms a complex with the host proteins IKKγ and Hsp90. Studies have shown that inhibiting Hsp90 blocked viral gene expression in primary infection and downregulated NF-κB activity [91,116]. Another study showed that treatment with a purine scaffold Hsp90 inhibitor, PU−H71, selectively binds the fraction of Hsp90 that is enriched in tumor cells. PU−H71 kills PEL cells and prolonged disease-free survival in a PEL xenograft mouse model [117]. Furthermore, PU−H71 and a pan-Bcl2 inhibitor, GX15−070, synergize in PEL cells to enhance PEL cell killing [117]. In a follow-up screen to identify additional PEL-dependencies, the nucleoside analog activated by adenosine kinase, 6−ETI, was a top hit. 6−ETI was validated to be effective in reducing tumors in a PEL xenograft model by inducing S phase arrest [118]. It has also been shown that PEL relies on the mitotic kinase NEK2 for survival [119]. Inhibition of NEK2 with the JH295 inhibitor resulted in the caspase-3 mediated apoptosis of PEL and prolonged survival and reduced the tumor burden in a PEL mouse model. JH295 also sensitizes lymphomas to other chemotherapeutic agents like rapamycin, enhancing cancer cell death [120,121]. Another group found that the inhibition of Ring finger protein 5 (RNF5), an ER-localized E3 ligase, decreases PEL xenograft tumor growth and downregulates both viral and cell cycle gene expression [122]. The RNF5 inhibitor INH2 works by increasing the levels of Ephrin receptors A3 and A4, reducing Erk and Akt activation as well as oncogenic KSHV lytic replication. Interestingly, a group found that capsaicin, a compound from chili pepper, can inhibit the growth of PEL cells by inhibiting ERK, p38 MAPK, and hIL-6 expression [123].

By inhibiting ERK and p38 MAPK phosphorylation, capsaicin can induce apoptosis in PEL cells in a caspase-9-dependent manner. This was confirmed in vitro in PEL cells and in PEL development in SCID mice. HIF-1α has been considered as a therapeutic target for PEL. HIF-1α is necessary for creating a metabolic state optimal for PEL growth, where HIF-1α knockdown or treatment with PX−478, a small-molecule inhibitor of HIF-1α, leads to a decreased expression of KSHV latent genes [124]. Thus, inhibition of HIF-1α could be a potential therapeutic strategy.

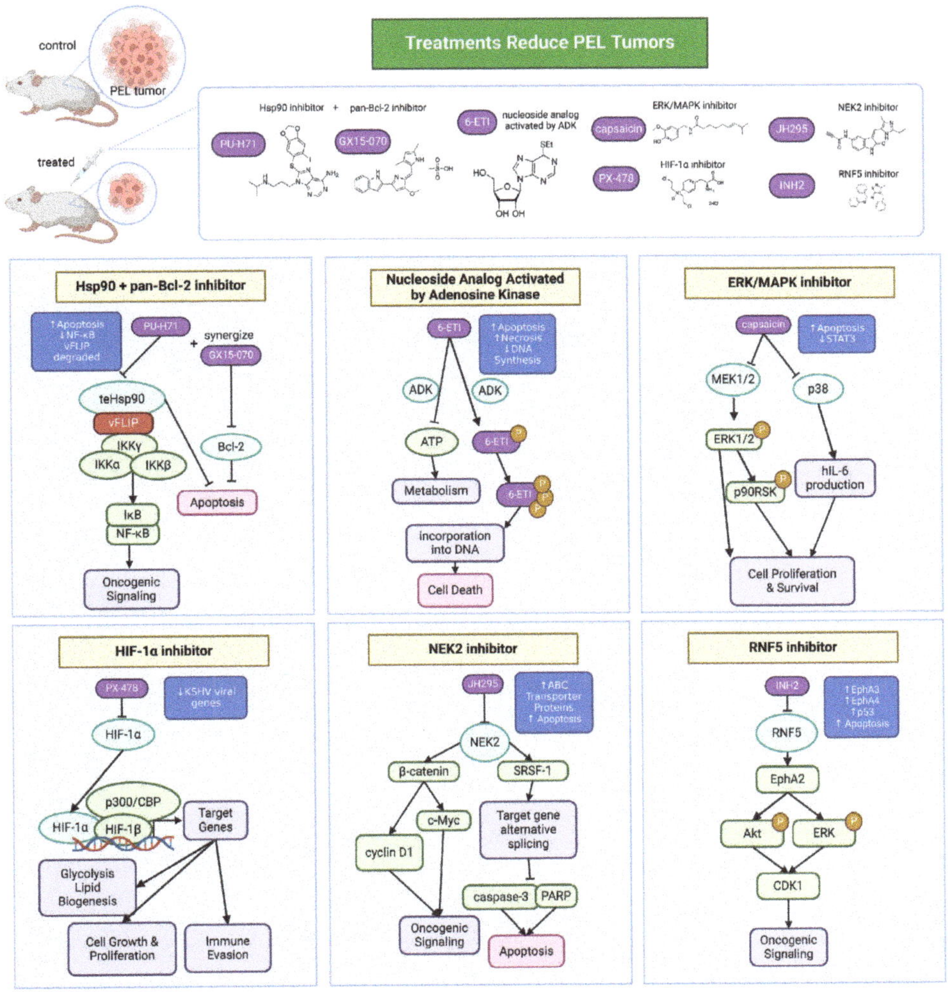

Figure 3. Therapeutics for PEL. KSHV co-opts host factors and machinery to promote its pathogenesis and persistence, dysregulating normal cell processes that then lead to oncogenic signaling. Development of therapeutics targeted towards host factors has shown efficacy in treating PEL-associated cell lines and mouse tumor models. Many of the therapeutics work to increase tumor cell death via apoptosis.

For KSHV-MCD, the monoclonal antibody drug rituximab has been used in patients, improving the survival of these patients and reducing the risk of lymphoma [125]. Additionally, a combination therapy with rituximab and the anthracycline doxorubicin has been used with efficacy in patients with MCD and KS [126]. Monoclonal antibodies targeting the

IL-6/IL-6 receptor have been approved for idiopathic MCD but are not treatment strategies for KSHV-MCD [127]. Given the role of vIL-6 in both MCD and PEL, there is compelling evidence supporting the utility of vIL-6 neutralization, and there are ongoing efforts to generate vIL-6-specific neutralizing antibodies [68].

Generally, for KSHV-associated malignancies, viral and host products contributing to KSHV lymphomagenesis are potential therapeutic targets. For example, therapies have been developed to inhibit viral oncogenes, like LANA [128], or to inhibit the signaling pathways involved, like the JAK/STAT pathway. JAK inhibitors have been studied to block the pathogenesis of both PEL and KSHV-MCD, given the importance of the IL-6/JAK/STAT3 axis in both diseases. From a panel of JAK inhibitors, pacritinib was found to most significantly inhibit the growth of PEL cell lines [129].

6. Conclusions

KSHV is the driver of the lymphoproliferative disorders PEL and MCD. KSHV encodes genes to commandeer cellular processes to benefit KSHV persistence, which gives way to lymphoproliferation through a dysregulated cell cycle and immune evasion. Furthering our understanding of the molecular mechanisms of KSHV-associated lymphomagenesis informs the development of more effective treatments. Active areas of investigation include studying the interactions between KSHV proteins and host cellular pathways, looking into ways to enhance the immune response against KSHV-infected cells, and developing drugs targeted to KSHV-induced epigenetic modifications.

Author Contributions: Writing—original draft preparation, C.J.Y.; writing—review and editing, B.D. All authors have read and agreed to the published version of the manuscript.

Funding: B.D. was supported by National Institutes of Health grants CA019014, CA291437, CA294800, and CA163217. C.J.Y. was supported by National Institutes of Health training grant T32 AI007001.

Acknowledgments: Figures were created with Biorender.com.

Conflicts of Interest: The authors declare no conflicts of interest.

References

1. Cesarman, E.; Damania, B.; Krown, S.E.; Martin, J.; Bower, M.; Whitby, D. Kaposi sarcoma. *Nat. Rev. Dis. Primers* **2019**, *5*, 9. [CrossRef]
2. Chang, Y.; Cesarman, E.; Pessin, M.S.; Lee, F.; Culpepper, J.; Knowles, D.M.; Moore, P.S. Identification of herpesvirus-like DNA sequences in AIDS-associated Kaposi's sarcoma. *Science* **1994**, *266*, 1865–1869. [CrossRef]
3. Soulier, J.; Grollet, L.; Oksenhendler, E.; Cacoub, P.; Cazals-Hatem, D.; Babinet, P.; d'Agay, M.F.; Clauvel, J.P.; Raphael, M.; Degos, L.; et al. Kaposi's sarcoma-associated herpesvirus-like DNA sequences in multicentric Castleman's disease. *Blood* **1995**, *86*, 1276–1280. [CrossRef]
4. Cesarman, E.; Chang, Y.; Moore, P.S.; Said, J.W.; Knowles, D.M. Kaposi's sarcoma-associated herpesvirus-like DNA sequences in AIDS-related body-cavity-based lymphomas. *N. Engl. J. Med.* **1995**, *332*, 1186–1191. [CrossRef]
5. Broussard, G.; Damania, B. Regulation of KSHV Latency and Lytic Reactivation. *Viruses* **2020**, *12*, 1034. [CrossRef]
6. Manners, O.; Murphy, J.C.; Coleman, A.; Hughes, D.J.; Whitehouse, A. Contribution of the KSHV and EBV lytic cycles to tumourigenesis. *Curr. Opin. Virol.* **2018**, *32*, 60–70. [CrossRef]
7. Chakraborty, S.; Veettil, M.V.; Chandran, B. Kaposi's Sarcoma Associated Herpesvirus Entry into Target Cells. *Front. Microbiol.* **2012**, *3*, 6. [CrossRef]
8. Thandra, K.C.; Barsouk, A.; Saginala, K.; Padala, S.A.; Barsouk, A.; Rawla, P. Epidemiology of Non-Hodgkin's Lymphoma. *Med. Sci.* **2021**, *9*, 5. [CrossRef]
9. Morton, L.M.; Wang, S.S.; Cozen, W.; Linet, M.S.; Chatterjee, N.; Davis, S.; Severson, R.K.; Colt, J.S.; Vasef, M.A.; Rothman, N.; et al. Etiologic heterogeneity among non-Hodgkin lymphoma subtypes. *Blood* **2008**, *112*, 5150–5160. [CrossRef]
10. Available online: https://www.lls.org/lymphoma/non-hodgkin-lymphoma/nhl-subtypes (accessed on 8 February 2024).
11. Jaffe, E.S.; Harris, N.L.; Stein, H.; Isaacson, P.G. Classification of lymphoid neoplasms: The microscope as a tool for disease discovery. *Blood* **2008**, *112*, 4384–4399. [CrossRef]
12. Chu, Y.; Liu, Y.; Fang, X.; Jiang, Y.; Ding, M.; Ge, X.; Yuan, D.; Lu, K.; Li, P.; Li, Y.; et al. The epidemiological patterns of non-Hodgkin lymphoma: Global estimates of disease burden, risk factors, and temporal trends. *Front. Oncol.* **2023**, *13*, 1059914. [CrossRef]
13. Giffin, L.; Damania, B. KSHV: Pathways to tumorigenesis and persistent infection. *Adv. Virus Res.* **2014**, *88*, 111–159. [CrossRef]

14. Nador, R.G.; Cesarman, E.; Chadburn, A.; Dawson, D.B.; Ansari, M.Q.; Sald, J.; Knowles, D.M. Primary effusion lymphoma: A distinct clinicopathologic entity associated with the Kaposi's sarcoma-associated herpes virus. *Blood* **1996**, *88*, 645–656. [CrossRef]
15. Jones, K.D.; Aoki, Y.; Chang, Y.; Moore, P.S.; Yarchoan, R.; Tosato, G. Involvement of interleukin-10 (IL-10) and viral IL-6 in the spontaneous growth of Kaposi's sarcoma herpesvirus-associated infected primary effusion lymphoma cells. *Blood* **1999**, *94*, 2871–2879. [CrossRef]
16. Lurain, K.; Polizzotto, M.N.; Aleman, K.; Bhutani, M.; Wyvill, K.M.; Goncalves, P.H.; Ramaswami, R.; Marshall, V.A.; Miley, W.; Steinberg, S.M.; et al. Viral, immunologic, and clinical features of primary effusion lymphoma. *Blood* **2019**, *133*, 1753–1761. [CrossRef]
17. Godfrey, A.; Anderson, J.; Papanastasiou, A.; Takeuchi, Y.; Boshoff, C. Inhibiting primary effusion lymphoma by lentiviral vectors encoding short hairpin RNA. *Blood* **2005**, *105*, 2510–2518. [CrossRef]
18. Wies, E.; Mori, Y.; Hahn, A.; Kremmer, E.; Sturzl, M.; Fleckenstein, B.; Neipel, F. The viral interferon-regulatory factor-3 is required for the survival of KSHV-infected primary effusion lymphoma cells. *Blood* **2008**, *111*, 320–327. [CrossRef]
19. Horenstein, M.G.; Nador, R.G.; Chadburn, A.; Hyjek, E.M.; Inghirami, G.; Knowles, D.M.; Cesarman, E. Epstein-Barr virus latent gene expression in primary effusion lymphomas containing Kaposi's sarcoma-associated herpesvirus/human herpesvirus-8. *Blood* **1997**, *90*, 1186–1191. [CrossRef]
20. Szekely, L.; Chen, F.; Teramoto, N.; Ehlin-Henriksson, B.; Pokrovskaja, K.; Szeles, A.; Manneborg-Sandlund, A.; Lowbeer, M.; Lennette, E.T.; Klein, G. Restricted expression of Epstein-Barr virus (EBV)-encoded, growth transformation-associated antigens in an EBV- and human herpesvirus type 8-carrying body cavity lymphoma line. *J. Gen. Virol.* **1998**, *79 Pt 6*, 1445–1452. [CrossRef]
21. Lange, P.T.; Damania, B. Modeling oncogenic herpesvirus infections in humanized mice. *Curr. Opin. Virol.* **2020**, *44*, 90–96. [CrossRef]
22. Bigi, R.; Landis, J.T.; An, H.; Caro-Vegas, C.; Raab-Traub, N.; Dittmer, D.P. Epstein-Barr virus enhances genome maintenance of Kaposi sarcoma-associated herpesvirus. *Proc. Natl. Acad. Sci. USA* **2018**, *115*, E11379–E11387. [CrossRef] [PubMed]
23. McHugh, D.; Caduff, N.; Barros, M.H.M.; Ramer, P.C.; Raykova, A.; Murer, A.; Landtwing, V.; Quast, I.; Styles, C.T.; Spohn, M.; et al. Persistent KSHV Infection Increases EBV-Associated Tumor Formation In Vivo via Enhanced EBV Lytic Gene Expression. *Cell Host Microbe* **2017**, *22*, 61–73.e67. [CrossRef] [PubMed]
24. Munz, C. The Role of Lytic Infection for Lymphomagenesis of Human gamma-Herpesviruses. *Front. Cell. Infect. Microbiol.* **2021**, *11*, 605258. [CrossRef]
25. Faure, A.; Hayes, M.; Sugden, B. How Kaposi's sarcoma-associated herpesvirus stably transforms peripheral B cells towards lymphomagenesis. *Proc. Natl. Acad. Sci. USA* **2019**, *116*, 16519–16528. [CrossRef]
26. Fajgenbaum, D.C.; van Rhee, F.; Nabel, C.S. HHV-8-negative, idiopathic multicentric Castleman disease: Novel insights into biology, pathogenesis, and therapy. *Blood* **2014**, *123*, 2924–2933. [CrossRef]
27. Zhou, T.; Wang, H.W.; Pittaluga, S.; Jaffe, E.S. Multicentric Castleman disease and the evolution of the concept. *Pathologica* **2021**, *113*, 339–353. [CrossRef]
28. Ramaswami, R.; Lurain, K.; Polizzotto, M.N.; Ekwede, I.; Waldon, K.; Steinberg, S.M.; Mangusan, R.; Widell, A.; Rupert, A.; George, J.; et al. Characteristics and outcomes of KSHV-associated multicentric Castleman disease with or without other KSHV diseases. *Blood Adv.* **2021**, *5*, 1660–1670. [CrossRef]
29. Oksenhendler, E.; Boutboul, D.; Fajgenbaum, D.; Mirouse, A.; Fieschi, C.; Malphettes, M.; Vercellino, L.; Meignin, V.; Gerard, L.; Galicier, L. The full spectrum of Castleman disease: 273 patients studied over 20 years. *Br. J. Haematol.* **2018**, *180*, 206–216. [CrossRef] [PubMed]
30. Vega, F.; Miranda, R.N.; Medeiros, L.J. KSHV/HHV8-positive large B-cell lymphomas and associated diseases: A heterogeneous group of lymphoproliferative processes with significant clinicopathological overlap. *Mod. Pathol.* **2020**, *33*, 18–28. [CrossRef]
31. Dossier, A.; Meignin, V.; Fieschi, C.; Boutboul, D.; Oksenhendler, E.; Galicier, L. Human herpesvirus 8-related Castleman disease in the absence of HIV infection. *Clin. Infect. Dis.* **2013**, *56*, 833–842. [CrossRef]
32. Aoki, Y.; Tosato, G.; Fonville, T.W.; Pittaluga, S. Serum viral interleukin-6 in AIDS-related multicentric Castleman disease. *Blood* **2001**, *97*, 2526–2527. [CrossRef] [PubMed]
33. Aoki, Y.; Jones, K.D.; Tosato, G. Kaposi's sarcoma-associated herpesvirus-encoded interleukin-6. *J. Hematother. Stem Cell Res.* **2000**, *9*, 137–145. [CrossRef] [PubMed]
34. Chadburn, A.; Hyjek, E.M.; Tam, W.; Liu, Y.; Rengifo, T.; Cesarman, E.; Knowles, D.M. Immunophenotypic analysis of the Kaposi sarcoma herpesvirus (KSHV; HHV-8)-infected B cells in HIV+ multicentric Castleman disease (MCD). *Histopathology* **2008**, *53*, 513–524. [CrossRef]
35. Suthaus, J.; Stuhlmann-Laeisz, C.; Tompkins, V.S.; Rosean, T.R.; Klapper, W.; Tosato, G.; Janz, S.; Scheller, J.; Rose-John, S. HHV-8-encoded viral IL-6 collaborates with mouse IL-6 in the development of multicentric Castleman disease in mice. *Blood* **2012**, *119*, 5173–5181. [CrossRef]
36. Rosean, T.R.; Holman, C.J.; Tompkins, V.S.; Jing, X.; Krasowski, M.D.; Rose-John, S.; Janz, S. KSHV-encoded vIL-6 collaborates with deregulated c-Myc to drive plasmablastic neoplasms in mice. *Blood Cancer J.* **2016**, *6*, e398. [CrossRef]
37. Sakakibara, S.; Tosato, G. Viral interleukin-6: Role in Kaposi's sarcoma-associated herpesvirus: Associated malignancies. *J. Interferon Cytokine Res.* **2011**, *31*, 791–801. [CrossRef]
38. Shaffer, A.L., 3rd; Young, R.M.; Staudt, L.M. Pathogenesis of human B cell lymphomas. *Annu. Rev. Immunol.* **2012**, *30*, 565–610. [CrossRef]

39. Keller, S.A.; Hernandez-Hopkins, D.; Vider, J.; Ponomarev, V.; Hyjek, E.; Schattner, E.J.; Cesarman, E. NF-kappaB is essential for the progression of KSHV- and EBV-infected lymphomas in vivo. *Blood* **2006**, *107*, 3295–3302. [CrossRef] [PubMed]
40. Bhatt, A.P.; Damania, B. AKTivation of PI3K/AKT/mTOR signaling pathway by KSHV. *Front. Immunol.* **2012**, *3*, 401. [CrossRef]
41. Xiang, Q.; Yang, Z.; Nicholas, J. STAT and Janus kinase targeting by human herpesvirus 8 interferon regulatory factor in the suppression of type-I interferon signaling. *PLoS Pathog.* **2022**, *18*, e1010676. [CrossRef]
42. Friborg, J., Jr.; Kong, W.; Hottiger, M.O.; Nabel, G.J. p53 inhibition by the LANA protein of KSHV protects against cell death. *Nature* **1999**, *402*, 889–894. [CrossRef] [PubMed]
43. Hume, A.J.; Kalejta, R.F. Regulation of the retinoblastoma proteins by the human herpesviruses. *Cell Div.* **2009**, *4*, 1. [CrossRef] [PubMed]
44. Bhatt, A.P.; Jacobs, S.R.; Freemerman, A.J.; Makowski, L.; Rathmell, J.C.; Dittmer, D.P.; Damania, B. Dysregulation of fatty acid synthesis and glycolysis in non-Hodgkin lymphoma. *Proc. Natl. Acad. Sci. USA* **2012**, *109*, 11818–11823. [CrossRef]
45. Gregory, S.M.; Davis, B.K.; West, J.A.; Taxman, D.J.; Matsuzawa, S.; Reed, J.C.; Ting, J.P.; Damania, B. Discovery of a viral NLR homolog that inhibits the inflammasome. *Science* **2011**, *331*, 330–334. [CrossRef]
46. West, J.; Damania, B. Upregulation of the TLR3 pathway by Kaposi's sarcoma-associated herpesvirus during primary infection. *J. Virol.* **2008**, *82*, 5440–5449. [CrossRef] [PubMed]
47. Lagos, D.; Vart, R.J.; Gratrix, F.; Westrop, S.J.; Emuss, V.; Wong, P.P.; Robey, R.; Imami, N.; Bower, M.; Gotch, F.; et al. Toll-like receptor 4 mediates innate immunity to Kaposi sarcoma herpesvirus. *Cell Host Microbe* **2008**, *4*, 470–483. [CrossRef]
48. West, J.A.; Wicks, M.; Gregory, S.M.; Chugh, P.; Jacobs, S.R.; Zhang, Z.; Host, K.M.; Dittmer, D.P.; Damania, B. An important role for mitochondrial antiviral signaling protein in the Kaposi's sarcoma-associated herpesvirus life cycle. *J. Virol.* **2014**, *88*, 5778–5787. [CrossRef] [PubMed]
49. Kerur, N.; Veettil, M.V.; Sharma-Walia, N.; Bottero, V.; Sadagopan, S.; Otageri, P.; Chandran, B. IFI16 acts as a nuclear pathogen sensor to induce the inflammasome in response to Kaposi Sarcoma-associated herpesvirus infection. *Cell Host Microbe* **2011**, *9*, 363–375. [CrossRef]
50. Ma, Z.; Jacobs, S.R.; West, J.A.; Stopford, C.; Zhang, Z.; Davis, Z.; Barber, G.N.; Glaunsinger, B.A.; Dittmer, D.P.; Damania, B. Modulation of the cGAS-STING DNA sensing pathway by gammaherpesviruses. *Proc. Natl. Acad. Sci. USA* **2015**, *112*, E4306–E4315. [CrossRef]
51. Wu, J.J.; Li, W.; Shao, Y.; Avey, D.; Fu, B.; Gillen, J.; Hand, T.; Ma, S.; Liu, X.; Miley, W.; et al. Inhibition of cGAS DNA Sensing by a Herpesvirus Virion Protein. *Cell Host Microbe* **2015**, *18*, 333–344. [CrossRef]
52. Inn, K.S.; Lee, S.H.; Rathbun, J.Y.; Wong, L.Y.; Toth, Z.; Machida, K.; Ou, J.H.; Jung, J.U. Inhibition of RIG-I-mediated signaling by Kaposi's sarcoma-associated herpesvirus-encoded deubiquitinase ORF64. *J. Virol.* **2011**, *85*, 10899–10904. [CrossRef] [PubMed]
53. Zhang, H.; Ni, G.; Damania, B. ADAR1 Facilitates KSHV Lytic Reactivation by Modulating the RLR-Dependent Signaling Pathway. *Cell Rep.* **2020**, *31*, 107564. [CrossRef]
54. Zhang, G.; Chan, B.; Samarina, N.; Abere, B.; Weidner-Glunde, M.; Buch, A.; Pich, A.; Brinkmann, M.M.; Schulz, T.F. Cytoplasmic isoforms of Kaposi sarcoma herpesvirus LANA recruit and antagonize the innate immune DNA sensor cGAS. *Proc. Natl. Acad. Sci. USA* **2016**, *113*, E1034–E1043. [CrossRef]
55. Broussard, G.; Ni, G.; Zhang, Z.; Li, Q.; Cano, P.; Dittmer, D.P.; Damania, B. Barrier-to-autointegration factor 1 promotes gammaherpesvirus reactivation from latency. *Nat. Commun.* **2023**, *14*, 434. [CrossRef] [PubMed]
56. Uppal, T.; Banerjee, S.; Sun, Z.; Verma, S.C.; Robertson, E.S. KSHV LANA—The master regulator of KSHV latency. *Viruses* **2014**, *6*, 4961–4998. [CrossRef]
57. Cotter, M.A., 2nd; Robertson, E.S. The latency-associated nuclear antigen tethers the Kaposi's sarcoma-associated herpesvirus genome to host chromosomes in body cavity-based lymphoma cells. *Virology* **1999**, *264*, 254–264. [CrossRef]
58. Ye, X.; Guerin, L.N.; Chen, Z.; Rajendren, S.; Dunker, W.; Zhao, Y.; Zhang, R.; Hodges, E.; Karijolich, J. Enhancer-promoter activation by the Kaposi sarcoma-associated herpesvirus episome maintenance protein LANA. *Cell Rep.* **2024**, *43*, 113888. [CrossRef]
59. Chen, W.; Hilton, I.B.; Staudt, M.R.; Burd, C.E.; Dittmer, D.P. Distinct p53, p53:LANA, and LANA complexes in Kaposi's Sarcoma--associated Herpesvirus Lymphomas. *J. Virol.* **2010**, *84*, 3898–3908. [CrossRef] [PubMed]
60. Platt, G.; Carbone, A.; Mittnacht, S. p16INK4a loss and sensitivity in KSHV associated primary effusion lymphoma. *Oncogene* **2002**, *21*, 1823–1831. [CrossRef]
61. Fujimuro, M.; Hayward, S.D. The latency-associated nuclear antigen of Kaposi's sarcoma-associated herpesvirus manipulates the activity of glycogen synthase kinase-3beta. *J. Virol.* **2003**, *77*, 8019–8030. [CrossRef]
62. Fujimuro, M.; Wu, F.Y.; ApRhys, C.; Kajumbula, H.; Young, D.B.; Hayward, G.S.; Hayward, S.D. A novel viral mechanism for dysregulation of beta-catenin in Kaposi's sarcoma-associated herpesvirus latency. *Nat. Med.* **2003**, *9*, 300–306. [CrossRef] [PubMed]
63. Lan, K.; Verma, S.C.; Murakami, M.; Bajaj, B.; Kaul, R.; Robertson, E.S. Kaposi's sarcoma herpesvirus-encoded latency-associated nuclear antigen stabilizes intracellular activated Notch by targeting the Sel10 protein. *Proc. Natl. Acad. Sci. USA* **2007**, *104*, 16287–16292. [CrossRef]
64. Verma, S.C.; Borah, S.; Robertson, E.S. Latency-associated nuclear antigen of Kaposi's sarcoma-associated herpesvirus up-regulates transcription of human telomerase reverse transcriptase promoter through interaction with transcription factor Sp1. *J. Virol.* **2004**, *78*, 10348–10359. [CrossRef]

65. Kwun, H.J.; da Silva, S.R.; Qin, H.; Ferris, R.L.; Tan, R.; Chang, Y.; Moore, P.S. The central repeat domain 1 of Kaposi's sarcoma-associated herpesvirus (KSHV) latency associated-nuclear antigen 1 (LANA1) prevents cis MHC class I peptide presentation. *Virology* **2011**, *412*, 357–365. [CrossRef] [PubMed]
66. Shamay, M.; Krithivas, A.; Zhang, J.; Hayward, S.D. Recruitment of the de novo DNA methyltransferase Dnmt3a by Kaposi's sarcoma-associated herpesvirus LANA. *Proc. Natl. Acad. Sci. USA* **2006**, *103*, 14554–14559. [CrossRef] [PubMed]
67. Mesri, M. Advances in Proteomic Technologies and Its Contribution to the Field of Cancer. *Adv. Med.* **2014**, *2014*, 238045. [CrossRef]
68. Suthaus, J.; Adam, N.; Grotzinger, J.; Scheller, J.; Rose-John, S. Viral Interleukin-6: Structure, pathophysiology and strategies of neutralization. *Eur. J. Cell Biol.* **2011**, *90*, 495–504. [CrossRef] [PubMed]
69. Aoki, Y.; Narazaki, M.; Kishimoto, T.; Tosato, G. Receptor engagement by viral interleukin-6 encoded by Kaposi sarcoma-associated herpesvirus. *Blood* **2001**, *98*, 3042–3049. [CrossRef]
70. Chen, D.; Cousins, E.; Sandford, G.; Nicholas, J. Human herpesvirus 8 viral interleukin-6 interacts with splice variant 2 of vitamin K epoxide reductase complex subunit 1. *J. Virol.* **2012**, *86*, 1577–1588. [CrossRef]
71. Li, Q.; Chen, D.; Xiang, Q.; Nicholas, J. Insulin-Like Growth Factor 2 Receptor Expression Is Promoted by Human Herpesvirus 8-Encoded Interleukin-6 and Contributes to Viral Latency and Productive Replication. *J. Virol.* **2019**, *93*. [CrossRef]
72. Aoki, Y.; Jaffe, E.S.; Chang, Y.; Jones, K.; Teruya-Feldstein, J.; Moore, P.S.; Tosato, G. Angiogenesis and hematopoiesis induced by Kaposi's sarcoma-associated herpesvirus-encoded interleukin-6. *Blood* **1999**, *93*, 4034–4043. [CrossRef] [PubMed]
73. Fullwood, R.A.; Low, G.M.; Chase, E.P.; Grasley, M.; Beal, S.S.; McCrary, I.M.; Daniels, C.W.; Ingersoll, K.; Berges, B.K. The Kaposi's sarcoma-associated herpesvirus viral interleukin 6 gene affects metastasis and expression of B cell markers in a murine xenograft model. *PLoS ONE* **2018**, *13*, e0204947. [CrossRef] [PubMed]
74. Li, W.; Wang, Q.; Qi, X.; Guo, Y.; Lu, H.; Chen, Y.; Lu, Z.; Yan, Q.; Zhu, X.; Jung, J.U.; et al. Viral interleukin-6 encoded by an oncogenic virus promotes angiogenesis and cellular transformation by enhancing STAT3-mediated epigenetic silencing of caveolin 1. *Oncogene* **2020**, *39*, 4603–4618. [CrossRef] [PubMed]
75. Zhou, J.; Wang, T.; Zhang, H.; Liu, J.; Wei, P.; Xu, R.; Yan, Q.; Chen, G.; Li, W.; Gao, S.J.; et al. KSHV vIL-6 promotes SIRT3-induced deacetylation of SERBP1 to inhibit ferroptosis and enhance cellular transformation by inducing lipoyltransferase 2 mRNA degradation. *PLoS Pathog.* **2024**, *20*, e1012082. [CrossRef] [PubMed]
76. Zhou, F.; Xue, M.; Qin, D.; Zhu, X.; Wang, C.; Zhu, J.; Hao, T.; Cheng, L.; Chen, X.; Bai, Z.; et al. HIV-1 Tat promotes Kaposi's sarcoma-associated herpesvirus (KSHV) vIL-6-induced angiogenesis and tumorigenesis by regulating PI3K/PTEN/AKT/GSK-3beta signaling pathway. *PLoS ONE* **2013**, *8*, e53145. [CrossRef]
77. Zhu, X.; Guo, Y.; Yao, S.; Yan, Q.; Xue, M.; Hao, T.; Zhou, F.; Zhu, J.; Qin, D.; Lu, C. Synergy between Kaposi's sarcoma-associated herpesvirus (KSHV) vIL-6 and HIV-1 Nef protein in promotion of angiogenesis and oncogenesis: Role of the AKT signaling pathway. *Oncogene* **2014**, *33*, 1986–1996. [CrossRef]
78. Jung, J.U.; Stager, M.; Desrosiers, R.C. Virus-encoded cyclin. *Mol. Cell. Biol.* **1994**, *14*, 7235–7244. [CrossRef]
79. Li, M.; Lee, H.; Yoon, D.W.; Albrecht, J.C.; Fleckenstein, B.; Neipel, F.; Jung, J.U. Kaposi's sarcoma-associated herpesvirus encodes a functional cyclin. *J. Virol.* **1997**, *71*, 1984–1991. [CrossRef]
80. Carbone, A.; Gloghini, A.; Bontempo, D.; Monini, P.; Tirelli, U.; Volpe, R.; Browning, P.J.; Gaidano, G. Proliferation in HHV-8-positive primary effusion lymphomas is associated with expression of HHV-8 cyclin but independent of p27(kip1). *Am. J. Pathol.* **2000**, *156*, 1209–1215. [CrossRef]
81. Verschuren, E.W.; Jones, N.; Evan, G.I. The cell cycle and how it is steered by Kaposi's sarcoma-associated herpesvirus cyclin. *J. Gen. Virol.* **2004**, *85*, 1347–1361. [CrossRef]
82. Van Dross, R.; Yao, S.; Asad, S.; Westlake, G.; Mays, D.J.; Barquero, L.; Duell, S.; Pietenpol, J.A.; Browning, P.J. Constitutively active K-cyclin/cdk6 kinase in Kaposi sarcoma-associated herpesvirus-infected cells. *J. Natl. Cancer Inst.* **2005**, *97*, 656–666. [CrossRef]
83. Zhi, H.; Zahoor, M.A.; Shudofsky, A.M.; Giam, C.Z. KSHV vCyclin counters the senescence/G1 arrest response triggered by NF-kappaB hyperactivation. *Oncogene* **2015**, *34*, 496–505. [CrossRef]
84. DiMaio, T.A.; Vogt, D.T.; Lagunoff, M. KSHV requires vCyclin to overcome replicative senescence in primary human lymphatic endothelial cells. *PLoS Pathog.* **2020**, *16*, e1008634. [CrossRef]
85. Davis, D.A.; Shrestha, P.; Yarchoan, R. Hypoxia and hypoxia-inducible factors in Kaposi sarcoma-associated herpesvirus infection and disease pathogenesis. *J. Med. Virol.* **2023**, *95*, e29071. [CrossRef]
86. Kumar Singh, R.; Pei, Y.; Bose, D.; Lamplugh, Z.L.; Sun, K.; Yuan, Y.; Lieberman, P.; You, J.; Robertson, E.S. KSHV-encoded vCyclin can modulate HIF1alpha levels to promote DNA replication in hypoxia. *eLife* **2021**, *10*, e57436. [CrossRef]
87. Wan, Q.; Tavakoli, L.; Wang, T.Y.; Tucker, A.J.; Zhou, R.; Liu, Q.; Feng, S.; Choi, D.; He, Z.; Gack, M.U.; et al. Hijacking of nucleotide biosynthesis and deamidation-mediated glycolysis by an oncogenic herpesvirus. *Nat. Commun.* **2024**, *15*, 1442. [CrossRef]
88. Chaudhary, P.M.; Jasmin, A.; Eby, M.T.; Hood, L. Modulation of the NF-kappa B pathway by virally encoded death effector domains-containing proteins. *Oncogene* **1999**, *18*, 5738–5746. [CrossRef]
89. Jacobs, S.R.; Damania, B. The viral interferon regulatory factors of KSHV: Immunosuppressors or oncogenes? *Front. Immunol.* **2011**, *2*, 19. [CrossRef]

90. Ballon, G.; Chen, K.; Perez, R.; Tam, W.; Cesarman, E. Kaposi sarcoma herpesvirus (KSHV) vFLIP oncoprotein induces B cell transdifferentiation and tumorigenesis in mice. *J. Clin. Investig.* **2011**, *121*, 1141–1153. [CrossRef]
91. Field, N.; Low, W.; Daniels, M.; Howell, S.; Daviet, L.; Boshoff, C.; Collins, M. KSHV vFLIP binds to IKK-gamma to activate IKK. *J. Cell Sci.* **2003**, *116*, 3721–3728. [CrossRef]
92. Hunte, R.; Alonso, P.; Thomas, R.; Bazile, C.A.; Ramos, J.C.; van der Weyden, L.; Dominguez-Bendala, J.; Khan, W.N.; Shembade, N. CADM1 is essential for KSHV-encoded vGPCR-and vFLIP-mediated chronic NF-kappaB activation. *PLoS Pathog.* **2018**, *14*, e1006968. [CrossRef]
93. Nakayama, R.; Ueno, Y.; Ueda, K.; Honda, T. Latent infection with Kaposi's sarcoma-associated herpesvirus enhances retrotransposition of long interspersed element-1. *Oncogene* **2019**, *38*, 4340–4351. [CrossRef]
94. Chen, L.; Ding, L.; Wang, X.; Huang, Y.; Gao, S.J. Activation of glucocorticoid receptor signaling inhibits KSHV-induced inflammation and tumorigenesis. *mBio* **2024**, *15*, e0301123. [CrossRef]
95. Russo, J.J.; Bohenzky, R.A.; Chien, M.C.; Chen, J.; Yan, M.; Maddalena, D.; Parry, J.P.; Peruzzi, D.; Edelman, I.S.; Chang, Y.; et al. Nucleotide sequence of the Kaposi sarcoma-associated herpesvirus (HHV8). *Proc. Natl. Acad. Sci. USA* **1996**, *93*, 14862–14867. [CrossRef]
96. Rivas, C.; Thlick, A.E.; Parravicini, C.; Moore, P.S.; Chang, Y. Kaposi's sarcoma-associated herpesvirus LANA2 is a B-cell-specific latent viral protein that inhibits p53. *J. Virol.* **2001**, *75*, 429–438. [CrossRef]
97. Baresova, P.; Musilova, J.; Pitha, P.M.; Lubyova, B. p53 tumor suppressor protein stability and transcriptional activity are targeted by Kaposi's sarcoma-associated herpesvirus-encoded viral interferon regulatory factor 3. *Mol. Cell. Biol.* **2014**, *34*, 386–399. [CrossRef]
98. Baresova, P.; Pitha, P.M.; Lubyova, B. Kaposi sarcoma-associated herpesvirus vIRF-3 protein binds to F-box of Skp2 protein and acts as a regulator of c-Myc protein function and stability. *J. Biol. Chem.* **2012**, *287*, 16199–16208. [CrossRef]
99. Pfeffer, S.; Sewer, A.; Lagos-Quintana, M.; Sheridan, R.; Sander, C.; Grasser, F.A.; van Dyk, L.F.; Ho, C.K.; Shuman, S.; Chien, M.; et al. Identification of microRNAs of the herpesvirus family. *Nat. Methods* **2005**, *2*, 269–276. [CrossRef]
100. Cai, X.; Lu, S.; Zhang, Z.; Gonzalez, C.M.; Damania, B.; Cullen, B.R. Kaposi's sarcoma-associated herpesvirus expresses an array of viral microRNAs in latently infected cells. *Proc. Natl. Acad. Sci. USA* **2005**, *102*, 5570–5575. [CrossRef] [PubMed]
101. Samols, M.A.; Hu, J.; Skalsky, R.L.; Renne, R. Cloning and identification of a microRNA cluster within the latency-associated region of Kaposi's sarcoma-associated herpesvirus. *J. Virol.* **2005**, *79*, 9301–9305. [CrossRef] [PubMed]
102. Yogev, O.; Henderson, S.; Hayes, M.J.; Marelli, S.S.; Ofir-Birin, Y.; Regev-Rudzki, N.; Herrero, J.; Enver, T. Herpesviruses shape tumour microenvironment through exosomal transfer of viral microRNAs. *PLoS Pathog.* **2017**, *13*, e1006524. [CrossRef]
103. Gottwein, E.; Corcoran, D.L.; Mukherjee, N.; Skalsky, R.L.; Hafner, M.; Nusbaum, J.D.; Shamulailatpam, P.; Love, C.L.; Dave, S.S.; Tuschl, T.; et al. Viral microRNA targetome of KSHV-infected primary effusion lymphoma cell lines. *Cell Host Microbe* **2011**, *10*, 515–526. [CrossRef]
104. Bartel, D.P. MicroRNAs: Target recognition and regulatory functions. *Cell* **2009**, *136*, 215–233. [CrossRef]
105. Bellare, P.; Ganem, D. Regulation of KSHV lytic switch protein expression by a virus-encoded microRNA: An evolutionary adaptation that fine-tunes lytic reactivation. *Cell Host Microbe* **2009**, *6*, 570–575. [CrossRef]
106. Lu, C.C.; Li, Z.; Chu, C.Y.; Feng, J.; Feng, J.; Sun, R.; Rana, T.M. MicroRNAs encoded by Kaposi's sarcoma-associated herpesvirus regulate viral life cycle. *EMBO Rep.* **2010**, *11*, 784–790. [CrossRef]
107. Lei, X.; Bai, Z.; Ye, F.; Xie, J.; Kim, C.G.; Huang, Y.; Gao, S.J. Regulation of NF-kappaB inhibitor IkappaBalpha and viral replication by a KSHV microRNA. *Nat. Cell Biol.* **2010**, *12*, 193–199. [CrossRef]
108. Liang, D.; Gao, Y.; Lin, X.; He, Z.; Zhao, Q.; Deng, Q.; Lan, K. A human herpesvirus miRNA attenuates interferon signaling and contributes to maintenance of viral latency by targeting IKKepsilon. *Cell Res.* **2011**, *21*, 793–806. [CrossRef]
109. Liang, Z.; Qin, Z.; Riker, A.I.; Xi, Y. CRISPR/Cas9 ablating viral microRNA promotes lytic reactivation of Kaposi's sarcoma-associated herpesvirus. *Biochem. Biophys. Res. Commun.* **2020**, *533*, 1400–1405. [CrossRef]
110. Gottwein, E.; Mukherjee, N.; Sachse, C.; Frenzel, C.; Majoros, W.H.; Chi, J.T.; Braich, R.; Manoharan, M.; Soutschek, J.; Ohler, U.; et al. A viral microRNA functions as an orthologue of cellular miR-155. *Nature* **2007**, *450*, 1096–1099. [CrossRef]
111. Skalsky, R.L.; Samols, M.A.; Plaisance, K.B.; Boss, I.W.; Riva, A.; Lopez, M.C.; Baker, H.V.; Renne, R. Kaposi's sarcoma-associated herpesvirus encodes an ortholog of miR-155. *J. Virol.* **2007**, *81*, 12836–12845. [CrossRef]
112. Qin, Z.; Freitas, E.; Sullivan, R.; Mohan, S.; Bacelieri, R.; Branch, D.; Romano, M.; Kearney, P.; Oates, J.; Plaisance, K.; et al. Upregulation of xCT by KSHV-encoded microRNAs facilitates KSHV dissemination and persistence in an environment of oxidative stress. *PLoS Pathog.* **2010**, *6*, e1000742. [CrossRef]
113. Lei, X.; Zhu, Y.; Jones, T.; Bai, Z.; Huang, Y.; Gao, S.J. A Kaposi's sarcoma-associated herpesvirus microRNA and its variants target the transforming growth factor beta pathway to promote cell survival. *J. Virol.* **2012**, *86*, 11698–11711. [CrossRef]
114. Suffert, G.; Malterer, G.; Hausser, J.; Viiliainen, J.; Fender, A.; Contrant, M.; Ivacevic, T.; Benes, V.; Gros, F.; Voinnet, O.; et al. Kaposi's sarcoma herpesvirus microRNAs target caspase 3 and regulate apoptosis. *PLoS Pathog.* **2011**, *7*, e1002405. [CrossRef]
115. Gottwein, E.; Cullen, B.R. A human herpesvirus microRNA inhibits p21 expression and attenuates p21-mediated cell cycle arrest. *J. Virol.* **2010**, *84*, 5229–5237. [CrossRef]
116. Qin, Z.; DeFee, M.; Isaacs, J.S.; Parsons, C. Extracellular Hsp90 serves as a co-factor for MAPK activation and latent viral gene expression during de novo infection by KSHV. *Virology* **2010**, *403*, 92–102. [CrossRef]

117. Nayar, U.; Lu, P.; Goldstein, R.L.; Vider, J.; Ballon, G.; Rodina, A.; Taldone, T.; Erdjument-Bromage, H.; Chomet, M.; Blasberg, R.; et al. Targeting the Hsp90-associated viral oncoproteome in gammaherpesvirus-associated malignancies. *Blood* **2013**, *122*, 2837–2847. [CrossRef]
118. Nayar, U.; Sadek, J.; Reichel, J.; Hernandez-Hopkins, D.; Akar, G.; Barelli, P.J.; Sahai, M.A.; Zhou, H.; Totonchy, J.; Jayabalan, D.; et al. Identification of a nucleoside analog active against adenosine kinase-expressing plasma cell malignancies. *J. Clin. Investig.* **2017**, *127*, 2066–2080. [CrossRef]
119. White, M.C.; Wong, J.P.; Damania, B. Inhibition of NEK2 Promotes Chemosensitivity and Reduces KSHV-positive Primary Effusion Lymphoma Burden. *Cancer Res. Commun.* **2024**, *4*, 1024–1040. [CrossRef]
120. Sin, S.H.; Roy, D.; Wang, L.; Staudt, M.R.; Fakhari, F.D.; Patel, D.D.; Henry, D.; Harrington, W.J., Jr.; Damania, B.A.; Dittmer, D.P. Rapamycin is efficacious against primary effusion lymphoma (PEL) cell lines in vivo by inhibiting autocrine signaling. *Blood* **2007**, *109*, 2165–2173. [CrossRef]
121. Bhatt, A.P.; Bhende, P.M.; Sin, S.H.; Roy, D.; Dittmer, D.P.; Damania, B. Dual inhibition of PI3K and mTOR inhibits autocrine and paracrine proliferative loops in PI3K/Akt/mTOR-addicted lymphomas. *Blood* **2010**, *115*, 4455–4463. [CrossRef]
122. Li, X.; Wang, F.; Zhang, X.; Sun, Q.; Kuang, E. Suppression of KSHV lytic replication and primary effusion lymphoma by selective RNF5 inhibition. *PLoS Pathog.* **2023**, *19*, e1011103. [CrossRef]
123. Moriguchi, M.; Watanabe, T.; Kadota, A.; Fujimuro, M. Capsaicin Induces Apoptosis in KSHV-Positive Primary Effusion Lymphoma by Suppressing ERK and p38 MAPK Signaling and IL-6 Expression. *Front. Oncol.* **2019**, *9*, 83. [CrossRef]
124. Shrestha, P.; Davis, D.A.; Veeranna, R.P.; Carey, R.F.; Viollet, C.; Yarchoan, R. Hypoxia-inducible factor-1 alpha as a therapeutic target for primary effusion lymphoma. *PLoS Pathog.* **2017**, *13*, e1006628. [CrossRef]
125. Pria, A.D.; Pinato, D.; Roe, J.; Naresh, K.; Nelson, M.; Bower, M. Relapse of HHV8-positive multicentric Castleman disease following rituximab-based therapy in HIV-positive patients. *Blood* **2017**, *129*, 2143–2147. [CrossRef]
126. Uldrick, T.S.; Polizzotto, M.N.; Aleman, K.; Wyvill, K.M.; Marshall, V.; Whitby, D.; Wang, V.; Pittaluga, S.; O'Mahony, D.; Steinberg, S.M.; et al. Rituximab plus liposomal doxorubicin in HIV-infected patients with KSHV-associated multicentric Castleman disease. *Blood* **2014**, *124*, 3544–3552. [CrossRef]
127. Nishimoto, N.; Kanakura, Y.; Aozasa, K.; Johkoh, T.; Nakamura, M.; Nakano, S.; Nakano, N.; Ikeda, Y.; Sasaki, T.; Nishioka, K.; et al. Humanized anti-interleukin-6 receptor antibody treatment of multicentric Castleman disease. *Blood* **2005**, *106*, 2627–2632. [CrossRef]
128. Berwanger, A.; Stein, S.C.; Kany, A.M.; Gartner, M.; Loretz, B.; Lehr, C.M.; Hirsch, A.K.H.; Schulz, T.F.; Empting, M. Disrupting Kaposi's Sarcoma-Associated Herpesvirus (KSHV) Latent Replication with a Small Molecule Inhibitor. *J. Med. Chem.* **2023**, *66*, 10782–10790. [CrossRef]
129. Wu, Y.; Wang, V.; Yarchoan, R. Pacritinib inhibits proliferation of primary effusion lymphoma cells and production of viral interleukin-6 induced cytokines. *Sci. Rep.* **2024**, *14*, 4125. [CrossRef]

cancers

MDPI

Review

Lymphoproliferations in People Living with HIV: Oncogenic Pathways, Diagnostic Challenges, and New Therapeutic Opportunities

Riccardo Dolcetti [1,2,3,*], Emanuela Vaccher [4] and Antonino Carbone [4,*]

[1] Peter MacCallum Cancer Centre, Melbourne, VIC 3000, Australia
[2] Sir Peter MacCallum Department of Oncology, The University of Melbourne, Melbourne, VIC 3000, Australia
[3] Department of Microbiology and Immunology, The University of Melbourne, Melbourne, VIC 3000, Australia
[4] Centro di Riferimento Oncologico, Istituto di Ricovero e Cura a Carattere Scientifico, National Cancer Institute, 33081 Aviano, Italy; evaccher@cro.it
* Correspondence: riccardo.dolcetti@petermac.org (R.D.); acarbone@cro.it (A.C.)

Simple Summary

People living with Human Immunodeficiency Virus (HIV) are at significantly increased risk of developing lymphoproliferative disorders. Although the advent of combination antiretroviral therapy (cART) has markedly improved survival, lymphomas remain a major cause of morbidity and mortality in this population. In this review, we discuss the key features of this heterogeneous group of tumors, highlight current diagnostic and therapeutic challenges, and explore emerging immunotherapeutic strategies aimed at improving lymphoma outcomes in individuals with HIV.

Abstract

Although efficiently managed by cART, chronic HIV infection remains associated with a high incidence of malignant lymphomas. This diverse group of tumors presents considerable challenges in research, diagnosis, and treatment due to their complex pathogenesis, heterogeneous tumor microenvironment, and frequently aggressive clinical behavior. In this review, we examine the multifactorial pathogenesis of lymphomas arising in people living with HIV (PLWH), encompassing both direct and indirect oncogenic mechanisms. We summarize the key histopathological features and microenvironmental characteristics that may influence therapeutic responses. Current treatment strategies approved for the treatment of lymphomas in PLWH are showing outcomes comparable with those observed in patients without HIV. Notably, the immune reconstitution achieved through cART has renewed interest in immunotherapeutic approaches for HIV-associated lymphomas, with several strategies under clinical investigation. However, progress in the diagnosis and management of these malignancies is hindered by fragmented research efforts and the frequent exclusion of PLWH from pivotal clinical trials. Coordinated efforts are essential to overcome these barriers, reduce lymphoma incidence, and improve survival outcomes in this vulnerable population.

Keywords: human immunodeficiency virus; lymphomas; Epstein–Barr virus; Kaposi's sarcoma-associated herpesvirus; lymphomagenesis; combined antiretroviral therapy; immunotherapy

https://doi.org/10.3390/cancers17132088

1. Introduction

With the widespread implementation of combination antiretroviral therapy (cART), lymphoma has become the most common malignancy among people living with HIV (PLWH) in many developed countries [1–4]. Over the last two decades, the incidence of certain subtypes, such as diffuse large B cell lymphoma (DLBCL) and primary central nervous system lymphomas (PCNSLs), has decreased, while the incidence of Burkitt lymphoma (BL) has remained stable, and Hodgkin lymphoma (HL) has increased [1]. These lymphomas are generally characterized by high-grade features, frequent extranodal involvement, advanced-stage presentation, and marked histopathologic heterogeneity, often posing significant diagnostic challenges [1]. The 5th edition of the World Health Organization (WHO) Classification of Haematolymphoid Tumors now includes a dedicated category for HIV-associated lymphomas [5]. In this review, we provide an overview of the current understanding of the complex pathogenesis of lymphoproliferation in PLWH, outline their main clinicopathologic features, discuss current therapeutic strategies, and explore emerging opportunities for immunotherapy aimed at improving outcomes in this population.

2. Pathogenesis of HIV-Associated Lymphoproliferative Disorders

The pathogenesis of lymphomas in PLWH is multifaceted and complex, sharing several features with lymphoproliferative disorders associated with other congenital or acquired immunodeficiencies [6]. Indirect factors related to HIV-induced immunosuppression, such as chronic antigenic stimulation, dysregulated cytokine production, and co-infection with oncogenic viruses, are recognized contributors to lymphomagenesis in this setting [7]. However, emerging evidence challenges the traditional view, suggesting that HIV itself may also play a more direct role in promoting lymphoma development [8].

2.1. Indirect Mechanisms

The pathogenesis of HIV-associated lymphoproliferative disorders involves multiple factors, primarily, though not exclusively, linked to the immunosuppressive effects of HIV infection.

HIV-induced immunosuppression and chronic antigenic stimulation. HIV-induced depletion of $CD4^+$ T cells leads to a progressive decline in immune competence, impairing the host's ability to control oncogenic infections and pre-malignant lymphocyte clones [9]. This immune dysfunction facilitates the unchecked expansion of lymphocytes harboring predisposing genetic alterations or oncogenic virus infections, thereby increasing lymphoma risk. Despite the virologic control achieved with cART, T cell dysfunction persists. Specifically, the sustained expression of inhibitory checkpoint receptors, such as PD-1, TIM-3, or TIGIT, on T cells limits their cytotoxic potential against emerging tumor cells [10]. Furthermore, chronic immune activation and inflammation—hallmarks of HIV infection—also contribute to persistent antigenic stimulation, which can drive lymphomagenesis [7,8]. HIV infection also disrupts cytokine homeostasis, promoting the overproduction of pro-inflammatory mediators such as IL-6, which support the survival and proliferation of neoplastic lymphocytes [11]. Macrophages are particularly affected: HIV infection impairs their function and contributes to both viral persistence and tumor-promoting [12]. For instance, the HIV viral protein U (Vpu) degrades the bone marrow stromal antigen 2 (BST2), a macrophage protein that restricts viral replication, thereby facilitating chronic inflammation and persistent infection [13,14]. Moreover, HIV-infected macrophages upregulate several microRNAs, including miRNA-99 and miRNA-146a, which stimulate the release of pro-inflammatory cytokines (e.g., TNFα, IL-6, and IL-1β), and dampen innate immune responses [15,16]. HIV-encoded Tat and Nef proteins further impair immunity by inducing

Indoleamine 2,3-dioxygenase (IDO), leading to Th17 cell depletion, regulatory T cell expansion, and reduced $CD8^+$ T cell proliferation—all of which foster immune escape and tumor development [17–19]. Natural killer (NK) cells are also affected. HIV infection impairs the activation of $CD56^{bright}$ NK cells and downregulates critical activating receptors (e.g., NCRs and NKG2D), largely due to chronic exposure to their ligands [20,21]. A dysfunctional $CD56^-CD16^+$ NK cell subset expands at the expense of the cytotoxic $CD56^{dim}CD16^+$ population, with reduced expression of NKp30 and NKp46 [22,23]. These altered NK cells exhibit impaired cytokine production, defective crosstalk with other immune cells, and diminished cytotoxicity toward tumor targets [22,23]. Although cART partially restores NK cell function [23], residual abnormalities persist and may contribute to ongoing lymphoma risk in PLWH.

Oncogenic virus co-infection. The role of co-infecting viruses in the development of HIV-associated malignancies has been extensively investigated, with Epstein–Barr virus (EBV) and Kaposi Sarcoma Herpesvirus (KSHV) being strongly implicated in specific HIV-related lymphoma subtypes [24–27]. In both cases, HIV-induced immunosuppression undermines immune surveillance, enabling the unrestricted proliferation of virus-infected cells and the aberrant expression of oncogenic viral proteins that directly drive lymphomagenesis. Notably, the degree of HIV-induced immunosuppression correlates with the type of EBV- or KSHV-associated lymphoma that arises in PLWH [24–27]. Before the widespread use of cART, severe immunodeficiency led to a high incidence of aggressive EBV^+ DLBCLs and immunoblastic lymphomas. These neoplasms typically exhibit the EBV latency III pattern, characterized by the expression of the full repertoire of EBV nuclear and membrane proteins [28]. The advent of cART significantly reduced the incidence of such lymphomas [1], reflecting improved immune control of EBV. In contrast, EBV-associated lymphomas arising in PLWHs with relatively preserved immune function—such as BL and HL—demonstrate more restricted patterns of EBV gene expression. BL is associated with EBV latency I (expression of EBNA1 only), while HL is linked to latency II (expression of EBNA1, LMP1, and LMP2) [25,26]. The incidence of these lymphoma types has not significantly changed following cART introduction [1], suggesting that their pathogenesis may be influenced by mechanisms beyond HIV-induced immunosuppression. Of note, both BL and HL originate in germinal centers—an anatomical niche where HIV replication can persist despite cART [1,5]. Within this microenvironment, the expression of EBV nuclear antigen 6 (EBNA6) by EBV-infected B cells may induce activation-induced cytidine deaminase (AICDA), a DNA-editing enzyme that can drive 2oncogogenic mutations [29,30]. The synergistic oncogenic potential of HIV and EBV has been demonstrated by elegant studies in humanized mouse models, where co-infection enhances systemic EBV replication, promotes immune activation, and facilitates EBV-driven lymphomagenesis [31]. Beyond its latent and lytic repertoire, EBV also contributes to lymphomagenesis via the expression of viral microRNAs. These miRNAs are encoded in two genomic regions—BamHI-A region rightward transcript (BART) and BamHI-H rightward fragment 1 (BHRF1)—and modulate host pathways involved in apoptosis, proliferation, immune evasion, and the tumor microenvironment [32,33]. BART miRNAs are broadly expressed across all latency types, particularly latency I and II, while BHRF1 miRNAs are largely restricted to latency III [32,33].

In addition to Kaposi's sarcoma, the oncogenic γ-herpesvirus KSHV is etiologically linked to certain B cell malignancies, notably primary effusion lymphoma (PEL) [34], and a subset of multicentric Castleman disease (MCD) [35]. KSHV appears to persist preferentially in immunoglobulin lambda-expressing B cells, which are enriched in both PEL and MCD tissues [36,37], although the precise reservoir of KSHV in asymptomatic carriers remains poorly defined. KSHV-driven lymphomagenesis is primarily attributed to latent infection. Most KSHV-infected tumor cells express a limited subset of latent viral proteins, with Latency-Associated Nuclear Antigen (LANA) being the most consistently expressed [24]. Functionally analogous to EBV's EBNA-1, LANA is essential for maintaining the viral episome during latency and contributes to oncogenesis by modulating both viral and host gene expression through interactions with cellular transcription factors [24]. Other latent KSHV proteins—such as viral FLICE inhibitory protein (vFLIP), v-cyclin, Kaposin, and viral interferon regulatory factor 3 (vIRF-3)—may also be expressed in PEL and MCD, though typically at lower frequencies [24]. KSHV lytic proteins also play roles in lymphomagenesis by disrupting apoptotic pathways, promoting angiogenesis, remodeling the tumor microenvironment, and enabling immune evasion. Among these, viral interleukin 6 (vIL-6) is particularly relevant, especially in MCD. vIL-6 promotes vascular endothelial growth factor (VEGF)-mediated vascular permeability, drives IL-6-dependent inflammation, and directly supports the growth of PEL cells [38–40]. Importantly, the vast majority of PEL cases are also co-infected with EBV, and there is growing evidence of a pathogenic interplay between the two viruses. KSHV has been shown to trigger lytic EBV replication, which appears to contribute to lymphomagenesis [24]. Notably, studies in humanized mouse models demonstrated that co-infection with KSHV and an EBV strain incapable of lytic replication led to a lower incidence of lymphoma, suggesting that abortive lytic EBV reactivation—rather than full viral replication—may be critical in promoting tumor development [41]. Complete EBV lytic replication likely reduces lymphoma formation due to cell lysis and subsequent clearance of infected cells.

Contribution of cART. While cART significantly improves overall immune function and reduces HIV-related mortality, its long-term use—particularly in individuals with advanced HIV infection—only partially reverses the metabolic alterations, chronic inflammation, and immune dysregulation that can indirectly contribute to the development of lymphoproliferative disorders [42,43]. In some cases, rapid immune restoration following cART initiation may lead to an exaggerated inflammatory response against latent viral infections, a condition known as Immune Reconstitution Inflammatory Syndrome (IRIS) [44]. IRIS has been associated with the emergence or unmasking of lymphoproliferative processes, although its precise role in HIV-related lymphomagenesis remains unclear. Further studies are needed to delineate the direct and indirect contribution of cART to lymphoma development in PLWH.

Microenvironmental factors. The tumor microenvironment (TME) of HIV-associated lymphomas is highly complex and plays a pivotal role in disease development by fostering immunosuppression, facilitating tumor immune evasion, and supporting tumor progression [7,45]. HIV-associated lymphomas often arise and evolve within a dynamic and interactive microenvironment, where malignant lymphocytes coexist and communicate with various immune and stromal components. This interaction mirrors the physiological relationship between normal B cells and their microenvironment [7,45]. In most cases, the lymphoma's growth remains partially dependent on signals from the TME. However, this interdependence may be lost when genetic alterations endow tumor cells with full autonomy. A prototypical example is BL, in which the t(8;14) chromosomal translocation leads to constitutive activation of the c-MYC oncogene, driving uncontrolled proliferation independent of microenvironmental cues [46].

2.2. Direct Mechanisms

Despite the partial restoration of immune function achieved with cART, PLWH continues to exhibit a disproportionately high incidence of lymphomas, including histological subtypes with distinctive cyto-histopathological features [1]. Emerging evidence suggests that, beyond immune suppression, HIV may also directly contribute to lymphomagenesis through mechanisms involving the biological activity of viral proteins [5]. Although HIV does not infect B lymphocytes—the primary target cells in most HIV-associated lymphomas—it can release viral proteins into the surrounding microenvironment. HIV-encoded proteins, such as gp120, Nef, Tat, and HIV matrix protein p17 (p17), can interact with or be internalized by bystander cells, including B lymphocytes, where they activate intracellular signaling pathways. These interactions may promote chronic antigenic stimulation, dysregulation of cellular processes, and ultimately contribute directly to the initiation and progression of lymphomagenesis [8,47,48].

Following its release in lymphoid tissues, the HIV Tat protein can functionally interact with and enter B lymphocytes, where it influences key cellular processes, including cell cycle regulation, apoptosis, redox homeostasis, and gene expression [43]. Tat has been shown to upregulate the production of pro-survival and proliferative cytokines such as IL-6 and IL-10, thereby enhancing B cell activation and proliferation [49,50]. Additionally, Tat induces the expression of the DNA methyltransferase 1 enzyme, implicating it in the epigenetic dysregulation of host gene expression and suggesting a potential role in lymphomagenesis [51]. Early cellular responses to Tat include oxidative stress, DNA damage, and increased genomic instability in B cells [52]. Tat also activates the Akt/mTORC1 signaling pathway and the expression of AICDA by suppressing its transcriptional repressors [53–55], further contributing to mutagenesis and chromosomal instability. More recently, chronic Tat expression has been shown to downregulate HLA-DR expression via the inhibition of the NF-κB pathway, impairing $CD4^+$ T cell-mediated immune recognition of EBV-transformed B cells and thus facilitating immune evasion [56]. In vivo studies have confirmed the oncogenic potential of Tat, with transgenic mice expressing Tat developing a range of different neoplasms, including lymphomas [57]. In addition to Tat, the HIV-encoded Nef protein has been implicated in lymphomagenesis by promoting AICDA and c-MYC expression, both of which are central to genomic instability and the formation of pathogenic chromosomal translocations [58].

Multiple lines of evidence support a potential lymphomagenic role for the HIV matrix protein p17. HIV-infected cells can release virion-free p17 independently of active viral protease activity, and this protein has been detected in the plasma and lymphoid tissues of PLWH [59,60]. Notably, p17 persists in lymphoid tissues even during cART and in the absence of active HIV replication [60]. This long-term presence within the tissue microenvironment suggests that p17 may serve as a chronic stimulus for B cell activation and proliferation. Supporting this hypothesis, the expression of p17 in mice transgenic for a defective HIV-1 provirus has been shown to result in lymphoma development [61]. Furthermore, specific p17 variants (vp17)—derived either from a Ugandan HIV-1 A1 strain or from individuals with HIV-associated non-Hodgkin lymphoma—demonstrate distinct biological activity compared to prototype clade B p17 or p17 proteins isolated from PLWH without lymphoma [62,63]. These vp17s exhibit unique amino acid insertions in their C-terminal region, resulting in marked conformational changes that expose a functional epitope in the N-terminal domain of the protein [62–64]. Recent studies have shown that this functional epitope interacts with the Protease-Activated Receptor 1, triggering downstream epithelial growth factor receptor (EGFR) transactivation and activation of the phosphoinositide 3-kinase (PI3K)-Akt signaling pathway—ultimately promoting B cell proliferation and clonogenicity [65]. Consistent with these mechanistic insights, retrospective cohort analyses have demonstrated a significantly higher prevalence of vp17s in PLWH with lymphoma compared to those without lymphoma [66].

The inflammatory microenvironment and the chronic B cell stimulation induced by HIV-encoded proteins and other exogenous factors may promote polyclonal or oligoclonal expansion of aberrant B lymphocytes. This proliferation is sustained by the dysregulated production of B cell growth-promoting cytokines, such as IL-6 and IL-10. In this context of persistent immune activation, the risk increases for monoclonal B cell expansions to acquire genetic lesions critical for lymphomagenesis. Indeed, recent studies have identified unique B cell subsets in PLWH that display pre-malignant features and may represent early stages in the evolution toward overt lymphoma [67]. Notably, HIV virions released from infected $CD4^+$ T cells can incorporate the costimulatory molecule CD40 ligand, thereby gaining the capacity to activate B lymphocytes in a manner that mimics physiological T–B interactions [68,69]. Regardless of the initiating stimulus, chronic B cell activation is closely associated with the induction of AICDA, a key enzyme in somatic hypermutation and class–switch recombination [70]. Elevated AICDA expression has been detected in circulating B lymphocytes of HIV^+ individuals prior to lymphoma diagnosis, particularly in those who later developed BL, a disease strongly associated with immunoglobulin gene translocations [70]. Furthermore, HIV-related dysregulation of B cell receptor (BCR) signaling has been proposed as an additional mechanism contributing to lymphomagenesis. It has been hypothesized that HIV may remodel the structure of the BCR through AICDA and recombinant-activating gene upregulation while simultaneously acting as a chronic antigen that activates downstream signaling pathways, including PI3K and MAPK, partly through reduced expression of the inhibitory receptor CD300a [71].

3. Pathologic Features of HIV-Associated Lymphoproliferations

Polymorphic lymphoproliferative disorders, which span a spectrum of malignant potential, are morphologically and clinically heterogeneous. They disrupt the normal tissue architecture and are frequently associated with necrosis [72–74]. Although classically associated with post-transplant immunosuppression, polymorphic lymphoproliferative disorders have also been documented in PLWH [74–80]. Histologically, these lesions are characterized morphologically by a heterogeneous, or polymorphic, cell population representing various stages of B cell differentiation—typically expressing CD20 and/or PAX5 and generally of non-germinal center origin—alongside plasma cells, other inflammatory cells, and Reed–Sternberg-like cells [81]. A variable proportion of the Reed–Sternberg-like cells are EBV^+ and express the EBV oncogenic latent membrane protein-1 (LMP1) [71,72,74,75,78–87]. These lesions typically contain monoclonal B cell populations and oligoclonal T cell infiltrates, the latter likely representing a reactive response to the expansion of EBV-infected B cells [70,71,74–77,79,81–83,88].

HIV-associated malignant lymphomas are classified using the same morphologic, immunophenotypic, and genotypic criteria applied to lymphomas in immunocompetent individuals [5,88,89] and, similarly, are broadly categorized into Hodgkin lymphomas (HLs) and non-Hodgkin lymphomas (NHLs). Compared to their counterparts in immunocompetent patients, HIV-associated lymphomas are more frequently EBV-positive. Moreover, certain lymphoma subtypes show a predilection for PLWH, including plasmablastic lymphoma (PBL) and primary effusion lymphoma (PEL), along with its solid variant, extra-cavitary PEL (EC-PEL) [89,90] (Table 1).

Table 1. Lymphoid proliferations associated with oncogenic virus infections in people with immunodeficiency [1,81,91,92].

Type of Lymphoproliferation	Viral Status (EBV and KSHV)
B-cell hyperplasias	EBV+; KSHV−
Polymorphic proliferations	EBV+ or −
Indolent B-cell lymphomas	EBV+/−; KSHV−

Table 1. *Cont.*

Type of Lymphoproliferation	Viral Status (EBV and KSHV)
B-cell lymphomas *	
Hodgkin lymphoma	EBV+; KSHV−
Burkitt lymphoma	EBV−/+; KSHV−
DLBCL	EBV+; KSHV−
PEL and its solid variants	EBV+; KSHV+
PBL of the oral cavity type	EBV+/−; KSHV−
MCD-associated LBCL	EBV−; KSHV+
PBL associated with MCD	EBV−; KSHV+

Immunodeficiency setting includes post-transplant (solid organ), iatrogenic (methotrexate), immune senescence, and people living with HIV/AIDS. * Must meet criteria for corresponding lymphoma in immunocompetent hosts.

3.1. *Diffuse Large B Cell Lymphoma (DLBCL)*

HIV-associated DLBCL is the most common subtype of HIV-associated lymphoma [1,6]. While HIV-associated DLBCLs share many histopathological and molecular features with their counterparts in immunocompetent individuals, they differ in their higher frequency of EBV positivity and distinct biological characteristics. These lymphomas can originate from either germinal center B cells (GCB subtype) or non-germinal center activated B cells (ABC subtype) [1,6]. The ABC-like subtype is frequently EBV-positive, while the GCB-like subtype is typically EBV-negative [93–95]. Both subtypes exhibit increased vascular density compared to DLBCL in HIV-negative individuals, reflecting an altered tumor microenvironment [96]. Cytogenetically, the prevalence of MYC and BCL6 rearrangements in HIV-associated DLBCL (14.9% and 27.7%, respectively) is comparable to that in HIV-negative patients, whereas BCL2 rearrangements are less frequently observed (4.3%) [97]. The TME in HIV-associated DLBCL includes a prominent population of tumor-associated macrophages (TAMs), which have been implicated in tumor progression [98]. These TAMs are often skewed toward an M2-like, pro-tumoral phenotype under the influence of cytokines such as IL-4 and IL-10 [99]. In addition to supporting tumor growth, M2-like macrophages promote angiogenesis in the tumor milieu of HIV-associated DLBCLs [7,45]. Importantly, both tumor cells and infiltrating macrophages in aggressive HIV-associated DLBCLs have been shown to express the checkpoint molecule programmed cell death ligand 1 (PD-L1) [100,101], highlighting the potential utility of immune checkpoint inhibitors targeting the PD-1/PD-L1 axis in this context.

3.2. *Classic Hodgkin Lymphoma (HL) in PLWH*

In high-income countries, PLWH have a 5- to 26-fold increased risk of developing HL compared to the general population [4]. The highest risk period for HL diagnosis often coincides with the months following initiation of cART, during which the inflammatory IRIS may occur [102,103]. Unlike NHL, the incidence of HL in PLWH has remained relatively stable or has declined only modestly in the cART era [3,4]. Notably, PLWH on effective cART—with suppressed viral load and restored $CD4^+$ T cell counts—still exhibit a ninefold higher risk of HL than HIV-negative individuals [104], further implicating IRIS and immune dysregulation as potential contributors to HIV-associated HL pathogenesis [103,105]. Classic Hodgkin lymphoma (cHL) in the context of HIV is EBV^+ in over 90% of cases [1,5,26,106–108] with frequent strong expression of the LMP-1 oncogenic EBV protein [109] (Figure 1). Although the immunophenotype of the neoplastic Hodgkin–Reed–Sternberg (HRS) cells closely resemble that observed in immunocompetent patients, the TME in PLWH is distinct, being characterized by a lower CD4 to CD8 T cell ratio and altered immune cell composition [26].

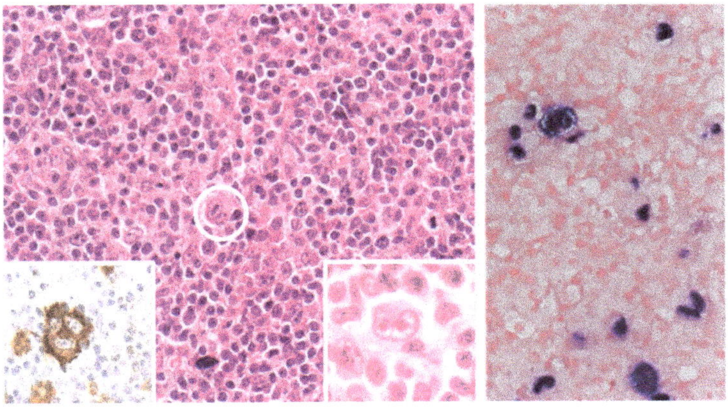

Figure 1. EBV-positive classic Hodgkin lymphoma (cHL) in a patient living with HIV. (**Left**) Hodgkin–Reed–Sternberg (HRS) cells (circled area) are seen in an inflammatory environment. A diagnostic HRS cell is encircled at the center, expresses CD30 antigen (in the inset to the left), and is rosetted by small lymphocytes (inset to the right). (**Right**) EBER in situ hybridization shows EBV positivity in sparse tumor cells. (**Left**) Hematoxliyn and Eosin stain, original magnification 40×, Inset to the left: immunohistochemistry, inset to the right: Hematoxliyn and Eosin; Inset: original magnification 100×; (**Right**) in situ hybridization, original magnification 40×.

At the time of diagnosis, HIV-associated cHL typically presents with a relatively high median CD4$^+$ T cell count is relatively high—ranging from approximately 275 to 306 cells/µL—indicating only moderate immunosuppression [109]. However, the number of functional and mature NK cells within HIV-associated HL tissues is reduced compared to HIV-negative HL cases, suggesting an impaired innate immune response [110]. The TME of cHL is complex and composed of various cellular and structural components, including B and T cells, fibroblasts, stromal cells, macrophages, mast cells, granulocytes, and an ECM [111,112]. The survival and proliferation of HRS cells are highly dependent on this supportive TME, particularly the presence of CD4^{high} T cells forming "rosettes" around HRS cells, which deliver essential growth-promoting signals [113].

Compared to HIV-unrelated EBV$^+$ cases, the TME of HIV$^+$EBV$^+$ cHL is characterized by significantly lower proportions of CD8$^+$ T cells co-expressing the inhibitory checkpoint molecules PD-1 and TIGIT. This reduction may reflect increased turnover, impaired T cell receptor (TCR) activation, or dysregulated PD-1/TIGIT expression. Notably, elevated levels of TIGIT ligands—CD155 and nectin 3—within the TME suggest that ligand/receptor interactions may contribute to the downregulation of TIGIT on CD8$^+$ T cells [114]. Recent transcriptomic analyses have shown that relative to HIV-unrelated cHL, the HIV$^+$EBV$^+$ cHL TME exhibits a reduced expression of genes involved in TCR signaling and T cell stemness. Concurrently, these cases display increased activation of Gα and G-protein-coupled receptor signaling pathways, which are known to suppress T cell proliferation and inhibit IL-2 production—key mechanisms of T cell activation [115]. Additionally, a significant upregulation of ECM remodeling pathways was observed in HIV$^+$EBV$^+$ cHL, potentially interfering with effective adaptive immune responses. These findings support the notion that HIV infection may impair anti-tumor immunity in cHL through both direct effects on TCR signaling and indirect reorganization of the ECM. EBV infection further contributes to immune evasion by promoting the accumulation of CD68^{high} and CD163^{high} macrophages, consistent with an immunosuppressive M2-like phenotype. This is likely mediated in part by the EBV-encoded LMP1, which modulates the TME to favor tumor immune escape [116].

3.3. Epstein–Barr Virus-Positive Marginal Zone Lymphoma (EBV-MZL)

EBV-positive marginal zone lymphoma (EBV-MZL) has been reported in transplant recipients and in other settings of immunodeficiency or immune dysregulation, including PLWH [73,116–118]. Similar to MZL in immunocompetent patients, EBV-MZL in PLWH can arise at various anatomical sites—such as the skin, lung, salivary gland, stomach, and lymph nodes—and typically displays comparable morphologic features [116–118].

3.4. HIV-Associated Burkitt Lymphoma (BL)

BL accounts for approximately 25–40% of lymphomas occurring in PLWH [1–3]. Despite the introduction of cART, the incidence of HIV-associated BL has not declined [1–4]. PLWH with moderate immunodeficiency carry a lifelong risk of 10–20% for developing BL, a risk that is more strongly associated with cumulative HIV viremia than with recent immune status [3,4]. Clinically, HIV-associated BL presents with distinct features compared to BL in HIV-negative individuals [119,120]. Patients often present with B symptoms, advanced-stage disease, and frequent extranodal involvement—including bone marrow, oral cavity, gastrointestinal tract, and central nervous system [1]. Notably, HIV-associated BLs occurring in the context of relatively preserved CD4$^+$ T cell counts are more likely to exhibit extranodal disease, B symptoms, frailty, and elevated lactate dehydrogenase levels [120]. EBV is detected in approximately 60% of HIV-associated BL cases, compared to about 20% in HIV-negative BL [121]. Histological features and EBV immunostaining are illustrated in Figure 2. Nearly all HIV-associated BLs harbor *MYC* gene rearrangements on chromosome 8. Importantly, the HIV Tat protein has been detected within BL tumor cells, where it enhances c-MYC expression through direct interaction with AP-1 sites on the c-MYC promoter [122]. In agreement with what we described above, these findings support a role for Tat in promoting oncogenic signaling within the BL TME, potentially contributing to the aggressiveness of this HIV-related lymphoma.

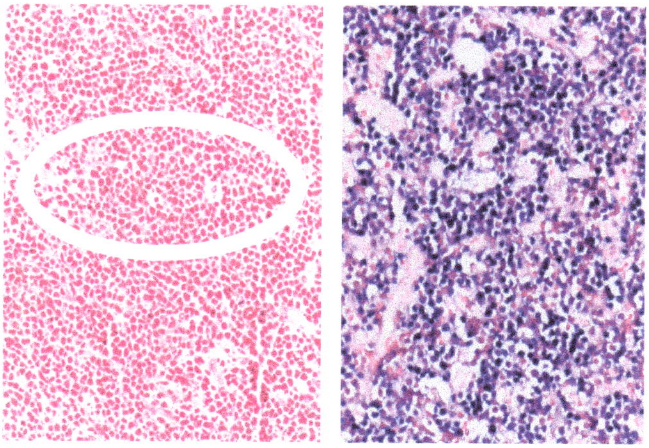

Figure 2. EBV-positive Burkitt lymphoma in a patient living with HIV. (**Left**) the tumor population is monotonous and displays small-intermediate size. Starry sky pattern is seen at the center of the microphotograph (indicated by an ellipse). (**Right**) EBER in situ hybridization shows EBV positivity in most tumor cells. (**Left**) Hematoxliyn and Eosin stain, original magnification 20×, (**Right**) in situ hybridization, original magnification 20×.

3.5. Other Rare HIV-Associated Lymphomas

Primary central nervous system lymphoma (PCNSL) is a rare but highly aggressive malignancy. In PLWH, the incidence of PCNSL is significantly increased, accounting for approximately 12–15% of HIV-associated lymphomas [123]. The vast majority of PCNSL cases (90–95%) are DLBCL, with a minority comprising BL or T-cell lymphomas. EBV infection is detected in 80–100% of PCNSL cases [123].

Plasmablastic lymphoma (PBL) and primary effusion lymphoma (PEL) are also rare and aggressive entities predominantly observed in PLWH [1,124,125]. While the exact incidence of HIV-associated PBL remains unclear, it accounts for approximately 2% of HIV-related lymphomas diagnosed in high-income countries during the cART era [1]. In contrast, in low- and middle-income countries (LMICs) with a high burden of HIV infection—such as South Africa—the incidence of PBL may be as high as 20% despite widespread cART availability [126]. However, high-quality epidemiological data on hematologic malignancies among PLWH in LMICs remain limited [4]. PBLs are consistently CD20-negative and EBV^{+} in about 75% of cases. It shows a marked male predominance and presents extranodally in over 95% of cases, with common sites including the oral cavity, gastrointestinal tract, skin, and bone [1]. The molecular pathogenesis of PBL involves dysregulation of *MYC* and activation of key oncogenic pathways such as JAK-STAT, RAS-RAF, and NOTCH, highlighting potential therapeutic targets for future investigation [124]. Despite this, clinical and biological insights into PBL remain limited, primarily based on case reports, underscoring the need for further research.

KSHV/HHV8 is an oncogenic virus linked to several HIV-associated malignancies: Kaposi sarcoma, NHL (PEL and DLBCL), MCD, and the Kaposi sarcoma inflammatory cytokine syndrome (KICS), and is classified as a human oncogenic virus [127,128]. The fifth edition of the WHO classification groups KSHV/HHV8-associated lymphoproliferative disorders into a single disease category, while KSHV/HHV8-associated MCD is classified under tumor-like lesions with B-cell predominance and immune dysregulation [5]. There are two clinical presentations of PEL: classic PEL, which involves serous effusions without tumor masses, and extra-cavitary PEL (EC-PEL), which manifests as nodal or extranodal masses with or without effusions. PELs are strongly associated with systemic inflammatory symptoms and frequently co-occur with other KSHV-associated diseases in over 80% of cases. These lymphomas most commonly arise in PLWH but can also be seen in post-transplant recipients and elderly patients with immune senescence [127,128]. While PELs in the HIV-infected patient population can be either EBV-positive or EBV-negative, cases in the elderly tend to be EBV-negative [92,129] (Figure 3).

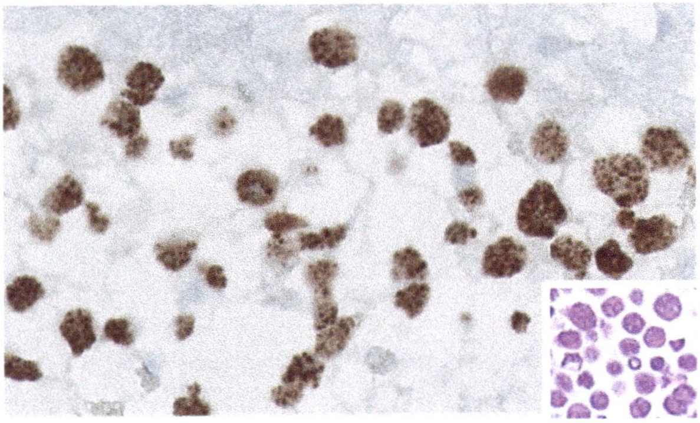

Figure 3. Kaposi sarcoma herpesvirus (KSHV)/human herpes virus 8 (HHV8)-positive primary effusion lymphoma (PEL) in a patient living with HIV. Sparse tumor cells displaying various sizes show nuclear positivity for latent nuclear antigen (LNA) of KSHV/HHV8. The staining is characteristically "speckled". The inset shows that tumor cells are anaplastic and usually display large size. Immunohistochemistry for LNA, original magnification 40×. Inset: Hematoxliyn and Eosin stain from a cell block of a pleural effusion, original magnification 40×.

There is a causal relationship between KSHV/Human Herpesvirus 8 (HHV8) and PEL, and this infection is a diagnostic requirement [128,129]. Another KSHV/HHV8-associated disorder, plasmablastic MCD, is characterized by polyclonal or oligoclonal B cells with unmutated immunoglobulin genes despite showing lambda light chain restriction. The disease typically arises in the context of immune dysregulation, and over 90% of cases occur in PLWH [130]. Given the shared pathogenesis, it is not uncommon for KSHV/HHV8-associated diseases to co-exist. PEL—often co-infected with EBV—along with PBL and KSHV/HHV8-positive DLBCL, have all been described in patients with MCD. These malignancies are generally associated with poor prognosis and often require aggressive systemic chemotherapy [131,132].

T-cell lymphomas are rare among PLWH, accounting for less than 5–10% of HIV-associated lymphomas. The most commonly reported subtypes include peripheral T-cell lymphoma, not otherwise specified [133], anaplastic large-cell lymphoma [134], and angioimmunoblastic T-cell lymphoma [135]. These lymphomas typically exhibit high-grade histologic features, a high proliferation index (Ki-67^{+}), and expression of markers such as CD30 and CD4. Prognosis is generally poor, with significantly shorter median survival compared to HIV-negative individuals with similar lymphoma subtypes [136]. A distinct T cell entity, adult T-cell leukemia/lymphoma (ATLL), is etiologically linked to Human T-cell Lymphotropic Virus type 1 (HTLV-1) infection. Co-infection with HIV, particularly in HTLV-1-endemic regions such as Japan, the Caribbean, and sub-Saharan Africa, may accelerate disease progression, although only about 5% of HTLV-1 carriers develop ATLL [137]. EBV has also been implicated in the pathogenesis of certain HIV-associated T-cell lymphomas, especially in extranodal NK/T-cell lymphoma, nasal type. This lymphoma subtype, which is strongly associated with EBV, is more prevalent in distinct geographic regions, including parts of Asia and Latin America [138].

4. Treatment Strategies

Modern cART has significantly improved life expectancy and cancer treatment outcomes in PLWH by ensuring sustained HIV suppression and partial recovery of immune function [130–132]. For DLBCL, BL, and cHL, survival rates in PLWH now closely resemble those of HIV-negative patients when treated with full-dose chemotherapy in conjunction with cART. However, the prognosis remains poor for PBL and KSHV-associated lymphomas [1,2,139,140]. Prognostic outcomes in HIV-associated lymphomas are more strongly influenced by lymphoma-specific factors—such as those captured by the age-adjusted International Prognostic Index (IPI), BL- or PEL-specific IPI scores, and failure to achieve complete response (CR) to therapy—rather than by HIV-specific factors or EBV-related biomarkers [1,141–144]. Despite advancements, population-based studies have shown that disparities in overall and cancer-specific survival persist between PLWH and the general cancer population, particularly among socially disadvantaged or medically underserved groups [142,145–149]. Importantly, HIV infection should not be an exclusion criterion for enrollment in cancer clinical trials, and HIV-associated malignancies should be treated according to established standards of care [1,150–152].

The management of these malignancies requires a multidisciplinary approach, with concurrent administration of cART and antineoplastic therapy being critical for immune restoration and long-term disease control [1,152–154]. Given the potential for significant drug–drug interactions (DDIs), careful evaluation of interactions between antineoplastic agents, cART, and supportive medications is essential. Ritonavir- and cobicistat-boosted regimens are associated with numerous DDIs that may compromise treatment efficacy or tolerability. In contrast, integrase strand transfer inhibitors such as raltegravir, dolutegravir, and bictegravir offer more favorable DDI profiles and lead to faster reductions in HIV viremia [155,156]. Supportive care, including prophylaxis against opportunistic infections and the use of hematopoietic growth factors, is particularly important in patients with low CD4 cell count [1,150,152]. Furthermore, the growing burden of comorbidities reduced comorbidity-free survival in aging PLWH [130,157], necessitating vigilant monitoring and coordinated, multidisciplinary care strategies [1,158]. Post-treatment cancer surveillance programs should also be implemented, as PLWH are at increased risk for secondary malignancies, including Kaposi sarcoma, Human Papilloma Virus-associated anogenital and head and neck cancers, and additional NHLs [159–163].

4.1. Current Treatment Strategies for HIV-Related NHL

Recent immunodeficiency and cumulative HIV viremia are significant risk factors for the development of NHL in PLWH, with an estimated 11- to 17-fold increased risk in the late cART era compared to the general population [3,4]. HIV-associated NHLs typically present with advanced-stage disease, extranodal involvement, B symptoms, and aggressive clinical behavior [1]. Despite improvements in HIV care, NHL remains a leading cause of mortality among PLWH, underscoring the ongoing need for evidence-based therapeutic strategies [147].

DLBCL accounts for 30–50% of NHL cases in PLWH and is particularly prevalent in individuals with severe immunodeficiency ($CD4^+$ T cell count < 200/μL) [4,164]. Prospective studies and pooled analysis have demonstrated that immunochemotherapeutic regimens incorporating cART and the anti-CD20 monoclonal antibody rituximab (R) are both safe and effective, yielding higher CR rates, longer progression-free survival (PFS), and improved overall survival (OS) in HIV-associated DLBCL [165–169]. Enhanced supportive care, particularly the prophylaxis of opportunistic infections, is critical in high-risk patients [1,150,152], especially in light of findings from a randomized phase 3 trial conducted in the early cART era, which identified safety concerns in patients with advanced HIV [170]. The most commonly used regimens include R-CHOP (cyclophosphamide, doxorubicin, vincristine, prednisone), infusional dose-adjusted (DA)-R-EPOCH (etoposide, prednisone, vincristine, doxorubicin, cyclophosphamide dose-adjusted to CD4 count), and infusional R-CDE (cyclophosphamide, doxorubicin, etoposide). In phase 2 trials, the 2-year OS rates were 75% for R-CHOP, 70% for DA-R-EPOCH, and 64% for R-CDE [165–169] (Table 2). A pooled analysis from two consecutive AIDS Malignancy Consortium studies indicated that infusional R-EPOCH may provide superior outcomes compared to bolus R-CHOP [171]. However, a randomized phase 3 trial in immunocompetent DLBCL patients showed that DA-R-EPOCH did not significantly improve outcomes compared to R-CHOP, suggesting that its benefit may be context-dependent [172].

In the randomized Phase 2 study AMC-075, the efficacy of dose-adjusted R-EPOCH alone was compared to its combination with the oral histone deacetylase inhibitor, oncolytic vorinostat, in patients with HIV-associated aggressive B-cell NHL (approximately 70% DLBCL. The addition of vorinostat conferred no significant benefit in terms of treatment outcomes or reduction of the HIV reservoir [173]. Notably, Myc protein expression was the only biomarker significantly associated with inferior outcomes: the 3-year event-free survival (EFS) of 44% in Myc-positive versus 83% in Myc-negative DLBCL, highlighting the urgent need for novel targeted therapies in this high-risk subgroup [173] (Table 2). A retrospective European study further suggested that Myc-driven HIV-associated DLBCL may benefit more from intensive chemotherapy regimens traditionally used for BL compared to standard R-CHOP, although these findings require validation in prospective clinical trials [174].

The prognostic significance of tumor histogenesis and molecular subtypes, such as double-hit and triple-hit B-cell lymphomas, remains controversial in HIV-associated

DLBCL [1]. This uncertainty likely reflects the heterogeneity of study populations and methodological differences across investigations [143]. In immunocompetent patients, recent advances—particularly the use of the antibody-drug conjugate polantuzumab vedotin—have demonstrated a survival advantage over R-CHOP [175]. However, data on the efficacy and safety of such novel agents in HIV-associated DLBCL are currently lacking, underscoring the need for dedicated studies in this population.

PLWH with moderate immunodeficiency have a lifelong risk of developing BL estimated at 10–20%, regardless of cART status. This risk correlates more strongly with cumulative HIV viremia than with recent immunodeficiency [3,176]. Outcomes for HIV-associated BL have improved in the late cART era with the use of immunochemotherapy regimens that include cART and intensive or infusional multi-agent therapies; notably, R-CHOP is considered inadequate for the treatment of BL [141,177,178]. A phase 2 trial of modified CODOX-M/IVAC-R (cyclophosphamide, doxorubicin, vincristine, methotrexate, etoposide, ifosfamide, cytarabine, rituximab) demonstrated a 2-year overall survival of 69%, with a favorable toxicity profile [177,179]. In patients without central nervous system involvement, risk-adjusted DA-EPOCH-R has shown high efficacy regardless of HIV status, achieving a 4-year EFS rate of 85% [179,180] (Table 2). Furthermore, a large international retrospective study reported that CODOX-M/IVAC-R was associated with significantly longer PFS (hazard ratio (HR) 0.45, p = 0.005), improved OS (HR 0.44, p = 0.007), and lower treatment-related mortality (TRM) (7% vs. 13–18%) compared to other regimens, supporting its role as a preferred option for HIV-associated BL [181].

Table 2. Major front-line clinical trials in HIV-associated non-Hodgkin Lymphoma (HIV-NHL) and classical Hodgkin Lymphoma (HIV-cHL).

Treatment	Study Design	Patients N°	Histology %	CR Rate %	PFS %	OS %	I.Death %	Ref.
HIV-NHL								
R-CHOP-R vs. CHOP	Phase 3	150	DLBCL 81	58 vs. 47	11.3 vs. 9.5 mos	28 vs. 35	14 vs. 2	[170]
R-CHOP	Phase 2	61	DLBCL 72	77	69 (2 yr)	75 (2 yr)	2	[166]
R-CHOP	Phase 2	95	DLBCL 81	69	NA	56 (3 yr)	7	[169]
R-CDE	Phase 2	74	DLBCL 72	70	EFS (2 yr) 52	64 (2 yr)	7	[165]
R-EPOCH vs. EPOCH-R	Rand. Phase 2	106	DLBCL 80	73 vs. 55	66 vs. 63 (2 yr)	70 vs. 67 (2 yr)	10 vs. 7	[167]
SC-EPOCH-RR	Phase 2	33	DLBCL 100	91	84 (5 yr)	68 (5 yr)		[168]
Vorinostat-R-EPOCH	Rand. Phase 2	90	DLBCL 100	58 vs. 74	EFS (3 yr) 63 vs. 69	70 vs. 77 (3 yr)	NA	[173]
R-CODOX-M/IVAC	Phase 2	34	BL 100	NA	69 (1 yr)	69 (2 yr)	3	[177]
DA-EPOCH-R	Phase 2	28 HIV	BL 100	NA	85 (4 yr)	NA	NA	[179]
HIV-cHL								
ABVD	Retrospective	93	cHL	74	EFS (5-yr) 59	81 (5 yr)	<1	[182]
(a) EF: ABVD (b) EU: BEACOPP (c) Adv: BEACOPP or ABVD	Phase 2	108	cHL	(a) EF: 96 (b) EU: 100 (c) Adv: 86	90 (2 yr)	91 (2 yr)	EF: 4 Adv: 7	[183]
BV-AVD	Phase 2	41	cHL	76	86 (2 yr)	92 (2 yr)	<1	[184]

To date, there is no universally accepted standard of care for PBL and PEL in PLWH. Prognosis remains poor in population-based HIV studies, with 5-year OS rates below 40% for PBL and below 30% for PEL [1,142,185]. Nevertheless, favorable outcomes have been reported in selected patients treated with anthracycline-based multiagent chemotherapy in combination with cART [1,145,185]. In the AMC-075 trial, 3-year EFS rates for PBL and PEL were 60% and 71%, respectively [173]. A large international cohort study of patients with

HIV-associated PEL treated with anthracycline-based chemotherapy, cART, and rituximab (when concurrent MCD was present) identified poor performance status and severe anemia as adverse prognostic factors. Notably, the median OS was 16.9 years in patients without these, compared to just 0.6 years in those with them—highlighting the urgent need for improved treatment strategies in HIV-associated PEL with poor prognosis [144]. Novel therapeutic approaches—including bortezomib, daratumumab, lenalidomide, nivolumab, and pomalidomide—are currently under investigation. Despite therapeutic advances, KSHV-associated large B-cell lymphomas remain particularly challenging, with persistently poor long-term outcomes [1].

Prospective and retrospective studies have shown that high-dose chemotherapy (HDC) followed by autologous stem cell transplantation (ASCT) is safe and effective in PLWH who have chemosensitive relapsed or refractory lymphoma, with TRM rates below 5% and 3-year overall survival ranging from 61% to 85% [186–188]. In contrast, data on allogeneic hematopoietic cell transplantation (allo-HCT) in PLWH are limited [189]. Nonetheless, for eligible PLWH, identifying a matched donor homozygous for the CCR5Δ32 mutation remains of particular interest, given reported cases of sustained HIV remission or virologic cure following allo-HCT with such donors [190,191]. Despite promising early data, PLWH continue to be excluded from most clinical trials investigating chimeric antigen receptor (CAR) T-cell therapy, representing an important unmet need in this population [192] (see below).

4.2. *Current Treatment Strategies for HIV-Related Classic Hodgkin Lymphoma*

PLWH who develop cHL typically present with moderate immunodeficiency, B symptoms, and advanced-stage disease involving the bone marrow, liver, and spleen. Primary bone marrow involvement occurs in approximately 3–14% of cases and is associated with particularly aggressive behavior [1,3]. As in the general population, stage-adapted therapy is recommended for HIV-associated cHL. The ABVD regimen (doxorubicin, bleomycin, vinblastine, dacarbazine) has demonstrated safety and efficacy in PLWH receiving cART, achieving a CR rate of 74% and a 5-year OS of 81% [183,193]. The BEACOPP regimen (bleomycin, etoposide, doxorubicin, cyclophosphamide, vincristine, procarbazine, and prednisolone) has also shown high efficacy, with a CR rate of 86%, but it is associated with considerable toxicity, including dose reductions or treatment delays in more than 50% of patients and a TRM of 6% [183] (Table 2).

Risk-adapted therapy guided by fluorodeoxyglucose positron emission tomography (FDG-PET) may serve as a standardized approach in HIV-associated cHL, mirroring strategies used in the general population [194]. However, caution is warranted, as non-malignant HIV-associated conditions can lead to false-positive PET-avid regions, particularly in patients with low $CD4^+$ T cell counts.

The phase 2 AMC-085 trial evaluated the combination of the anti-CD30 antibody–drug conjugate brentuximab vedotin (BV) with AVD (doxorubicin, vincristine, dacarbazine) in newly diagnosed HIV-associated cHL. The study demonstrated promising outcomes, with 2-year PFS and 2-year OS rates of 86% and 92%, respectively, and manageable toxicity [176] (Table 2). Programmed death receptor (PD)-1 blockade has shown efficacy in cHL generally [195], but the historical exclusion of PLWH from oncology clinical trials has delayed the integration of immune checkpoint inhibitors (ICIs) into HIV-associated cHL treatment. Emerging data support the safety and efficacy of ICIs in PLWH with advanced malignancies [196–198]. Notably, the multicenter phase 3 SWOG 1826 trial, which included a small cohort of PLWH with well-controlled HIV infection, demonstrated that nivolumab, an anti-PD-1 monoclonal antibody, combined with AVD in newly diagnosed advanced cHL significantly improved PFS compared to BV-AVD in newly diagnosed advanced-stage cHL (HR 0.48, 95% CI 0.27–0.87, $p < 0.001$) and was associated with better tolerability [199]. Although these data need to be confirmed in a large series of HIV-cHL, they could accelerate the harmonization of treatment between HIV-infected and uninfected patients with cHL.

HDC, followed by ASCT, remains the standard of care for relapsed/refractory cHL, even in PLWH. For patients who experience relapse after HDC-ASCT, treatment with ICIs, such as nivolumab or pembrolizumab, is recommended, in line with standard practice in the general population [195]. However, the use of ICIs in PLWH may be complicated by the reactivation or exacerbation of latent infections, including Mycobacterium tuberculosis, hepatitis B virus, and cytomegalovirus [200,201]. Therefore, comprehensive microbiological screening, along with appropriate prophylaxis or treatment for opportunistic infections, is critical to mitigate the risk of infection-related adverse events in this population.

4.3. Current Treatment Strategies for HIV-Related Multicentric Castleman Disease

MCD is a relapsing and remitting lymphoproliferative disorder characterized by systemic inflammatory symptoms, edema or effusions, cytopenias, generalized lymphadenopathy, and splenomegaly. Active disease is associated with lytic KSHV replication and dysregulation of cytokines, with elevated levels of vIL-6, human IL-6, and IL-10 [92]. A recently described "fluid form" of MCD has been identified using multiparametric cytometry, where the presence of LANA-positive, lambda-restricted plasmablasts in effusions or peripheral blood samples may be indicative of MCD [202,203]. Nonetheless, lymph node biopsy remains essential for a definitive diagnosis of MCD. More than half of PLWH diagnosed with MCD also have concurrent Kaposi Sarcoma, and 20–35% may develop PEL or KSHV-associated large B cell lymphoma at diagnosis or during the disease course [144].

If left untreated, MCD is typically fatal. However, several prospective and retrospective multicenter studies have demonstrated that rituximab significantly improves survival, with reported 5-year OS rates between 80% and 95% [1,204]. In cases of severe disease, the combination of rituximab with cytotoxic agents such as etoposide or liposomal doxorubicin (particularly when Kaposi sarcoma is present) is recommended. In a U.S. cohort, PLWH with MCD treated with cART, rituximab, and liposomal doxorubicin had 5- and 10-year OS rates of 90% and 73%, respectively [129]. Antiviral therapy and IL-6 blockade have demonstrated limited clinical benefit in this setting [1,205]. Novel targeted therapies, including the anti-CD38 monoclonal antibody daratumumab and the JAK-STAT pathway inhibitor pacritinib, are currently under investigation [144]. Early recognition is critical, as misdiagnosis and delayed diagnosis of KSHV-associated diseases can significantly impair timely, disease-specific treatment and negatively impact patient outcomes [206].

5. Immunotherapeutic Perspectives for Lymphomas in PLWH

Despite the complex spectrum of immune dysfunctions associated with HIV infection, accumulating evidence indicates that a variety of immune-based therapies can be safely and effectively used to treat cancer in PLWH. These include monoclonal antibodies, immunomodulatory agents and cytokines, allo-HSCT, and cell-based therapies. Notably, cART itself serves as a critical form of immunotherapy for certain HIV-associated cancers such as Kaposi sarcoma and PCNSL. The immune reconstitution achieved through cART provides a foundational prerequisite for the success of any type of immunotherapy and remains a cornerstone of HIV-related cancer management. However, access to novel immunotherapies for lymphoma in PLWH has been limited, largely due to socioeconomic disparities and a historical lack of data on safety and efficacy in this population. Increasing evidence now supports the safety and clinical benefit of various immunotherapeutic approaches in PLWH, leading to growing advocacy for their inclusion in cancer clinical trials. These efforts are crucial for ensuring equitable access to innovative treatments and improving outcomes in this historically underrepresented group.

ICIs, particularly those targeting the PD-1/PD-L1 and CTLA-4 pathways, have demonstrated substantial clinical benefit across a range of malignancies in the general population, particularly in tumors characterized by high mutation burden and pre-existing anti-tumor immunity. These agents are particularly promising for PLWH, as they function by reversing immune exhaustion rather than augmenting immune suppression. PLWH typically exhibit elevated levels of exhausted $CD8^+$ and $CD4^+$ T cells that express multiple inhibitory checkpoint molecules, suggesting a potential heightened responsiveness to ICI [207]. Beyond their anticancer effects, ICIs may influence HIV persistence. Several studies have shown that PD-1 blockade can enhance HIV transcription and replication from latently infected cells, potentially facilitating immune recognition and clearance of HIV-infected cells that persist on cART. In a study of 32 PLWH with cancer receiving cART, anti-PD-1 treatment was associated with in vivo reversal of HIV latency, supporting a rationale for combining ICIs with latency-reversing agents or other HIV-curative strategies [208]. However, results have been variable: the AIDS Malignancy Consortium 095 Study found that anti-PD-1 monotherapy had no significant impact on HIV latency or reservoir size. In contrast, dual checkpoint blockade with anti-PD-1 and anti-CTLA-4 modestly increased levels of cell-associated unspliced HIV RNA, suggesting potential synergy in targeting the replication-competent HIV reservoir [209].

However, the use of ICIs in patients with HIV-related lymphomas has historically been limited due to safety concerns, particularly the potential for disrupting immune homeostasis and triggering immune-related adverse events (irAEs) or chronic inflammatory syndromes. However, emerging evidence has provided reassurance regarding the safety profile of ICIs in PLWH. Recent studies report that severe irAEs (grade $\geq$ 3) occur in approximately 0–20% of treated individuals, with most patients maintaining stable HIV viral loads and $CD4^+$ T-cell counts throughout treatment [210–215]. Notably, the prospective, real-world ANRS CO24 OncoVIHAC cohort study, which evaluated ICI therapy in PLWH with cancer, reported a 1-year incidence of a first grade $\geq$ 3 irAE episode of 15.0%, with a cumulative incidence of all severe irAE episodes of 26.9 per 100 person-years. The study also identified several risk factors for the development of severe irAEs, including low $CD4^+$ T-cell count, longer duration since HIV diagnosis, history of cancer surgery, and positive cytomegalovirus serology at the initiation of ICI therapy [215].

The assessment of the therapeutic efficacy of ICIs in PLWH and cancer remains limited by the small number of patients enrolled in clinical trials. While ICIs have demonstrated modest benefits in immunocompetent patients with aggressive NHLs, the expression of viral antigens encoded by EBV and/or KSHV in certain HIV-related lymphomas may enhance their immunogenicity, potentially increasing responsiveness to ICI therapy [213,214]. This aligns with the broader observation that ICI efficacy is generally higher in patients with more immunogenic tumors that elicit pre-existing anti-tumor immune responses. As noted above, available clinical data suggest that anti-PD-1 therapy can be effective in HIV cHL [215], as well as in NHLs driven by oncogenic viruses in PLWH [213,214]. Despite these encouraging findings, the proportion of PLWH and cancer who respond to ICI therapy remains limited, mirroring outcomes in the general population. This underscores the need for novel therapeutic strategies to enhance ICI efficacy and broaden their applicability in this setting. One promising approach involves the use of rational drug combinations that synergize with ICIs. A retrospective study of a small cohort of PLWH with lymphomas evaluated pembrolizumab alone or in combination with pomalidomide, an immunomodulatory agent with both direct anti-lymphoma activity and the ability to enhance T and NK cell activation. Pomalidomide also reverses EBV- and KSHV-mediated immune evasion by restoring the expression of MHC I, ICAM-1, and CD86 on tumor cells [216,217]. A response rate of 50% was observed, with notable activity against PEL, suggesting that this combination merits further investigation [217,218]. Advances in understanding the immune landscape of HIV-related lymphomas are also guiding the development of mechanism-based ICI combinations. For example, the upregulation of the TIGIT ligand CD155 in neoplastic cells of HIV-associated cHL, along with increased expression of nectin-3, provides a compelling rationale for targeting the TIGIT/CD155/Nectin-3 axis. Clinical trials are currently underway to evaluate the therapeutic potential of disrupting this pathway in cHL and other malignancies [219].

Chimeric antigen receptor (CAR) T cell therapy has represented a major advancement in the treatment of relapsed or refractory DLBCL in the general population [220]. While five generations of CAR T cell products have been developed, all current FDA-approved products for hematologic malignancies are second-generation CAR T cells [221]. Given the persistently poor outcomes in PLWH and relapsed/refractory DLBCL despite therapeutic advances [164], CAR T cell therapy emerges as an especially attractive option for this population. However, PLWH were excluded from pivotal licensing trials of CAR T cell therapies, thereby delaying the application of this potentially effective approach to HIV-infected patients with lymphoma [222–224]. A recent review detailed the clinical characteristics and outcomes of the limited number of PLWH with DLBCL treated with CAR T cells [224]. In a small cohort, treatment with axicabtagene ciloleucel resulted in clinical responses in four of six patients (three complete remissions and one partial remission). Adverse events included cytokine release syndrome (CRS) of grade ≤ 2 and immune effector cell-associated neurotoxicity syndrome (ICANS) of grade 3–4 in four patients. Although preliminary, these findings suggest that CAR T cell therapy in PLWH may have an acceptable safety profile and therapeutic efficacy, with no compelling clinical contraindications identified to date. One of the key barriers to the broader application of CAR T cell therapy in PLWH is the challenge of achieving sufficient numbers of functional, particularly concerning $CD4^+$ T cells. This lymphocyte subset is not only critical for the manufacturing of a clinical-grade product but also contributes to the therapeutic efficacy and efficacy of CAR T cells [225–228]. This issue may be particularly relevant in patients with advanced HIV disease or low CD4 counts at diagnosis. Nonetheless, several case reports have demonstrated that CAR T cell products can be successfully manufactured from patients with controlled HIV infection, even when $CD4^+$ T cell counts are relatively low [226]. Additional concerns include the presence of low-level viremia or transient increases in viral load during cART, which is relatively common and generally does not reflect the loss of virologic control. These should not, therefore, be used as exclusion criteria for CAR T cell treatment in PLWH. Another consideration is the use of lentiviral vectors in CAR T cell manufacturing. In the context of HIV infection, this raises potential concerns regarding enhanced CAR transgene expression during HIV replication, as well as the risk of false-positive HIV test results due to vector components [229]. Consequently, gamma-retroviral vectors have thus far been preferred for manufacturing CAR T cell products in PLWH with lymphoma. In summary, although data remain limited, current evidence supports the feasibility, safety, and potential efficacy of CAR T cell therapy in PLWH with relapsed/refractory lymphoma. Future clinical applications in this setting may benefit from advances in next-generation CAR T cell technologies, including strategies to enhance T cell potency, overcome tumor heterogeneity and antigen escape, and improve safety and specificity [230]. Looking ahead, the rational combination of these innovations may allow CAR T cells to not only treat HIV-related malignancies but also potentially contribute to HIV cure strategies. This could be achieved through the development of multi-specific CAR T cells targeting tumor- and virus-associated antigens and engineered to be resistant to HIV infection via gene editing techniques [231].

Hematopoietic stem cell transplantation (HSCT) has demonstrated encouraging outcomes in the clinical management of lymphomas. In the context of HIV infection, particular attention has been given to allo-HSCT using donor stem cells carrying the CCR5Δ32 mutation, which confers resistance to HIV by preventing viral entry into host cells. This approach has successfully led to long-term HIV remission or apparent cure in five individuals with HIV-associated hematologic malignancies [232]. Notably, a recent case report showed that sustained HIV remission can also be achieved following allogeneic HSCT from a donor without the CCR5Δ32 mutation, indicating that this genetic deletion is not an absolute requirement for achieving long-term viral control or potential cure [233]. These findings are highly significant, as they suggest a broader applicability of allo-HSCT to a greater number of HIV-infected patients with cancer, including lymphoma, potentially improving their clinical outcomes.

Adoptive cell therapy with EBV-specific T cells is also being explored as a potential strategy for treating EBV-associated malignancies in PLWH. This approach has already demonstrated clinical benefit in managing post-transplant lymphoproliferative disease (PTLD) and controlling EBV reactivation in immunocompromised hosts [234]. In HIV-negative individuals with relapsed or refractory NHL or cHL, treatment with autologous cytotoxic T lymphocytes specific for EBV antigens LMP1/LMP2 or LMP2 alone yielded encouraging results, with most patients achieving 2-year EFS with minimal toxicity [235]. Similarly, infusion of autologous EBNA-1-specific T cells has been shown to restore EBV-specific immunity and control EBV-infected B cell expansion in both adult and pediatric patients with EBV viremia or PTLD following HSCT [236]. More recently, EBV-specific T cells engineered to resist the immunosuppressive activity of Transforming Growth Factor-β have been shown to safely expand and persist in patients with cHL, even in the absence of lymphodepleting chemotherapy, and have induced CR even in patients with resistant disease [237]. Despite these promising results, the application of adoptive EBV-specific T cell therapy in PLWH remains investigational. Challenges include T cell exhaustion and low precursor frequencies, which may hinder the generation of effective autologous EBV-specific T cells in HIV-positive individuals. To overcome these limitations, the use of allogeneic or third-party EBV-specific T cells is being considered [238].

6. Conclusions

Lymphomas in PLWH represent a highly heterogeneous group of malignancies that pose substantial challenges in cancer research, diagnosis, and treatment. These challenges stem from their complex pathogenesis, immunologically composite TME, and frequently aggressive clinical course. Despite notable progress in recent years, advancements in the diagnostic evaluation and therapeutic management of lymphomas in PLWH remain constrained by fragmented research efforts and the frequent exclusion of this population from pivotal clinical trials. Collaborative, multidisciplinary strategies are urgently needed to address these gaps, reduce the incidence of HIV-associated lymphoproliferative disorders, and ultimately improve clinical outcomes and survival for these patients.

Author Contributions: R.D., E.V. and A.C. have equally contributed to the conceptualization, writing, and reviewing of the manuscript. All authors have read and agreed to the published version of the manuscript.

Funding: This research received no external funding.

Conflicts of Interest: The authors declare no conflicts of interest.

Abbreviations

The following abbreviations are used in this manuscript:

ABC	Activated B-cell Like
AICDA	Activation-Induced Cytidine Deaminase
ASCT	Autologous stem cell transplantation
Allo-HCT	Allogeneic Hematopoietic Cell Transplantation
ASCT	Autologous Stem Cell Transplantation
ATLL	Adult T-cell leukemia/lymphoma
BART	BamHI-A region rightward transcript
BCR	B Cell Receptor
BHRF1	BamHI-H rightward fragment 1
BST2	Bone marrow stromal antigen 2
BL	Burkitt Lymphoma
cART	Combined Anti-Retroviral Therapy
CAR	Chimeric Antigen Receptor
CR	Complete Response
DA	Dose-adjusted
DDI	Drug–Drug Interactions
DLBCL	Diffuse Large B Cell Lymphoma
EFS	Event-Free Survival
EBNA6	EBV nuclear antigen 6
EBV	Epstein–Barr Virus
EC	Extra cavitary
EGFR	Epithelial growth factor receptor
EFS	Event-free survival
GCB	Germinal Center B-cell Like
HDC	High-Dose Chemotherapy
HHV-8	Human Herpesvirus-8
HIV	Human Immunodeficiency Virus
HL	Hodgkin Lymphoma
HR	Hazard ratio
HSCT	Hematopoietic Stem Cell Transplantation
HTLV-1	Human T-cell Lymphotropic Virus type 1
KICS	Kaposi's Sarcoma Inflammatory Cytokine Syndrome
KSHV	Kaposi's Sarcoma-associated Herpes Virus
ICIs	Immune Checkpoint Inhibitors
IDO	Indoleamine 2,3-dioxygenase
IPI	International Prognostic Index
irAEs	Immune-Related Adverse Events
IRIS	Immune Reconstitution Inflammatory Syndrome
LANA	Latency-Associated Nuclear Antigen
LMP1	Latent Membrane Protein 1
MCD	Multicentric Castleman Disease
MZL	Marginal Zone Lymphoma
NK	Natural killer
OS	Overall survival
P17	HIV Matrix Protein p17
PFS	Progression-Free Survival
PBL	Plasmablastic Lymphoma
PD-1	Programmed Cell Death 1
PD-L1	Programmed Cell Death Ligand 1
PEL	Primary Effusion Lymphoma
PI3K	Phosphoinositide 3-kinase

PLWH	People Living With HIV
PCNSL	Primary central Nervous System Lymphoma
TAM	Tumor-Associated Macrophage
TCR	T cell receptor
TME	Tumor Microenvironment
TRM	Treatment-related mortality
VEGF	Vascular endothelial growth factor
vFLIP	Viral FLICE inhibitory protein
vIL-6	Viral interleukin 6
vIRF-3	Viral interferon regulatory factor 3
Vpu	HIV viral protein U
WHO	World Health Organization

References

1. Carbone, A.; Vaccher, E.; Gloghini, A. Hematologic cancers in individuals infected by HIV. *Blood* **2022**, *139*, 995–1012. [CrossRef] [PubMed]
2. Ramaswami, R.; Chia, G.; Dalla Pria, A.; Pinato, D.J.; Parker, K.; Nelson, M.; Bower, M. Evolution of HIV-Associated Lymphoma Over 3 Decades. *J. Acquir. Immune Defic. Syndr.* **2016**, *72*, 177–183. [CrossRef] [PubMed]
3. Hernández-Ramírez, R.U.; Shiels, M.S.; Dubrow, R.; Engels, E.A. Cancer risk in HIV-infected people in the USA from 1996 to 2012: A population-based, registry-linkage study. *Lancet HIV* **2017**, *4*, e495–e504. [CrossRef]
4. Kimani, S.M.; Painschab, M.S.; Horner, M.J.; Muchengeti, M.; Fedoriw, Y.; Shiels, M.S.; Gopal, S. Epidemiology of haematological malignancies in people living with HIV. *Lancet HIV* **2020**, *7*, e641–e651. [CrossRef]
5. Alaggio, R.; Amador, C.; Anagnostopoulos, I.; Attygalle, A.D.; Araujo, I.B.O.; Berti, E.; Bhagat, G.; Borges, A.M.; Boyer, D.; Calaminici, M.; et al. The 5th edition of the World Health Organization Classification of Haematolymphoid Tumours: Lymphoid Neoplasms. *Leukemia* **2022**, *36*, 1720–1748. [CrossRef]
6. Grulich, A.E.; Vajdic, C.M. The epidemiology of cancers in human immunodeficiency virus infection and after organ transplantation. *Semin. Oncol.* **2015**, *42*, 247–257. [CrossRef]
7. Pantanowitz, L.; Carbone, A.; Dolcetti, R. Microenvironment and HIV-related lymphomagenesis. *Semin. Cancer Biol.* **2015**, *34*, 52–57. [CrossRef]
8. Dolcetti, R.; Gloghini, A.; Caruso, A.; Carbone, A. A lymphomagenic role for HIV beyond immune suppression? *Blood* **2016**, *127*, 1403–1409. [CrossRef]
9. Armani-Tourret, M.; Bone, B.; Tan, T.S.; Sun, W.; Bellefroid, M.; Struyve, T.; Louella, M.; Yu, X.G.; Lichterfeld, M. Immune targeting of HIV-1 reservoir cells: A path to elimination strategies and cure. *Nat. Rev. Microbiol.* **2024**, *22*, 328–344. [CrossRef]
10. Benito, J.M.; Restrepo, C.; García-Foncillas, J.; Rallón, N. Immune checkpoint inhibitors as potential therapy for reverting T-cell exhaustion and reverting HIV latency in people living with HIV. *Front. Immunol.* **2023**, *14*, 1270881. [CrossRef]
11. Makgoeng, S.B.; Bolanos, R.S.; Jeon, C.Y.; Weiss, R.E.; Arah, O.A.; Breen, E.C.; Martínez-Maza, O.; Hussain, S.K. Markers of Immune Activation and Inflammation, and Non-Hodgkin Lymphoma: A Meta-Analysis of Prospective Studies. *JNCI Cancer Spectr.* **2018**, *2*, pky082. [CrossRef] [PubMed]
12. Veenhuis, R.T.; Abreu, C.M.; Shirk, E.N.; Gama, L.; Clements, J.E. HIV replication and latency in monocytes and macrophages. *Semin. Immunol.* **2021**, *51*, 101472. [CrossRef] [PubMed]
13. Mitchell, R.S.; Katsura, C.; Skasko, M.A.; Fitzpatrick, K.; Lau, D.; Ruiz, A.; Stephens, E.B.; Margottin-Goguet, F.; Benarous, R.; Guatelli, J.C. Vpu antagonizes BST-2-mediated restriction of HIV-1 release via beta-TrCP and endo-lysosomal trafficking. *PLoS Pathog.* **2009**, *5*, e1000450. [CrossRef] [PubMed]
14. Singh, H.; Jadhav, S.; Arif Khan, A.; Aggarwal, S.K.; Choudhari, R.; Verma, S.; Aggarwal, S.; Gupta, V.; Singh, A.; Nain, S.; et al. APOBEC3, TRIM5α, and BST2 polymorphisms in healthy individuals of various populations with special references to its impact on HIV transmission. *Microb. Pathog.* **2022**, *162*, 105326. [CrossRef]
15. Lê-Bury, G.; Niedergang, F. Defective Phagocytic Properties of HIV-Infected Macrophages: How Might They Be Implicated in the Development of Invasive Salmonella Typhimurium? *Front. Immunol.* **2018**, *9*, 531. [CrossRef]
16. Wang, L.; Li, G.; Yao, Z.Q.; Moorman, J.P.; Ning, S. MicroRNA regulation of viral immunity, latency, and carcinogenesis of selected tumor viruses and HIV. *Rev. Med. Virol.* **2015**, *25*, 320–341. [CrossRef]
17. Planès, R.; Bahraoui, E. HIV-1 Tat protein induces the production of IDO in human monocyte derived-dendritic cells through a direct mechanism: Effect on T cells proliferation. *PLoS ONE* **2013**, *8*, e74551. [CrossRef]
18. Favre, D.; Mold, J.; Hunt, P.W.; Kanwar, B.; Loke, P.; Seu, L.; Barbour, J.D.; Lowe, M.M.; Jayawardene, A.; Aweeka, F.; et al. Tryptophan catabolism by indoleamine 2,3-dioxygenase 1 alters the balance of TH17 to regulatory T cells in HIV disease. *Sci. Transl. Med.* **2010**, *2*, 32ra36. [CrossRef]

19. Terness, P.; Bauer, T.M.; Röse, L.; Dufter, C.; Watzlik, A.; Simon, H.; Opelz, G. Inhibition of allogeneic T cell proliferation by indoleamine 2,3-dioxygenase-expressing dendritic cells: Mediation of suppression by tryptophan metabolites. *J. Exp. Med.* **2002**, *196*, 447–457. [CrossRef]
20. Judge, C.J.; Kostadinova, L.; Sherman, K.E.; Butt, A.A.; Falck-Ytter, Y.; Funderburg, N.T.; Landay, A.L.; Lederman, M.M.; Sieg, S.F.; Sandberg, J.K.; et al. CD56^{bright}NK IL-7Rα expression negatively associates with HCV level, and IL-7-induced NK function is impaired during HCV and HIV infections. *J. Leukoc. Biol.* **2017**, *102*, 171–184. [CrossRef]
21. Lucar, O.; Reeves, R.K.; Jost, S. A Natural Impact: NK Cells at the Intersection of Cancer and HIV Disease. *Front. Immunol.* **2019**, *10*, 1850. [CrossRef] [PubMed]
22. Cocker, A.T.H.; Liu, F.; Djaoud, Z.; Guethlein, L.A.; Parham, P. CD56-negative NK cells: Frequency in peripheral blood, expansion during HIV-1 infection, functional capacity, and KIR expression. *Front. Immunol.* **2022**, *13*, 992723. [CrossRef] [PubMed]
23. Zhang, Z.; Zhou, Y.; Lu, J.; Chen, Y.F.; Hu, H.Y.; Xu, X.Q.; Fu, G.F. Changes in NK Cell Subsets and Receptor Expressions in HIV-1 Infected Chronic Patients and HIV Controllers. *Front. Immunol.* **2021**, *12*, 792775. [CrossRef] [PubMed]
24. Lurain, K.A.; Ramaswami, R.; Krug, L.T.; Whitby, D.; Ziegelbauer, J.M.; Wang, H.-W.; Yarchoan, R. HIV-associated cancers and lymphoproliferative disorders caused by Kaposi sarcoma herpesvirus and Epstein-Barr virus. *Clin. Microbiol. Rev.* **2024**, *37*, e0002223. [CrossRef]
25. Shindiapina, P.; Ahmed, E.H.; Mozhenkova, A.; Abebe, T.; Baiocchi, R.A. Immunology of EBV-Related Lymphoproliferative Disease in HIV-Positive Individuals. *Front. Oncol.* **2020**, *10*, 1723. [CrossRef]
26. Carbone, A.; Gloghini, A.; Caruso, A.; De Paoli, P.; Dolcetti, R. The impact of EBV and HIV infection on the microenvironmental niche underlying Hodgkin lymphoma pathogenesis. *Int. J. Cancer* **2017**, *140*, 1233–1245. [CrossRef]
27. Yarchoan, R.; Uldrick, T.S. HIV-Associated Cancers and Related Diseases. *N. Engl. J. Med.* **2018**, *378*, 1029–1041. [CrossRef]
28. Banerjee, A.; Dass, D.; Mukherjee, S.; Kaul, M.; Harshithkumar, R.; Bagchi, P.; Mukherjee, A. The 'Oma's of the Gammas-Cancerogenesis by γ-Herpesviruses. *Viruses* **2024**, *16*, 1928. [CrossRef]
29. Kalchschmidt, J.S.; Bashford-Rogers, R.; Paschos, K.; Gillman, A.C.; Styles, C.T.; Kellam, P.; Allday, M.J. Epstein-Barr virus nuclear protein EBNA3C directly induces expression of AID and somatic mutations in B cells. *J. Exp. Med.* **2016**, *213*, 921–928. [CrossRef]
30. Shponka, V.; Reveles, C.Y.; Alam, S.; Jaramillo, M.; Maguire, A.; Rimsza, L.M.; Kendrick, S. Frequent expression of activation-induced cytidine deaminase in diffuse large B-cell lymphoma tissues from persons living with HIV. *AIDS* **2020**, *34*, 2025–2035. [CrossRef]
31. Whitehurst, C.B.; Rizk, M.; Teklezghi, A.; Spagnuolo, R.A.; Pagano, J.S.; Wahl, A. HIV co-infection augments EBV-induced tumorigenesis in vivo. *Front. Virol.* **2022**, *2*, 861628. [CrossRef] [PubMed]
32. Notarte, K.I.; Senanayake, S.; Macaranas, I.; Albano, P.M.; Mundo, L.; Fennell, E.; Leoncini, L.; Murray, P. MicroRNA and Other Non-Coding RNAs in Epstein-Barr Virus-Associated Cancers. *Cancers* **2021**, *13*, 3909. [CrossRef] [PubMed]
33. Jalilian, S.; Bastani, M.N. From virus to cancer: Epstein-Barr virus miRNA connection in Burkitt's lymphoma. *Infect. Agent. Cancer* **2024**, *19*, 54. [CrossRef] [PubMed]
34. Cesarman, E.; Chang, Y.; Moore, P.S.; Said, J.W.; Knowles, D.M. Kaposi's sarcoma-associated herpesvirus-like DNA sequences in AIDS-related body-cavity-based lymphomas. *N. Engl. J. Med.* **1995**, *332*, 1186–1191. [CrossRef]
35. Soulier, J.; Grollet, L.; Oksenhendler, E.; Cacoub, P.; Cazals-Hatem, D.; Babinet, P.; d'Agay, M.F.; Clauvel, J.P.; Raphael, M.; Degos, L.; et al. Kaposi's sarcoma-associated herpesvirus-like DNA sequences in multicentric Castleman's disease. *Blood* **1995**, *86*, 1276–1280. [CrossRef] [PubMed]
36. Chadburn, A.; Hyjek, E.M.; Tam, W.; Liu, Y.; Rengifo, T.; Cesarman, E.; Knowles, D.M. Immunophenotypic analysis of the Kaposi sarcoma herpesvirus (KSHV.; HHV-8)-infected B cells in HIV+ multicentric Castleman disease (MCD). *Histopathology* **2008**, *53*, 513–524. [CrossRef]
37. Du, M.Q.; Liu, H.; Diss, T.C.; Ye, H.; Hamoudi, R.A.; Dupin, N.; Meignin, V.; Oksenhendler, E.; Boshoff, C.; Isaacson, P.G. Kaposi sarcoma-associated herpesvirus infects monotypic (IgM lambda) but polyclonal naive B cells in Castleman disease and associated lymphoproliferative disorders. *Blood* **2001**, *97*, 2130–2136. [CrossRef]
38. Hu, D.; Wang, V.; Yang, M.; Abdullah, S.; Davis, D.A.; Uldrick, T.S.; Polizzotto, M.N.; Veeranna, R.P.; Pittaluga, S.; Tosato, G.; et al. Induction of Kaposi's Sarcoma-Associated Herpesvirus-Encoded Viral Interleukin-6 by X-Box Binding Protein 1. *J. Virol.* **2015**, *90*, 368–378. [CrossRef]
39. Aoki, Y.; Yarchoan, R.; Wyvill, K.; Okamoto, S.; Little, R.F.; Tosato, G. Detection of viral interleukin-6 in Kaposi sarcoma-associated herpesvirus-linked disorders. *Blood* **2001**, *97*, 2173–2176. [CrossRef]
40. Liu, C.; Okruzhnov, Y.; Li, H.; Nicholas, J. Human herpesvirus 8 (HHV-8)-encoded cytokines induce expression of and autocrine signaling by vascular endothelial growth factor (VEGF) in HHV-8-infected primary-effusion lymphoma cell lines and mediate VEGF-independent antiapoptotic effects. *J. Virol.* **2001**, *75*, 10933–10940. [CrossRef]
41. Münz, C. Recent advances in animal models of lymphomagenesis caused by human γ-herpesviruses. *Curr. Opin. Virol.* **2025**, *71*, 101461. [CrossRef] [PubMed]
42. Deeks, S.G. HIV infection, inflammation, immunosenescence, and aging. *Annu. Rev. Med.* **2011**, *62*, 141–155. [CrossRef] [PubMed]

43. Straista, M.; Caccuri, F.; Arnaut, N.; Caruso, A.; Slevin, M. Pathological Mechanisms Involved in HIV-Associated Lymphomagenesis: Novel Targeted Therapeutic Approaches. *Cells* **2025**, *14*, 705. [CrossRef] [PubMed]
44. Novak, R.M.; Richardson, J.T.; Buchacz, K.; Chmiel, J.S.; Durham, M.D.; Palella, F.J.; Wendrow, A.; Wood, K.; Young, B.; Brooks, J.T.; et al. Investigators Immune reconstitution inflammatory syndrome: Incidence implications for mortality. *AIDS* **2012**, *26*, 721–730. [CrossRef]
45. Liu, Y.; Zhou, X.; Wang, X. Targeting the tumor microenvironment in B-cell lymphoma: Challenges and opportunities. *J. Hematol. Oncol.* **2021**, *14*, 125. [CrossRef]
46. Alsharif, R.; Dunleavy, K. Burkitt Lymphoma and Other High-Grade B-Cell Lymphomas with or without MYC, BCL2, and/or BCL6 Rearrangements. *Hematol. Oncol. Clin. North. Am.* **2019**, *33*, 587–596. [CrossRef]
47. Feng, Y.; Wang, Z.; Zeng, D.; Song, S.; Yang, Y.; Wang, A.; Xu, J.; Guo, W.; Wu, M.; Shi, Y.; et al. High expression of HIV-1 matrix protein p17 in both lymphoma and lymph node tissues of AIDS patients. *Pathol. Res. Pract.* **2022**, *237*, 154061. [CrossRef]
48. Lazzi, S.; Bellan, C.; De Falco, G.; Cinti, C.; Ferrari, F.; Nyongo, A.; Claudio, P.P.; Tosi, G.M.; Vatti, R.; Gloghini, A.; et al. Expression of RB2/p130 tumor-suppressor gene in AIDS-related non-Hodgkin's lymphomas: Implications for disease pathogenesis. *Hum. Pathol.* **2002**, *33*, 723–731. [CrossRef]
49. Scala, G.; Ruocco, M.R.; Ambrosino, C.; Mallardo, M.; Giordano, V.; Baldassarre, F.; Dragonetti, E.; Quinto, I.; Venuta, S. The expression of the interleukin 6 gene is induced by the human immunodeficiency virus 1 TAT protein. *J. Exp. Med.* **1994**, *179*, 961–971. [CrossRef]
50. Blazevic, V.; Heino, M.; Lagerstedt, A.; Ranki, A.; Krohn, K.J. Interleukin-10 gene expression induced by HIV-1 Tat and Rev in the cells of HIV-1 infected individuals. *J. Acquir. Immune Defic. Syndr. Hum. Retrovirol.* **1996**, *13*, 208–214. [CrossRef] [PubMed]
51. Luzzi, A.; Morettini, F.; Gazaneo, S.; Mundo, L.; Onnis, A.; Mannucci, S.; Rogena, E.A.; Bellan, C.; Leoncini, L.; De Falco, G. HIV-1 Tat induces DNMT over-expression through microRNA dysregulation in HIV-related non-Hodgkin lymphomas. *Infect. Agents Cancer* **2014**, *9*, 41. [CrossRef] [PubMed]
52. Valyaeva, A.A.; Tikhomirova, M.A.; Potashnikova, D.M.; Bogomazova, A.N.; Snigiryova, G.P.; Penin, A.A.; Logacheva, M.D.; Arifulin, E.A.; Shmakova, A.A.; Germini, D.; et al. Ectopic expression of HIV-1 Tat modifies gene expression in cultured B cells: Implications for the development of B-cell lymphomas in HIV-1-infected patients. *PeerJ* **2022**, *10*, e13986. [CrossRef] [PubMed]
53. Akbay, B.; Germini, D.; Bissenbaev, A.K.; Musinova, Y.R.; Sheval, E.V.; Vassetzky, Y.; Dokudovskaya, S. HIV-1 Tat Activates Akt/mTORC1 Pathway and AICDA Expression by Downregulating Its Transcriptional Inhibitors in B Cells. *Int. J. Mol. Sci.* **2021**, *22*, 1588. [CrossRef] [PubMed]
54. Sall, F.B.; El Amine, R.; Markozashvili, D.; Tsfasman, T.; Oksenhendler, E.; Lipinski, M.; Vassetzky, Y.; Germini, D. HIV-1 Tat protein induces aberrant activation of AICDA in human B-lymphocytes from peripheral blood. *J. Cell. Physiol.* **2019**, *234*, 15678–15685. [CrossRef]
55. Germini, D.; Tsfasman, T.; Klibi, M.; El-Amine, R.; Pichugin, A.; Iarovaia, O.V.; Bilhou-Nabera, C.; Subra, F.; Bou Saada, Y.; Sukhanova, A.; et al. HIV Tat induces a prolonged MYC relocalization next to IGH in circulating B-cells. *Leukemia* **2017**, *31*, 2515–2522. [CrossRef]
56. Shmakova, A.; Hugot, C.; Kozhevnikova, Y.; Schwager Karpukhina, A.; Tsimailo, I.; Gérard, L.; Boutboul, D.; Oksenhendler, E.; Szewczyk-Roszczenko, O.; Roszczenko, P.; et al. Chronic HIV-1 Tat action induces HLA-DR downregulation in B cells: A mechanism for lymphoma immune escape in people living with HIV. *J. Med. Virol.* **2024**, *96*, e29423. [CrossRef] [PubMed]
57. Kundu, R.K.; Sangiorgi, F.; Wu, L.Y.; Pattengale, P.K.; Hinton, D.R.; Gill, P.S.; Maxson, R. Expression of the human immunodeficiency virus-Tat gene in lymphoid tissues of transgenic mice is associated with B-cell lymphoma. *Blood* **1999**, *94*, 275–282. [CrossRef] [PubMed]
58. Mdletshe, N.; Nel, A.; Shires, K.; Mowla, S. HIV Nef enhances the expression of oncogenic c-MYC and activation-induced cytidine deaminase in Burkitt lymphoma cells, promoting genomic instability. *Infect. Agent. Cancer* **2020**, *15*, 54. [CrossRef]
59. Caccuri, F.; Iaria, M.L.; Campilongo, F.; Varney, K.; Rossi, A.; Mitola, S.; Schiarea, S.; Bugatti, A.; Mazzuca, P.; Giagulli, C.; et al. Cellular aspartyl proteases promote the unconventional secretion of biologically active HIV-1 matrix protein p17. *Sci. Rep.* **2016**, *6*, 38027. [CrossRef]
60. Popovic, M.; Tenner-Racz, K.; Pelser, C.; Stellbrink, H.J.; van Lunzen, J.; Lewis, G.; Kalyanaraman, V.S.; Gallo, R.C.; Racz, P. Persistence of HIV-1 structural proteins and glycoproteins in lymph nodes of patients under highly active antiretroviral therapy. *Proc. Natl. Acad. Sci. USA* **2005**, *102*, 14807–14812. [CrossRef]
61. Curreli, S.; Krishnan, S.; Reitz, M.; Lunardi-Iskandar, Y.; Lafferty, M.K.; Garzino-Demo, A.; Zella, D.; Gallo, R.C.; Bryant, J. B cell lymphoma in HIV transgenic mice. *Retrovirology* **2013**, *10*, 92. [CrossRef]
62. Dolcetti, R.; Giagulli, C.; He, W.; Selleri, M.; Caccuri, F.; Eyzaguirre, L.M.; Mazzuca, P.; Corbellini, S.; Campilongo, F.; Marsico, S.; et al. Role of HIV-1 matrix protein p17 variants in lymphoma pathogenesis. *Proc. Natl. Acad. Sci. USA* **2015**, *112*, 14331–14336. [CrossRef] [PubMed]
63. Giagulli, C.; Marsico, S.; Magiera, A.K.; Bruno, R.; Caccuri, F.; Barone, I.; Fiorentini, S.; Andò, S.; Caruso, A. Opposite effects of HIV-1 p17 variants on PTEN activation and cell growth in B cells. *PLoS ONE* **2011**, *6*, e17831. [CrossRef] [PubMed]

64. Giagulli, C.; D'Ursi, P.; He, W.; Zorzan, S.; Caccuri, F.; Varney, K.; Orro, A.; Marsico, S.; Otjacques, B.; Laudanna, C.; et al. A single amino acid substitution confers B-cell clonogenic activity to the HIV-1 matrix protein p17. *Sci. Rep.* **2017**, *7*, 6555. [CrossRef] [PubMed]
65. Giagulli, C.; Caccuri, F.; Zorzan, S.; Bugatti, A.; Zani, A.; Filippini, F.; Manocha, E.; D'Ursi, P.; Orro, A.; Dolcetti, R.; et al. B-cell clonogenic activity of HIV-1 p17 variants is driven by PAR1-mediated EGF transactivation. *Cancer Gene Ther.* **2021**, *28*, 649–666. [CrossRef]
66. Caccuri, F.; Messali, S.; Zani, A.; Campisi, G.; Giovanetti, M.; Zanussi, S.; Vaccher, E.; Fabris, S.; Bugatti, A.; Focà, E.; et al. HIV-1 mutants expressing B cell clonogenic matrix protein p17 variants are increasing their prevalence worldwide. *Proc. Natl. Acad. Sci USA* **2022**, *119*, e2122050119. [CrossRef]
67. Martínez, L.E.; Comin-Anduix, B.; Güemes-Aragon, M.; Ibarrondo, J.; Detels, R.; Mimiaga, M.J.; Epeldegui, M. Characterization of unique B-cell populations in the circulation of people living with HIV prior to non-Hodgkin lymphoma diagnosis. *Front. Immunol.* **2024**, *15*, 1441994. [CrossRef]
68. Martin, G.; Roy, J.; Barat, C.; Ouellet, M.; Gilbert, C.; Tremblay, M.J. Human immunodeficiency virus type 1-associated CD40 ligand transactivates B lymphocytes and promotes infection of CD4+ T cells. *J. Virol.* **2007**, *81*, 5872–5881. [CrossRef]
69. Imbeault, M.; Ouellet, M.; Giguère, K.; Bertin, J.; Bélanger, D.; Martin, G.; Tremblay, M.J. Acquisition of host-derived CD40L by HIV-1 in vivo and its functional consequences in the B-cell compartment. *J. Virol.* **2011**, *85*, 2189–2200. [CrossRef]
70. Robbiani, D.F.; Bunting, S.; Feldhahn, N.; Bothmer, A.; Camps, J.; Deroubaix, S.; McBride, K.M.; Klein, I.A.; Stone, G.; Eisenreich, T.R.; et al. AID produces DNA double-strand breaks in non-Ig genes and mature B cell lymphomas with reciprocal chromosome translocations. *Mol. Cell* **2009**, *36*, 631–641. [CrossRef]
71. Liang, Y.; Chen, X.; Zhang, X.; Guo, C.; Zhang, Y. Virus-driven dysregulation of the BCR pathway: A potential mechanism for the high prevalence of HIV related B-cell lymphoma. *Ann. Hematol.* **2024**, *103*, 4839–4849. [CrossRef] [PubMed]
72. Knowles, D.M.; Cesarman, E.; Chadburn, A.; Frizzera, G.; Chen, J.; Rose, E.A.; Michler, R.E. Correlative morphologic and molecular genetic analysis demonstrates three distinct categories of posttransplantation lymphoproliferative disorders. *Blood* **1995**, *85*, 552–565. [CrossRef] [PubMed]
73. Swerdlow, S.H.W.S.; Chadburn, A.; Ferry, J.A. Post-transplant lymphoproliferative disorders. In *WHO Classification of Tumours of Haematopoietic and Lymphoid Tissues*, 4th ed.; Swerdlow, S.C., Harris, N.L., Jaffe, E.S., Pileri, S.A., Stein, H., Thiele, J., Arber, D.A., Hasserjian, R.P., Le Beau, M.M., Orazi, A., et al., Eds.; IARC: Lyon, France, 2017; pp. 453–462.
74. Natkunam, Y.; Goodlad, J.R.; Chadburn, A.; de Jong, D.; Gratzinger, D.; Chan, J.K.; Said, J.; Jaffe, E.S. EBV-Positive B-Cell Proliferations of Varied Malignant Potential: 2015 SH/EAHP Workshop Report-Part 1. *Am. J. Clin. Pathol.* **2017**, *147*, 129–152. [CrossRef] [PubMed]
75. Gratzinger, D.; Jaffe, E.S.; Chadburn, A.; Chan, J.K.; de Jong, D.; Goodlad, J.R.; Said, J.; Natkunam, Y. Primary/Congenital Immunodeficiency: 2015 SH/EAHP Workshop Report-Part 5. *Am. J. Clin. Pathol.* **2017**, *147*, 204–216. [CrossRef]
76. Nador, R.G.; Chadburn, A.; Gundappa, G.; Cesarman, E.; Said, J.W.; Knowles, D.M. Human immunodeficiency virus (HIV)-associated polymorphic lymphoproliferative disorders. *Am. J. Surg. Pathol.* **2003**, *27*, 293–302. [CrossRef]
77. Dojcinov, S.D.; Venkataraman, G.; Pittaluga, S.; Wlodarska, I.; Schrager, J.A.; Raffeld, M.; Hills, R.K.; Jaffe, E.S. Age-related EBV-associated lymphoproliferative disorders in the Western population: A spectrum of reactive lymphoid hyperplasia and lymphoma. *Blood* **2011**, *117*, 4726–4735. [CrossRef]
78. Dojcinov, S.D.; Venkataraman, G.; Raffeld, M.; Pittaluga, S.; Jaffe, E.S. EBV positive mucocutaneous ulcer—A study of 26 cases associated with various sources of immunosuppression. *Am. J. Surg. Pathol.* **2010**, *34*, 405–417. [CrossRef]
79. Hart, M.; Thakral, B.; Yohe, S.; Balfour, H.H., Jr.; Singh, C.; Spears, M.; McKenna, R.W. EBV-positive mucocutaneous ulcer in organ transplant recipients: A localized indolent posttransplant lymphoproliferative disorder. *Am. J. Surg. Pathol.* **2014**, *38*, 1522–1529. [CrossRef]
80. Pina-Oviedo, S.; Miranda, R.N.; Medeiros, L.J. Cancer Therapy-associated Lymphoproliferative Disorders: An Under-recognized Type of Immunodeficiency-associated Lymphoproliferative Disorder. *Am. J. Surg. Pathol.* **2018**, *42*, 116–129. [CrossRef]
81. Carbone, A.; Chadburn, A.; Gloghini, A.; Vaccher, E.; Bower, M. Immune deficiency/dysregulation-associated lymphoproliferative disorders. Revised classification and management. *Blood Rev.* **2024**, *64*, 101167. [CrossRef]
82. Novoa-Takara, L.; Perkins, S.L.; Qi, D.; Shidham, V.B.; Vesole, D.H.; Hariharan, S.; Luo, Y.; Ewton, A.; Chang, C.C. Histogenetic phenotypes of B cells in posttransplant lymphoproliferative disorders by immunohistochemical analysis correlate with transplant type: Solid organ vs hematopoietic stem cell transplantation. *Am. J. Clin. Pathol.* **2005**, *123*, 104–112. [CrossRef] [PubMed]
83. Kaleem, Z.; Hassan, A.; Pathan, M.H.; White, G. Flow cytometric evaluation of posttransplant B-cell lymphoproliferative disorders. *Arch. Pathol. Lab. Med.* **2004**, *128*, 181–186. [CrossRef] [PubMed]
84. Ramalingam, P.; Rybicki, L.; Smith, M.D.; Abrahams, N.A.; Tubbs, R.R.; Pettay, J.; Farver, C.F.; Hsi, E.D. Posttransplant lymphoproliferative disorders in lung transplant patients: The Cleveland Clinic experience. *Mod. Pathol.* **2002**, *15*, 647–656. [CrossRef] [PubMed]

85. Ichikawa, A.; Arakawa, F.; Kiyasu, J.; Sato, K.; Miyoshi, H.; Niino, D.; Kimura, Y.; Takeuchi, M.; Yoshida, M.; Ishibashi, Y.; et al. Methotrexate/iatrogenic lymphoproliferative disorders in rheumatoid arthritis: Histology, Epstein-Barr virus, and clonality are important predictors of disease progression and regression. *Eur. J. Haematol.* **2013**, *91*, 20–28. [CrossRef]
86. Bräuninger, A.; Spieker, T.; Mottok, A.; Baur, A.S.; Küppers, R.; Hansmann, M.L. Epstein-Barr virus (EBV)-positive lymphoproliferations in post-transplant patients show immunoglobulin V gene mutation patterns suggesting interference of EBV with normal B cell differentiation processes. *Eur. J. Immunol.* **2003**, *33*, 1593–1602. [CrossRef]
87. Tokuhira, M.; Tamaru, J.I.; Kizaki, M. Clinical management for other iatrogenic immunodeficiency-associated lymphoproliferative disorders. *J. Clin. Exp. Hematop.* **2019**, *59*, 72–92. [CrossRef]
88. Chadburn, A.; Gloghini, A.; Carbone, A. Classification of B-Cell Lymphomas and Immunodeficiency-Related Lymphoproliferations: What's New? *Hemato* **2023**, *4*, 26–41. [CrossRef]
89. Said, J.C.E.; Rosenwald, A.; Harris, N.L. Lymphomas associated with HIV infection. In *WHO Classification of Tumours of Haematopoietic and Lymphoid Tissues*, 4th ed.; Swerdlow, S.C., Harris, N.L., Jaffe, E.S., Pileri, S.A., Stein, H., Thiele, J., Arber, D.A., Hasserjian, R.P., Le Beau, M.M., Orazi, A., et al., Eds.; IARC: Lyon, France, 2017; pp. 449–452.
90. Chadburn, A.; Abdul-Nabi, A.M.; Teruya, B.S.; Lo, A.A. Lymphoid proliferations associated with human immunodeficiency virus infection. *Arch. Pathol. Lab. Med.* **2013**, *137*, 360–370. [CrossRef]
91. IARC. *Monographs on the Evaluation of Carcinogenic Risks to Humans. A Review of Human Carcinogens. Part B: Biological Agents*; IARC Working Group on the Evaluation of Carcinogenic Risks to Humans: Lyon, France, 2012; Volume 100.
92. Carbone, A.; Borok, M.; Damania, B.; Gloghini, A.; Polizzotto, M.N.; Jayanthan, R.K.; Fajgenbaum, D.C.; Bower, M. Castleman disease. *Nat. Rev. Dis. Primers* **2021**, *7*, 84. [CrossRef]
93. Chapman, J.R.; Bouska, A.C.; Zhang, W.; Alderuccio, J.P.; Lossos, I.S.; Rimsza, L.M.; Maguire, A.; Yi, S.; Chan, W.C.; Vega, F.; et al. EBV-positive HIV-associated diffuse large B cell lymphomas are characterized by JAK/STAT (STAT3) pathway mutations and unique clinicopathologic features. *Br. J. Haematol.* **2021**, *194*, 870–878. [CrossRef]
94. Besson, C.; Lancar, R.; Prevot, S.; Algarte-Genin, M.; Delobel, P.; Bonnet, F.; Meyohas, M.C.; Partisani, M.; Oberic, L.; Gabarre, J.; et al. Outcomes for HIV-associated diffuse large B-cell lymphoma in the modern combined antiretroviral therapy era. *AIDS* **2017**, *31*, 2493–2501. [CrossRef] [PubMed]
95. Baptista, M.J.; Tapia, G.; Morgades, M.; Muncunill, J.; Muñoz-Marmol, A.M.; Montoto, S.; Gribben, J.G.; Calaminici, M.; Martinez, A.; Gonzalez-Farre, B.; et al. Using the Lymph2Cx assay for assessing cell-of-origin subtypes of HIV-related diffuse large B-cell lymphoma. *Leuk. Lymphoma* **2019**, *60*, 1087–1091. [CrossRef] [PubMed]
96. Liapis, K.; Clear, A.; Owen, A.; Coutinho, R.; Greaves, P.; Lee, A.M.; Montoto, S.; Calaminici, M.; Gribben, J.G. The microenvironment of AIDS-related diffuse large B-cell lymphoma provides insight into the pathophysiology and indicates possible therapeutic strategies. *Blood* **2013**, *122*, 424–433. [CrossRef]
97. Baptista, M.J.; Tapia, G.; Muñoz-Marmol, A.M.; Muncunill, J.; Garcia, O.; Montoto, S.; Gribben, J.G.; Calaminici, M.; Martinez, A.; Veloza, L.; et al. Genetic and phenotypic characterisation of HIV-associated aggressive B-cell non-Hodgkin lymphomas, which do not occur specifically in this population: Diagnostic and prognostic implications. *Histopathology* **2022**, *81*, 826–840. [CrossRef] [PubMed]
98. Vaughan, J.; Wiggill, T.; Mia, Z.; Patel, M. Tumour-associated macrophages in diffuse large B-cell lymphoma: The prognostic and therapeutic impact in a South African centre with high HIV seroprevalence. *Immunol. Res.* **2024**, *72*, 1393–1403. [CrossRef]
99. Martinez, F.O.; Gordon, S. The M1 and M2 paradigm of macrophage activation: Time for reassessment. *F1000Prime Rep.* **2014**, *6*, 13. [CrossRef]
100. Epeldegui, M.; Conti, D.V.; Guo, Y.; Cozen, W.; Penichet, M.L.; Martínez-Maza, O. Elevated numbers of PD-L1 expressing B cells are associated with the development of AIDS-NHL. *Sci. Rep.* **2019**, *9*, 9371. [CrossRef]
101. Goodman, A.; Patel, S.P.; Kurzrock, R. PD-1-PD-L1 immune-checkpoint blockade in B-cell lymphomas. *Nat. Rev. Clin. Oncol.* **2017**, *14*, 203–220. [CrossRef]
102. Kowalkowski, M.A.; Mims, M.P.; Amiran, E.S.; Lulla, P.; Chiao, E.Y. Effect of immune reconstitution on the incidence of HIV-related Hodgkin lymphoma. *PLoS ONE* **2013**, *8*, e77409. [CrossRef]
103. Carroll, V.; Garzino-Demo, A. HIV-associated lymphoma in the era of combination antiretroviral therapy: Shifting the immunological landscape. *Pathog. Dis.* **2015**, *73*, ftv044. [CrossRef]
104. Hleyhel, M.; Bouvier, A.M.; Belot, A.; Tattevin, P.; Pacanowski, J.; Genet, P.; De Castro, N.; Berger, J.L.; Dupont, C.; Lavolé, A.; et al. Risk of non-AIDS-defining cancers among HIV-1-infected individuals in France between 1997 and 2009: Results from a French cohort. *AIDS* **2014**, *28*, 2109–2118. [CrossRef] [PubMed]
105. Kowalkowski, M.A.; Mims, M.A.; Day, R.S.; Du, X.L.; Chan, W.; Chiao, E.Y. Longer duration of combination antiretroviral therapy reduces the risk of Hodgkin lymphoma: A cohort study of HIV-infected male veterans. *Cancer Epidemiol.* **2014**, *38*, 386–392. [CrossRef] [PubMed]
106. Said, J.W. Immunodeficiency-related Hodgkin lymphoma and its mimics. *Adv. Anat. Pathol.* **2007**, *14*, 189–194. [CrossRef]

107. Carbone, A.; Spina, M.; Gloghini, A.; Tirelli, U. Classical Hodgkin's lymphoma arising in different host's conditions: Pathobiology parameters, therapeutic options, and outcome. *Am. J. Hematol.* **2011**, *86*, 170–179. [CrossRef]
108. De Paoli, P.; Carbone, A. Microenvironmental abnormalities induced by viral cooperation: Impact on lymphomagenesis. *Semin. Cancer Biol.* **2015**, *34*, 70–80. [CrossRef]
109. Carbone, A.; Gloghini, A.; Serraino, D.; Spina, M. HIV-associated Hodgkin lymphoma. *Curr. Opin. HIV AIDS* **2009**, *4*, 3–10. [CrossRef]
110. Koulis, A.; Trivedi, P.; Ibrahim, H.; Bower, M.; Naresh, K.N. The role of the microenvironment in human immunodeficiency virus-associated classical Hodgkin lymphoma. *Histopathology* **2014**, *65*, 749–756. [CrossRef]
111. Weniger, M.A.; Küppers, R. Molecular biology of Hodgkin lymphoma. *Leukemia* **2021**, *35*, 968–981. [CrossRef]
112. Alfano, M.; Locatelli, I.; D'Arrigo, C.; Mora, M.; Vozzi, G.; De Acutis, A.; Pece, R.; Tavella, S.; Costa, D.; Poggi, A.; et al. Lysyl-Oxidase Dependent Extracellular Matrix Stiffness in Hodgkin Lymphomas: Mechanical and Topographical Evidence. *Cancers* **2022**, *14*, 259. [CrossRef]
113. Veldman, J.; Visser, L.; Huberts-Kregel, M.; Muller, N.; Hepkema, B.; van den Berg, A.; Diepstra, A. Rosetting T cells in Hodgkin lymphoma are activated by immunological synapse components HLA class II and CD58. *Blood* **2020**, *136*, 2437–2441. [CrossRef]
114. Chantziou, A.; Brenna, C.; Ioannidou, K.; Chen, O.Y.; Korkolopoulou, P.; Antoniadou, A.; Psichogiou, M.; Papaioannou, M.; Tsirigotis, P.; Foukas, P.G.; et al. HIV infection is associated with compromised tumor microenvironment adaptive immune reactivity in Hodgkin lymphoma. *Blood Adv.* **2024**, *8*, 6215–6231. [CrossRef] [PubMed]
115. Dukers, D.F.; Meij, P.; Vervoort, M.B.; Vos, W.; Scheper, R.J.; Meijer, C.J.; Bloemena, E.; Middeldorp, J.M. Direct immunosuppressive effects of EBV-encoded latent membrane protein 1. *J. Immunol.* **2000**, *165*, 663–670. [CrossRef] [PubMed]
116. de Jong, D.; Roemer, M.G.; Chan, J.K.; Goodlad, J.; Gratzinger, D.; Chadburn, A.; Jaffe, E.S.; Said, J.; Natkunam, Y. B-Cell and Classical Hodgkin Lymphomas Associated With Immunodeficiency: 2015 SH/EAHP Workshop Report-Part 2. *Am. J. Clin. Pathol.* **2017**, *147*, 153–170. [CrossRef] [PubMed]
117. Gibson, S.E.; Swerdlow, S.H.; Craig, F.E.; Surti, U.; Cook, J.R.; Nalesnik, M.A.; Lowe, C.; Wood, K.M.; Bacon, C.M. EBV-positive extranodal marginal zone lymphoma of mucosa-associated lymphoid tissue in the posttransplant setting: A distinct type of posttransplant lymphoproliferative disorder? *Am. J. Surg. Pathol.* **2011**, *35*, 807–815. [CrossRef]
118. Gong, S.; Crane, G.M.; McCall, C.M.; Xiao, W.; Ganapathi, K.A.; Cuka, N.; Davies-Hill, T.; Xi, L.; Raffeld, M.; Pittaluga, S.; et al. Expanding the Spectrum of EBV-positive Marginal Zone Lymphomas: A Lesion Associated With Diverse Immunodeficiency Settings. *Am. J. Surg. Pathol.* **2018**, *42*, 1306–1316. [CrossRef]
119. Guech-Ongey, M.; Simard, E.P.; Anderson, W.F.; Engels, E.A.; Bhatia, K.; Devesa, S.S.; Mbulaiteye, S.M. AIDS-related Burkitt lymphoma in the United States: What do age and CD4 lymphocyte patterns tell us about etiology and/or biology? *Blood* **2010**, *116*, 5600–5604. [CrossRef]
120. Han, X.; Jemal, A.; Hulland, E.; Simard, E.P.; Nastoupil, L.; Ward, E.; Flowers, C.R. HIV Infection and Survival of Lymphoma Patients in the Era of Highly Active Antiretroviral Therapy. *Cancer Epidemiol. Biomark. Prev.* **2017**, *26*, 303–311. [CrossRef]
121. Mbulaiteye, S.M.; Pullarkat, S.T.; Nathwani, B.N.; Weiss, L.M.; Rao, N.; Emmanuel, B.; Lynch, C.F.; Hernandez, B.; Neppalli, V.; Hawes, D.; et al. Epstein-Barr virus patterns in US Burkitt lymphoma tumors from the SEER residual tissue repository during 1979-2009. *APMIS* **2014**, *122*, 5–15. [CrossRef]
122. Alves de Souza Rios, L.; Mapekula, L.; Mdletshe, N.; Chetty, D.; Mowla, S. HIV-1 Transactivator of Transcription (Tat) Co-operates With AP-1 Factors to Enhance c-MYC Transcription. *Front. Cell Dev. Biol.* **2021**, *9*, 693706. [CrossRef]
123. D'Angelo, C.R. Diagnostic, Pathologic, and Therapeutic Considerations for Primary CNS Lymphoma. *JCO Oncol. Pract.* **2024**, *20*, 195–202. [CrossRef]
124. Li, J.W.; Peng, H.L.; Zhou, X.Y.; Wang, J.J. Plasmablastic lymphoma: Current knowledge and future directions. *Front. Immunol.* **2024**, *15*, 1354604. [CrossRef] [PubMed]
125. Xiao, Q.; Liu, Y.; Shu, X.; Li, Y.; Zhang, X.; Wang, C.; He, S.; Li, J.; Li, T.; Liu, T.; et al. Molecular mechanisms of viral oncogenesis in haematological malignancies: Perspectives from metabolic reprogramming, epigenetic regulation and immune microenvironment remodeling. *Exp. Hematol. Oncol.* **2025**, *14*, 69. [CrossRef] [PubMed]
126. Cesarman, E.; Chadburn, A.; Rubinstein, P.G. KSHV/HHV8-mediated hematologic diseases. *Blood* **2022**, *139*, 1013–1025. [CrossRef] [PubMed]
127. Ascoli, V.; Scalzo, C.C.; Danese, C.; Vacca, K.; Pistilli, A.; Lo Coco, F. Human herpes virus-8 associated primary effusion lymphoma of the pleural cavity in HIV-negative elderly men. *Eur. Respir. J.* **1999**, *14*, 1231–1234. [CrossRef]
128. Carbone, A.; De Paoli, P.; Gloghini, A.; Vaccher, E. KSHV-associated multicentric Castleman disease: A tangle of different entities requiring multitarget treatment strategies. *Int. J. Cancer* **2015**, *137*, 251–261. [CrossRef]
129. Ramaswami, R.; Lurain, K.; Polizzotto, M.N.; Ekwede, I.; Waldon, K.; Steinberg, S.M.; Mangusan, R.; Widell, A.; Rupert, A.; George, J.; et al. Characteristics and outcomes of KSHV-associated multicentric Castleman disease with or without other KSHV diseases. *Blood Adv.* **2021**, *5*, 1660–1670. [CrossRef]

130. Marcus, J.L.; Leyden, W.A.; Alexeeff, S.E.; Anderson, A.N.; Hechter, R.C.; Hu, H.; Lam, J.O.; Towner, W.J.; Yuan, Q.; Horberg, M.A.; et al. Comparison of Overall and Comorbidity-Free Life Expectancy Between Insured Adults With and Without HIV Infection, 2000-2016. *JAMA Netw. Open* **2020**, *3*, e207954. [CrossRef]
131. Trickey, A.; McGinnis, K.; Gill, M.J.; Abgrall, S.; Berenguer, J.; Wyen, C.; Hessamfar, M.; Reiss, P.; Kusejko, K.; Silverberg, M.J.; et al. Longitudinal trends in causes of death among adults with HIV on antiretroviral therapy in Europe and North America from 1996 to 2020: A collaboration of cohort studies. *Lancet HIV* **2024**, *11*, e176–e185. [CrossRef]
132. Tusch, E.; Ryom, L.; Pelchen-Matthews, A.; Mocroft, A.; Elbirt, D.; Oprea, C.; Günthard, H.F.; Staehelin, C.; Zangerle, R.; Suarez, I.; et al. Trends in Mortality in People With HIV From 1999 through 2020: A Multicohort Collaboration. *Clin. Infect. Dis.* **2024**, *79*, 1242–1257. [CrossRef]
133. Castillo, J.; Perez, K.; Milani, C.; Dezube, B.J.; Pantanowitz, L. Peripheral T-cell lymphomas in HIV-infected individuals: A comprehensive review. *J. HIV Ther.* **2009**, *14*, 34–40. [PubMed]
134. Perez, K.; Castillo, J.; Dezube, B.J.; Pantanowitz, L. Human Immunodeficiency Virus-associated anaplastic large cell lymphoma. *Leuk. Lymphoma* **2010**, *51*, 430–438. [CrossRef] [PubMed]
135. Arzoo, K.K.; Bu, X.; Espina, B.M.; Seneviratne, L.; Nathwani, B.; Levine, A.M. T-cell lymphoma in HIV-infected patients. *J. Acquir. Immune Defic. Syndr.* **2004**, *36*, 1020–1027. [CrossRef] [PubMed]
136. Biggar, R.J.; Chaturvedi, A.K.; Goedert, J.J.; Engels, E.A. AIDS-related cancer severity of immunosuppression in persons with AIDS. *J. Natl. Cancer Inst.* **2007**, *99*, 962–972. [CrossRef] [PubMed]
137. Dhasmana, D.; Taylor, G.P. Human T-lymphotropic virus/HIV co-infection: A clinical review. *Curr. Opin. Infect. Dis.* **2014**, *27*, 16–28. [CrossRef]
138. Barros, M.H.M.; Alves, P.D.S. Contribution of the Epstein-Barr virus to the oncogenesis of mature T-cell lymphoproliferative neoplasms. *Front. Oncol.* **2023**, *13*, 1240359. [CrossRef]
139. Olszewski, A.J.; Jakobsen, L.H.; Collins, G.P.; Cwynarski, K.; Bachanova, V.; Blum, K.A.; Boughan, K.M.; Bower, M.; Dalla Pria, A.; Danilov, A.; et al. Burkitt Lymphoma International Prognostic Index. *J. Clin. Oncol.* **2021**, *39*, 1129–1138. [CrossRef]
140. Lipscomb, J.; Switchenko, J.M.; Flowers, C.R.; Gillespie, T.W.; Wortley, P.M.; Bayakly, A.R.; Almon, L.; Ward, K.C. Impact of multi-agent systemic therapy on all-cause and disease-specific survival for people living with HIV who are diagnosed with non-Hodgkin lymphoma: Population-based analyses from the state of Georgia. *Leuk. Lymphoma* **2023**, *64*, 151–160. [CrossRef]
141. Atallah-Yunes, S.A.; Murphy, D.J.; Noy, A. HIV-associated Burkitt lymphoma. *Lancet Haematol.* **2020**, *7*, e594–e600. [CrossRef]
142. Di Ciaccio, P.R.; Polizzotto, M.N.; Cwynarski, K.; Gerrie, A.S.; Burton, C.; Bower, M.; Kuruvilla, J.; Montoto, S.; McKay, P.; Fox, C.P.; et al. The influence of immunodeficiency, disease features, and patient characteristics on survival in plasmablastic lymphoma. *Blood* **2024**, *143*, 152–165. [CrossRef]
143. Manyau, M.C.P.; Zambuko, B.; Chatambudza, M.; Zhou, D.T.; Manasa, J. Still Far to Go With Characterisation of Molecular and Genetic Features of Diffuse Large B-Cell Lymphoma in People Living With HIV: A Scoping Review. *Oncol. Rev.* **2024**, *18*, 1375291. [CrossRef]
144. Lurain, K.; Ramaswami, R.; Oksenhendler, E.; Boutboul, D.; Dalla Pria, A.; Ulrich, L.; Shanmugasundaram, K.; Uldrick, T.S.; Bower, M.; Yarchoan, R.; et al. Primary Effusion Lymphoma Prognostic Score (PEL-PS): A Validated International Prognostic Score in HIV-Associated Primary Effusion Lymphoma. *Am. J. Hematol.* **2025**, *100*, 393–401. [CrossRef] [PubMed]
145. Coghill, A.E.; Shiels, M.S.; Suneja, G.; Engels, E.A. Elevated Cancer-Specific Mortality Among HIV-Infected Patients in the United States. *J. Clin. Oncol.* **2015**, *33*, 2376–2383. [CrossRef] [PubMed]
146. Clark, E.; Royse, K.E.; Dong, Y.; Chang, E.; Raychaudhury, S.; Kramer, J.; White, D.L.; Chiao, E. Stable Incidence and Poor Survival for HIV-Related Burkitt Lymphoma Among the US Veteran Population During the Antiretroviral Era. *J. Acquir. Immune Defic. Syndr.* **2020**, *84*, 18–25. [CrossRef] [PubMed]
147. Horner, M.J.; Shiels, M.S.; Pfeiffer, R.M.; Engels, E.A. Deaths Attributable to Cancer in the US Human Immunodeficiency Virus Population During 2001-2015. *Clin. Infect. Dis.* **2021**, *72*, e224–e231. [CrossRef]
148. McGee-Avila, J.K.; Suneja, G.; Engels, E.A.; Rositch, A.F.; Horner, M.J.; Luo, Q.; Shiels, M.S.; Islam, J.Y. Cancer Treatment Disparities in People With HIV in the United States, 2001-2019. *J. Clin. Oncol.* **2024**, *42*, 1810–1820. [CrossRef]
149. Wieland, C.M.; Tuin, A.M.; Dort, E.J.; Hall, A.G.; Krishnan, M.; Velagapudi, M. Long-Term Survival Rates and Treatment Trends of Burkitt Lymphoma in Patients with HIV-A National Cancer Database (NCDB) Study. *Cancers* **2024**, *16*, 1397. [CrossRef]
150. Noy, A. Optimizing treatment of HIV-associated lymphoma. *Blood* **2019**, *134*, 1385–1394. [CrossRef]
151. Vaccher, E.; Gloghini, A.; Carbone, A. HIV-related lymphomas. *Curr. Opin. Oncol.* **2022**, *34*, 439–445. [CrossRef]
152. Lurain, K. Treating Cancer in People With HIV. *J. Clin. Oncol.* **2023**, *41*, 3682–3688. [CrossRef]
153. Barta, S.K.; Xue, X.; Wang, D.; Tamari, R.; Lee, J.Y.; Mounier, N.; Kaplan, L.D.; Ribera, J.M.; Spina, M.; Tirelli, U.; et al. Treatment factors affecting outcomes in HIV-associated non-Hodgkin lymphomas: A pooled analysis of 1546 patients. *Blood* **2013**, *122*, 3251–3262. [CrossRef]

154. Tan, C.R.C.; Barta, S.K.; Lee, J.; Rudek, M.A.; Sparano, J.A.; Noy, A. Combination antiretroviral therapy accelerates immune recovery in patients with HIV-related lymphoma treated with EPOCH: A comparison within one prospective trial AMC034. *Leuk. Lymphoma* **2018**, *59*, 1851–1860. [CrossRef] [PubMed]
155. Olin, J.L.; Klibanov, O.; Chan, A.; Spooner, L.M. Managing Pharmacotherapy in People Living With HIV and Concomitant Malignancy. *Ann. Pharmacother.* **2019**, *53*, 812–832. [CrossRef] [PubMed]
156. Okoli, C.; Schwenk, A.; Radford, M.; Myland, M.; Taylor, S.; Darley, A.; Barnes, J.; Fox, A.; Grimson, F.; Reeves, I.; et al. Polypharmacy and potential drug-drug interactions for people with HIV in the UK from the Climate-HIV database. *HIV Med.* **2020**, *21*, 471–480. [CrossRef]
157. d'Arminio Monforte, A.; Bonnet, F.; Bucher, H.C.; Pourcher, V.; Pantazis, N.; Pelchen-Matthews, A.; Touloumi, G.; Wolf, E. What do the changing patterns of comorbidity burden in people living with HIV mean for long-term management? Perspectives from European HIV cohorts. *HIV Med.* **2020**, *21* (Suppl. S2), 3–16. [CrossRef]
158. Dalla Pria, A.; Merchant, S.; Bower, M. Oncological challenges for an ageing population living with HIV. *AIDS* **2017**, *31* (Suppl. S2), S185–S189. [CrossRef]
159. Mukhtar, F.; Ilozumba, M.; Utuama, O.; Cimenler, O. Change in Pattern of Secondary Cancers After Kaposi Sarcoma in the Era of Antiretroviral Therapy. *JAMA Oncol.* **2018**, *4*, 48–53. [CrossRef]
160. Hessol, N.A.; Whittemore, H.; Vittinghoff, E.; Hsu, L.C.; Ma, D.; Scheer, S.; Schwarcz, S.K. Incidence of first and second primary cancers diagnosed among people with HIV, 1985–2013: A population-based, registry linkage study. *Lancet HIV* **2018**, *5*, e647–e655. [CrossRef]
161. Mahale, P.; Ugoji, C.; Engels, E.A.; Shiels, M.S.; Peprah, S.; Morton, L.M. Cancer risk following lymphoid malignancies among HIV-infected people. *AIDS* **2020**, *34*, 1237–1245. [CrossRef]
162. Abrahão, R.; Li, Q.W.; Malogolowkin, M.H.; Alvarez, E.M.; Ribeiro, R.C.; Wun, T.; Keegan, T.H.M. Chronic medical conditions and late effects following non-Hodgkin lymphoma in HIV-uninfected and HIV-infected adolescents and young adults: A population-based study. *Br. J. Haematol.* **2020**, *190*, 371–384. [CrossRef]
163. Poizot-Martin, I.; Lions, C.; Delpierre, C.; Makinson, A.; Allavena, C.; Fresard, A.; Brégigeon, S.; Rojas Rojas, T.; Delobel, P.; Group The Dat'AIDS Study. Prevalence and Spectrum of Second Primary Malignancies among People Living with HIV in the French Dat'AIDS Cohort. *Cancers* **2022**, *14*, 401. [CrossRef]
164. Schommers, P.; Hentrich, M.; Hoffmann, C.; Gillor, D.; Zoufaly, A.; Jensen, B.; Bogner, J.R.; Thoden, J.; Wasmuth, J.C.; Wolf, T.; et al. Survival of AIDS-related diffuse large B-cell lymphoma, Burkitt lymphoma, and plasmablastic lymphoma in the German HIV Lymphoma Cohort. *Br. J. Haematol.* **2015**, *168*, 806–810. [CrossRef] [PubMed]
165. Spina, M.; Jaeger, U.; Sparano, J.A.; Talamini, R.; Simonelli, C.; Michieli, M.; Rossi, G.; Nigra, E.; Berretta, M.; Cattaneo, C.; et al. Rituximab plus infusional cyclophosphamide, doxorubicin, and etoposide in HIV-associated non-Hodgkin lymphoma: Pooled results from 3 phase 2 trials. *Blood* **2005**, *105*, 1891–1897. [CrossRef] [PubMed]
166. Boué, F.; Gabarre, J.; Gisselbrecht, C.; Reynes, J.; Cheret, A.; Bonnet, F.; Billaud, E.; Raphael, M.; Lancar, R.; Costagliola, D. Phase II trial of CHOP plus rituximab in patients with HIV-associated non-Hodgkin's lymphoma. *J. Clin. Oncol.* **2006**, *24*, 4123–4128. [CrossRef] [PubMed]
167. Sparano, J.A.; Lee, J.Y.; Kaplan, L.D.; Levine, A.M.; Ramos, J.C.; Ambinder, R.F.; Wachsman, W.; Aboulafia, D.; Noy, A.; Henry, D.H.; et al. Rituximab plus concurrent infusional EPOCH chemotherapy is highly effective in HIV-associated B-cell non-Hodgkin lymphoma. *Blood* **2010**, *115*, 3008–3016. [CrossRef]
168. Dunleavy, K.; Little, R.F.; Pittaluga, S.; Grant, N.; Wayne, A.S.; Carrasquillo, J.A.; Steinberg, S.M.; Yarchoan, R.; Jaffe, E.S.; Wilson, W.H. The role of tumor histogenesis, FDG-PET, and short-course EPOCH with dose-dense rituximab (SC-EPOCH-RR) in HIV-associated diffuse large B-cell lymphoma. *Blood* **2010**, *115*, 3017–3024. [CrossRef]
169. Ribera, J.M.; Morgades, M.; González-Barca, E.; Miralles, P.; López-Guillermo, A.; Gardella, S.; López, A.; Abella, E.; García, M.; Navarro, J.T.; et al. Long-term follow-up of patients with HIV-related diffuse large B-cell lymphomas treated in a phase II study with rituximab and CHOP. *Br. J. Haematol.* **2012**, *157*, 637–639. [CrossRef]
170. Kaplan, L.D.; Lee, J.Y.; Ambinder, R.F.; Sparano, J.A.; Cesarman, E.; Chadburn, A.; Levine, A.M.; Scadden, D.T. Rituximab does not improve clinical outcome in a randomized phase 3 trial of CHOP with or without rituximab in patients with HIV-associated non-Hodgkin lymphoma: AIDS-Malignancies Consortium Trial 010. *Blood* **2005**, *106*, 1538–1543. [CrossRef]
171. Barta, S.K.; Lee, J.Y.; Kaplan, L.D.; Noy, A.; Sparano, J.A. Pooled analysis of AIDS malignancy consortium trials evaluating rituximab plus CHOP or infusional EPOCH chemotherapy in HIV-associated non-Hodgkin lymphoma. *Cancer* **2012**, *118*, 3977–3983. [CrossRef]
172. Bartlett, N.L.; Wilson, W.H.; Jung, S.H.; Hsi, E.D.; Maurer, M.J.; Pederson, L.D.; Polley, M.C.; Pitcher, B.N.; Cheson, B.D.; Kahl, B.S.; et al. Dose-Adjusted EPOCH-R Compared With R-CHOP as Frontline Therapy for Diffuse Large B-Cell Lymphoma: Clinical Outcomes of the Phase III Intergroup Trial Alliance/CALGB 50303. *J. Clin. Oncol.* **2019**, *37*, 1790–1799. [CrossRef]

173. Ramos, J.C.; Sparano, J.A.; Chadburn, A.; Reid, E.G.; Ambinder, R.F.; Siegel, E.R.; Moore, P.C.; Rubinstein, P.G.; Durand, C.M.; Cesarman, E.; et al. Impact of Myc in HIV-associated non-Hodgkin lymphomas treated with EPOCH and outcomes with vorinostat (AMC-075 trial). *Blood* **2020**, *136*, 1284–1297. [CrossRef]
174. Pagani, C.; Rusconi, C.; Dalla Pria, A.; Ravano, E.; Schommers, P.; Bastos-Oreiro, M.; Verga, L.; Gini, G.; Spina, M.; Arcaini, L.; et al. MYC rearrangements in HIV-associated large B-cell lymphomas: EUROMYC, a European retrospective study. *Blood Adv.* **2024**, *8*, 968–977. [CrossRef] [PubMed]
175. Tilly, H.; Morschhauser, F.; Sehn, L.H.; Friedberg, J.W.; Trněný, M.; Sharman, J.P.; Herbaux, C.; Burke, J.M.; Matasar, M.; Rai, S.; et al. Polatuzumab Vedotin in Previously Untreated Diffuse Large B-Cell Lymphoma. *N. Engl. J. Med.* **2022**, *386*, 351–363. [CrossRef] [PubMed]
176. Gibson, T.M.; Morton, L.M.; Shiels, M.S.; Clarke, C.A.; Engels, E.A. Risk of non-Hodgkin lymphoma subtypes in HIV-infected people during the HAART era: A population-based study. *AIDS* **2014**, *28*, 2313–2318. [CrossRef] [PubMed]
177. Noy, A.; Lee, J.Y.; Cesarman, E.; Ambinder, R.; Baiocchi, R.; Reid, E.; Ratner, L.; Wagner-Johnston, N.; Kaplan, L. AIDS Malignancy Consortium. AMC 048: Modified CODOX-M/IVAC-rituximab is safe and effective for HIV-associated Burkitt lymphoma. *Blood* **2015**, *126*, 160–166. [CrossRef]
178. Alwan, F.; He, A.; Montoto, S.; Kassam, S.; Mee, M.; Burns, F.; Edwards, S.; Wilson, A.; Tenant-Flowers, M.; Marcus, R.; et al. Adding rituximab to CODOX-M/IVAC chemotherapy in the treatment of HIV-associated Burkitt lymphoma is safe when used with concurrent combination antiretroviral therapy. *AIDS* **2015**, *29*, 903–910. [CrossRef]
179. Roschewski, M.; Dunleavy, K.; Abramson, J.S.; Powell, B.L.; Link, B.K.; Patel, P.; Bierman, P.J.; Jagadeesh, D.; Mitsuyasu, R.T.; Peace, D.; et al. Multicenter Study of Risk-Adapted Therapy With Dose-Adjusted EPOCH-R in Adults With Untreated Burkitt Lymphoma. *J. Clin. Oncol.* **2020**, *38*, 2519–2529. [CrossRef]
180. Dunleavy, K.; Pittaluga, S.; Shovlin, M.; Steinberg, S.M.; Cole, D.; Grant, C.; Widemann, B.; Staudt, L.M.; Jaffe, E.S.; Little, R.F.; et al. Low-intensity therapy in adults with Burkitt's lymphoma. *N. Engl. J. Med.* **2013**, *369*, 1915–1925. [CrossRef]
181. Alderuccio, J.P.; Olszewski, A.J.; Evens, A.M.; Collins, G.P.; Danilov, A.V.; Bower, M.; Jagadeesh, D.; Zhu, C.; Sperling, A.; Kim, S.H.; et al. HIV-associated Burkitt lymphoma: Outcomes from a US-UK collaborative analysis. *Blood Adv.* **2021**, *5*, 2852–2862. [CrossRef]
182. Montoto, S.; Shaw, K.; Okosun, J.; Gandhi, S.; Fields, P.; Wilson, A.; Shanyinde, M.; Cwynarski, K.; Marcus, R.; de Vos, J.; et al. HIV status does not influence outcome in patients with classical Hodgkin lymphoma treated with chemotherapy using doxorubicin, bleomycin, vinblastine, and dacarbazine in the highly active antiretroviral therapy era. *J. Clin. Oncol.* **2012**, *30*, 4111–4116. [CrossRef]
183. Hentrich, M.; Berger, M.; Wyen, C.; Siehl, J.; Rockstroh, J.K.; Müller, M.; Fätkenheuer, G.; Seidel, E.; Nickelsen, M.; Wolf, T.; et al. Stage-adapted treatment of HIV-associated Hodgkin lymphoma: Results of a prospective multicenter study. *J. Clin. Oncol.* **2012**, *30*, 4117–4123. [CrossRef]
184. Rubinstein, P.G.; Moore, P.C.; Bimali, M.; Lee, J.Y.; Rudek, M.A.; Chadburn, A.; Ratner, L.; Henry, D.H.; Cesarman, E.; DeMarco, C.E.; et al. Brentuximab vedotin with AVD for stage II-IV HIV-related Hodgkin lymphoma (AMC 085): Phase 2 results from an open-label, single arm, multicentre phase 1/2 trial. *Lancet Haematol.* **2023**, *10*, e624–e632. [CrossRef] [PubMed]
185. Lurain, K.; Polizzotto, M.N.; Aleman, K.; Bhutani, M.; Wyvill, K.M.; Gonçalves, P.H.; Ramaswami, R.; Marshall, V.A.; Miley, W.; Steinberg, S.M.; et al. Viral, immunologic, and clinical features of primary effusion lymphoma. *Blood* **2019**, *133*, 1753–1761. [CrossRef] [PubMed]
186. Gabarre, J.; Azar, N.; Autran, B.; Katlama, C.; Leblond, V. High-dose therapy and autologous haematopoietic stem-cell transplantation for HIV-1-associated lymphoma. *Lancet* **2000**, *355*, 1071–1072. [CrossRef] [PubMed]
187. Re, A.; Cattaneo, C.; Michieli, M.; Casari, S.; Spina, M.; Rupolo, M.; Allione, B.; Nosari, A.; Schiantarelli, C.; Vigano, M.; et al. High-dose therapy and autologous peripheral-blood stem-cell transplantation as salvage treatment for HIV-associated lymphoma in patients receiving highly active antiretroviral therapy. *J. Clin. Oncol.* **2003**, *21*, 4423–4427. [CrossRef]
188. Alvarnas, J.C.; Le Rademacher, J.; Wang, Y.; Little, R.F.; Akpek, G.; Ayala, E.; Devine, S.; Baiocchi, R.; Lozanski, G.; Kaplan, L.; et al. Autologous hematopoietic cell transplantation for HIV-related lymphoma: Results of the BMT CTN 0803/AMC 071 trial. *Blood* **2016**, *128*, 1050–1058. [CrossRef]
189. Ambinder, R.F.; Wu, J.; Logan, B.; Durand, C.M.; Shields, R.; Popat, U.R.; Little, R.F.; McMahon, D.K.; Cyktor, J.; Mellors, J.W.; et al. Allogeneic Hematopoietic Cell Transplant for HIV Patients with Hematologic Malignancies: The BMT CTN-0903/AMC-080 Trial. *Biol. Blood Marrow Transplant.* **2019**, *25*, 2160–2166. [CrossRef]
190. Allers, K.; Hütter, G.; Hofmann, J.; Loddenkemper, C.; Rieger, K.; Thiel, E.; Schneider, T. Evidence for the cure of HIV infection by CCR5Δ32/Δ32 stem cell transplantation. *Blood* **2011**, *117*, 2791–2799. [CrossRef]
191. Gupta, R.K.; Abdul-Jawad, S.; McCoy, L.E.; Mok, H.P.; Peppa, D.; Salgado, M.; Martinez-Picado, J.; Nijhuis, M.; Wensing, A.M.J.; Lee, H.; et al. HIV-1 remission following CCR5Δ32/Δ32 haematopoietic stem-cell transplantation. *Nature* **2019**, *568*, 244–248. [CrossRef]

192. Rust, B.J.; Kiem, H.P.; Uldrick, T.S. CAR T-cell therapy for cancer and HIV through novel approaches to HIV-associated haematological malignancies. *Lancet Haematol.* **2020**, *7*, e690–e696. [CrossRef]
193. Carbone, A.; Gloghini, A.; Serraino, D.; Spina, M.; Tirelli, U.; Vaccher, E. Immunodeficiency-associated Hodgkin lymphoma. *Expert Rev. Hematol.* **2021**, *14*, 547–559. [CrossRef]
194. Danilov, A.V.; Li, H.; Press, O.W.; Shapira, I.; Swinnen, L.J.; Noy, A.; Reid, E.; Smith, S.M.; Friedberg, J.W. Feasibility of interim positron emission tomography (PET)-adapted therapy in HIV-positive patients with advanced Hodgkin lymphoma (HL): A sub-analysis of SWOG S0816 Phase 2 trial. *Leuk. Lymphoma* **2017**, *58*, 461–465. [CrossRef]
195. Serrao, A.; Canichella, M.; De Luca, M.L.; Tartaglia, G.; Annechini, G.; D'Elia, G.M.; Pulsoni, A. Nivolumab as a safe and effective treatment in an HIV patient with refractory Hodgkin lymphoma. *Ann. Hematol.* **2019**, *98*, 1505–1506. [CrossRef] [PubMed]
196. Uldrick, T.S.; Gonçalves, P.H.; Abdul-Hay, M.; Claeys, A.J.; Emu, B.; Ernstoff, M.S.; Fling, S.P.; Fong, L.; Kaiser, J.C.; Lacroix, A.M.; et al. Assessment of the Safety of Pembrolizumab in Patients With HIV and Advanced Cancer-A Phase 1 Study. *JAMA Oncol.* **2019**, *5*, 1332–1339. [CrossRef] [PubMed]
197. Gonzalez-Cao, M.; Morán, T.; Dalmau, J.; Garcia-Corbacho, J.; Bracht, J.W.P.; Bernabe, R.; Juan, O.; de Castro, J.; Blanco, R.; Drozdowskyj, A.; et al. Assessment of the Feasibility and Safety of Durvalumab for Treatment of Solid Tumors in Patients With HIV-1 Infection: The Phase 2 DURVAST Study. *JAMA Oncol.* **2020**, *6*, 1063–1067. [CrossRef] [PubMed]
198. El Zarif, T.; Nassar, A.H.; Adib, E.; Fitzgerald, B.G.; Huang, J.; Mouhieddine, T.H.; Rubinstein, P.G.; Nonato, T.; McKay, R.R.; Li, M.; et al. Safety and Activity of Immune Checkpoint Inhibitors in People Living With HIV and Cancer: A Real-World Report From the Cancer Therapy Using Checkpoint Inhibitors in People Living With HIV-International (CATCH-IT) Consortium. *J. Clin. Oncol.* **2023**, *41*, 3712–3723. [CrossRef]
199. Herrera, A.F.; LeBlanc, M.; Castellino, S.M.; Li, H.; Rutherford, S.C.; Evens, A.M.; Davison, K.; Punnett, A.; Parsons, S.K.; Ahmed, S.; et al. Nivolumab+AVD in Advanced-Stage Classic Hodgkin's Lymphoma. *N. Engl. J. Med.* **2024**, *391*, 1379–1389. [CrossRef]
200. Langan, E.A.; Graetz, V.; Allerheiligen, J.; Zillikens, D.; Rupp, J.; Terheyden, P. Immune checkpoint inhibitors and tuberculosis: An old disease in a new context. *Lancet Oncol.* **2020**, *21*, e55–e65. [CrossRef]
201. Anidi, I.U.; Sakai, S.; Brooks, K.; Fling, S.P.; Wagner, M.J.; Lurain, K.; Lindestam Arlehamn, C.S.; Sette, A.; Knox, K.S.; Brenchley, J.M.; et al. Exacerbation of CMV and Nontuberculous Mycobacterial Infections Following PD-1 Blockade for HIV-Associated Kaposi Sarcoma. *Open Forum Infect. Dis.* **2024**, *11*, ofae183. [CrossRef]
202. Zhou, T.; Yuan, C.M.; Lurain, K.; Rous, C.; Weaver, L.; Raffeld, M.; Stetler-Stevenson, M.; Uldrick, T.S.; Filie, A.C.; Pittaluga, S.; et al. A novel approach for characterization of KSHV-associated multicentric Castleman disease from effusions. *Br. J. Haematol.* **2023**, *200*, 462–475. [CrossRef]
203. Martin de Frémont, G.; Vanjak, A.; Sbihi, Z.; Knapp, S.; Garzaro, M.; Chbihi, M.; Fournier, B.; Poirot, J.; Dossier, A.; Silvestrini, M.A.; et al. Characteristics of circulating KSHV-infected viroblasts during active KSHV+ multicentric Castleman disease. *Blood Adv.* **2023**, *7*, 1682–1691. [CrossRef]
204. Uldrick, T.S.; Polizzotto, M.N.; Aleman, K.; Wyvill, K.M.; Marshall, V.; Whitby, D.; Wang, V.; Pittaluga, S.; O'Mahony, D.; Steinberg, S.M.; et al. Rituximab plus liposomal doxorubicin in HIV-infected patients with KSHV-associated multicentric Castleman disease. *Blood* **2014**, *124*, 3544–3552. [CrossRef] [PubMed]
205. Ramaswami, R.; Lurain, K.; Peer, C.J.; Serquiña, A.; Wang, V.; Widell, A.; Goncalves, P.; Steinberg, S.M.; Marshall, V.; George, J.; et al. Tocilizumab in patients with symptomatic Kaposi sarcoma herpesvirus-associated multicentric Castleman disease. *Blood* **2020**, *135*, 2316–2319. [CrossRef] [PubMed]
206. Carbone, A.; Vaccher, E.; Gloghini, A.; Pria, A.D.; Bower, M. Management of KSHV/HHV8-Positive Multicentric Castleman Disease Concurrent With Other KSHV/HHV8-Positive Disorders: An Emergent Challenge in PLWH. *J. Acquir. Immune Defic. Syndr.* **2025**, *99*, e1–e2. [CrossRef]
207. Trautmann, L.; Janbazian, L.; Chomont, N.; Said, E.A.; Gimmig, S.; Bessette, B.; Boulassel, M.R.; Delwart, E.; Sepulveda, H.; Balderas, R.S.; et al. Upregulation of PD-1 expression on HIV-specific CD8+ T cells leads to reversible immune dysfunction. *Nat. Med.* **2006**, *12*, 1198–1202. [CrossRef]
208. Uldrick, T.S.; Adams, S.V.; Fromentin, R.; Roche, M.; Fling, S.P.; Gonçalves, P.H.; Lurain, K.; Ramaswami, R.; Wang, C.J.; Gorelick, R.J.; et al. Pembrolizumab induces HIV latency reversal in people living with HIV and cancer on antiretroviral therapy. *Sci. Transl. Med.* **2022**, *14*, eabl3836. [CrossRef]
209. Rasmussen, T.A.; Rajdev, L.; Rhodes, A.; Dantanarayana, A.; Tennakoon, S.; Chea, S.; Spelman, T.; Lensing, S.; Rutishauser, R.; Bakkour, S.; et al. Impact of Anti-PD-1 and Anti-CTLA-4 on the Human Immunodeficiency Virus (HIV) Reservoir in People Living With HIV With Cancer on Antiretroviral Therapy: The AIDS Malignancy Consortium 095 Study. *Clin. Infect. Dis.* **2021**, *73*, e1973–e1981. [CrossRef]
210. Abbar, B.; Baron, M.; Katlama, C.; Marcelin, A.G.; Veyri, M.; Autran, B.; Guihot, A.; Spano, J.P. Immune checkpoint inhibitors in people living with HIV: What about anti-HIV effects? *AIDS* **2020**, *34*, 167–175. [CrossRef]

211. Lavole, A.; Mazieres, J.; Schneider, S.; Brosseau, S.; Kiakouama, L.; Greillier, L.; Guihot, A.; Abbar, B.; Baron, M.; Makinson, A.; et al. Assessment of nivolumab in HIV-Infected patients with advanced non-small cell lung cancer after prior chemotherapy. The IFCT-1602 CHIVA2 phase 2 clinical trial. *Lung Cancer* **2021**, *158*, 146–150. [CrossRef]
212. Assoumou, L.; Baldé, R.; Katlama, C.; Abbar, B.; Delobel, P.; Allegre, T.; Lavole, A.; Makinson, A.; Zaegel-Faucher, O.; Greillier, L.; et al. Safety and tolerability of immune checkpoint inhibitors in people with HIV infection and cancer: Insights from the national prospective real-world OncoVIHAC ANRS CO24 cohort study. *J. Immunother. Cancer* **2024**, *12*, e009728. [CrossRef]
213. Lesokhin, A.M.; Ansell, S.M.; Armand, P.; Scott, E.C.; Halwani, A.; Gutierrez, M.; Millenson, M.M.; Cohen, A.D.; Schuster, S.J.; Lebovic, D.; et al. Nivolumab in Patients With Relapsed or Refractory Hematologic Malignancy: Preliminary Results of a Phase Ib Study. *J. Clin. Oncol.* **2016**, *34*, 2698–2704. [CrossRef]
214. Ansell, S.M.; Minnema, M.C.; Johnson, P.; Timmerman, J.M.; Armand, P.; Shipp, M.A.; Rodig, S.J.; Ligon, A.H.; Roemer, M.G.M.; Reddy, N.; et al. Nivolumab for Relapsed/Refractory Diffuse Large B-Cell Lymphoma in Patients Ineligible for or Having Failed Autologous Transplantation: A Single-Arm, Phase II Study. *J. Clin. Oncol.* **2019**, *37*, 481–489. [CrossRef] [PubMed]
215. Lurain, K.; Zarif, T.E.; Ramaswami, R.; Nassar, A.H.; Adib, E.; Abdel-Wahab, N.; Chintapally, N.; Drolen, C.E.; Feldman, T.; Haykal, T.; et al. Real-World Multicenter Study of PD-1 Blockade in HIV-Associated Classical Hodgkin Lymphoma Across the United States. *Clin. Lymphoma Myeloma Leuk.* **2024**, *24*, 523–530. [CrossRef] [PubMed]
216. Lurain, K.; Ramaswami, R.; Mangusan, R.; Widell, A.; Ekwede, I.; George, J.; Ambinder, R.; Cheever, M.; Gulley, J.L.; Goncalves, P.H.; et al. Use of pembrolizumab with or without pomalidomide in HIV-associated non-Hodgkin's lymphoma. *J. Immunother. Cancer* **2021**, *9*, e002097. [CrossRef] [PubMed]
217. Gopalakrishnan, R.; Matta, H.; Tolani, B.; Triche TJr Chaudhary, P.M. Immunomodulatory drugs target IKZF1-IRF4-MYC axis in primary effusion lymphoma in a cereblon-dependent manner and display synergistic cytotoxicity with BRD4 inhibitors. *Oncogene* **2016**, *35*, 1797–1810. [CrossRef]
218. Davis, D.A.; Shrestha, P.; Aisabor, A.I.; Stream, A.; Galli, V.; Pise-Masison, C.A.; Tagawa, T.; Ziegelbauer, J.M.; Franchini, G.; Yarchoan, R. Pomalidomide increases immune surface marker expression and immune recognition of oncovirus-infected cells. *Oncoimmunology* **2018**, *8*, e1546544. [CrossRef]
219. Jin, S.; Zhang, Y.; Zhou, F.; Chen, X.; Sheng, J.; Zhang, J. TIGIT: A promising target to overcome the barrier of immunotherapy in hematological malignancies. *Front. Oncol.* **2022**, *12*, 1091782. [CrossRef]
220. Milunović, V.; Dragčević, D.; Bogeljić Patekar, M.; Mandac Smoljanović, I.; Gašparov, S. The Improving Outcomes in Relapsed-Refractory Diffuse Large B Cell Lymphoma: The Role of CAR T-Cell Therapy. *Curr. Treat. Options Oncol.* **2025**, *26*, 445–464. [CrossRef]
221. Khodke, P.; Kumbhar, B.V. Engineered CAR-T cells: An immunotherapeutic approach for cancer treatment and beyond. *Adv. Protein Chem. Struct. Biol.* **2024**, *140*, 157–198. [CrossRef]
222. Neelapu, S.S.; Locke, F.L.; Bartlett, N.L.; Lekakis, L.J.; Miklos, D.B.; Jacobson, C.A.; Braunschweig, I.; Oluwole, O.O.; Siddiqi, T.; Lin, Y.; et al. Axicabtagene Ciloleucel CAR T-Cell Therapy in Refractory Large B-Cell Lymphoma. *N. Engl. J. Med.* **2017**, *377*, 2531–2544. [CrossRef]
223. Schuster, S.J.; Bishop, M.R.; Tam, C.S.; Waller, E.K.; Borchmann, P.; McGuirk, J.P.; Jäger, U.; Jaglowski, S.; Andreadis, C.; Westin, J.R.; et al. Tisagenlecleucel in Adult Relapsed or Refractory Diffuse Large B-Cell Lymphoma. *N. Engl. J. Med.* **2019**, *380*, 45–56. [CrossRef]
224. Kamdar, M.; Solomon, S.R.; Arnason, J.; Johnston, P.B.; Glass, B.; Bachanova, V.; Ibrahimi, S.; Mielke, S.; Mutsaers, P.; Hernandez-Ilizaliturri, F.; et al. Lisocabtagene maraleucel versus standard of care with salvage chemotherapy followed by autologous stem cell transplantation as second-line treatment in patients with relapsed or refractory large B-cell lymphoma (TRANSFORM): Results from an interim analysis of an open-label, randomised, phase 3 trial. *Lancet* **2022**, *399*, 2294–2308. [CrossRef] [PubMed]
225. Olivieri, D.J.; Gopal, A.K.; Uldrick, T.S.; Menon, M.P. Exclusion of People Living with HIV in Aggressive B-Cell Non-Hodgkin Lymphoma Studies: A Cross-Sectional Analysis of Clinical Trials from 2014 to 2024. *Cancer Investig.* **2025**, *43*, 141–148. [CrossRef] [PubMed]
226. Hattenhauer, S.T.; Mispelbaum, R.; Hentrich, M.; Boesecke, C.; Monin, M.B. Enabling CAR T-cell therapies for HIV-positive lymphoma patients—A call for action. *HIV Med.* **2023**, *24*, 957–964. [CrossRef] [PubMed]
227. Tao, Z.; Chyra, Z.; Kotulová, J.; Celichowski, P.; Mihályová, J.; Charvátová, S.; Hájek, R. Impact of T cell characteristics on CAR-T cell therapy in hematological malignancies. *Blood Cancer J.* **2024**, *14*, 213. [CrossRef]
228. Abbasi, A.; Peeke, S.; Shah, N.; Mustafa, J.; Khatun, F.; Lombardo, A.; Abreu, M.; Elkind, R.; Fehn, K.; de Castro, A.; et al. Axicabtagene ciloleucel CD19 CAR-T cell therapy results in high rates of systemic and neurologic remissions in ten patients with refractory large B cell lymphoma including two with HIV and viral hepatitis. *J. Hematol. Oncol.* **2020**, *13*, 1. [CrossRef]
229. Villalba, J.A.; Maus, M.V.; Frigault, M.J.; Zaffini, R.; Gandhi, R.T.; Rosenberg, E.S.; Siedner, M.J. False-Positive Human Immunodeficiency Virus Test Results in Patients Receiving Lentivirus-Based Chimeric Antigen Receptor T-Cell Therapy: Case Report, Review of the Literature, and Proposed Recommendations. *J. Infect. Dis.* **2022**, *225*, 1933–1936. [CrossRef]

230. Patel, K.K.; Tariveranmoshabad, M.; Kadu, S.; Shobaki, N.; June, C. From concept to cure: The evolution of CAR-T cell therapy. *Mol. Ther.* **2025**, *33*, 2123–2140. [CrossRef]
231. Urak, R.; Pahlavanneshan, S.; Gittins, B.; Nakamura, R.; Zaia, J.A.; Baird, J.H.; Clark, M.C.; Forman, S.J.; Wang, X. Simultaneous targeting of B-cell malignancies and human immunodeficiency virus with bispecific chimeric antigen receptor T cells. *Haematologica* **2024**, *109*, 3053–3058. [CrossRef]
232. Ambinder, R.F.; Capoferri, A.A.; Durand, C.M. Haemopoietic cell transplantation in patients living with HIV. *Lancet HIV* **2020**, *7*, e652–e660. [CrossRef]
233. Sáez-Cirión, A.; Mamez, A.C.; Avettand-Fenoel, V.; Nabergoj, M.; Passaes, C.; Thoueille, P.; Decosterd, L.; Hentzien, M.; Perdomo-Celis, F.; Salgado, M.; et al. Sustained HIV remission after allogeneic hematopoietic stem cell transplantation with wild-type CCR5 donor cells. *Nat. Med.* **2024**, *30*, 3544–3554. [CrossRef]
234. McLaughlin, L.P.; Gottschalk, S.; Rooney, C.M.; Bollard, C.M. EBV-Directed T Cell Therapeutics for EBV-Associated Lymphomas. *Methods Mol. Biol.* **2017**, *1532*, 255–265. [CrossRef] [PubMed]
235. Bollard, C.M.; Gottschalk, S.; Torrano, V.; Diouf, O.; Ku, S.; Hazrat, Y.; Carrum, G.; Ramos, C.; Fayad, L.; Shpall, E.J.; et al. Sustained complete responses in patients with lymphoma receiving autologous cytotoxic T lymphocytes targeting Epstein-Barr virus latent membrane proteins. *J. Clin. Oncol.* **2014**, *32*, 798–808. [CrossRef] [PubMed]
236. Icheva, V.; Kayser, S.; Wolff, D.; Tuve, S.; Kyzirakos, C.; Bethge, W.; Greil, J.; Albert, M.H.; Schwinger, W.; Nathrath, M.; et al. Adoptive transfer of epstein-barr virus (EBV) nuclear antigen 1-specific t cells as treatment for EBV reactivation and lymphoproliferative disorders after allogeneic stem-cell transplantation. *J. Clin. Oncol.* **2013**, *31*, 39–48. [CrossRef] [PubMed]
237. Bollard, C.M.; Tripic, T.; Cruz, C.R.; Dotti, G.; Gottschalk, S.; Torrano, V.; Dakhova, O.; Carrum, G.; Ramos, C.A.; Liu, H.; et al. Tumor-Specific T-Cells Engineered to Overcome Tumor Immune Evasion Induce Clinical Responses in Patients With Relapsed Hodgkin Lymphoma. *J. Clin. Oncol.* **2018**, *36*, 1128–1139. [CrossRef]
238. Khoury, R.; Grimley, M.S.; Nelson, A.S.; Leemhuis, T.; Cancelas, J.A.; Cook, E.; Wang, Y.; Heyenbruch, D.; Bollard, C.M.; Keller, M.D.; et al. Third-party virus-specific T cells for the treatment of double-stranded DNA viral reactivation and posttransplant lymphoproliferative disease after solid organ transplant. *Am. J. Transplant.* **2024**, *24*, 1634–1643. [CrossRef]

cancers

MDPI

Review

Molecular Mechanisms in the Transformation from Indolent to Aggressive B Cell Malignancies

Nawar Maher [1,†], Samir Mouhssine [1,†], Bassam Francis Matti [2], Alaa Fadhil Alwan [3] and Gianluca Gaidano [1,*]

1 Division of Hematology, Department of Translational Medicine, Università del Piemonte Orientale and Azienda Ospedaliero-Universitaria Maggiore della Carità, 28100 Novara, Italy; nawar.maher@uniupo.it (N.M.); samir.mouhssine@uniupo.it (S.M.)
2 Department of Hematology and Bone Marrow Transplant, Hematology and Bone Marrow Transplant Center, Medical City, Baghdad 00964, Iraq; bassam_francis@yahoo.com
3 Department of Clinical Hematology, The National Center of Hematology, Mustansiriyah University, Baghdad 10001, Iraq; ala_sh73@uomustansiriyah.edu.iq
* Correspondence: gianluca.gaidano@uniupo.it; Tel.: +39-339-3322688
† These authors contributed equally to this work.

Simple Summary: A significant fraction of patients affected by indolent B cell malignancies, such as chronic lymphocytic leukemia, follicular lymphoma, marginal zone lymphoma and lymphoplasmacytic lymphoma, may experience a transformation into a more aggressive form of lymphoma, often referred to as diffuse large B cell lymphoma. This transformation poses significant treatment challenges due to the increased aggressiveness of the disease and its resistance to available therapies. This review aims to explore the molecular mechanisms underlying this transformation, identifying genetic alterations and immune system changes that contribute to the progression of the disease. These changes affect several processes such as cell growth, DNA repair, and immune response. Understanding these mechanisms is crucial for developing targeted treatments and improving patient outcomes. This work provides an overview of the complex processes behind the transformation of these tumors, highlighting the need for continued investigation to enhance treatment strategies and patient care.

Abstract: Histological transformation (HT) into aggressive lymphoma is a turning point in a significant fraction of patients affected by indolent lymphoproliferative neoplasms, namely, chronic lymphocytic leukemia (CLL), follicular lymphoma (FL), marginal zone lymphomas (MZLs), and lymphoplasmacytic lymphoma (LPL). The most common histologic subtype resulting from HT is diffuse large B cell lymphoma (DLBCL). Consistently, DLBCL arisen from indolent non-Hodgkin lymphomas still represents an unmet clinical need, due to disease aggressiveness and resistance to available treatment approaches. Numerous molecular alterations associated with the occurrence of the aggressive HT of CLL (known as Richter transformation) have been identified, namely, genetic lesions of *TP53*, *CDKN2A*, *NOTCH1*, and *c-MYC*, dysregulation of immune checkpoints, and hyperactivation of the B cell receptor pathway. HT of FL involves a combination of genetic and epigenetic alterations affecting genes involved in cell-cycle control (e.g., *CDKN2A/B*), DNA damage response (e.g., *TP53* and *BCL6*), proliferation including the NF-κB pathway, and microenvironmental interactions (namely, *TNFSRF14*). The molecular pathways driving MZL transformation include copy number alterations, mutations of *NOTCH2*, *KLF2*, *TNFAIP3*, *KMT2D*, and *FOXP1*, as well as epigenetic dysregulation. The molecular alterations underlying the aggressive transformation of LPL are poorly understood, with *CXCR4* mutations being the primary factor involved. Current knowledge of the molecular mechanisms of HT has led to promising therapeutic strategies targeting immune checkpoints, BCL2, and BTK, and exploiting bispecific monoclonal antibodies and chimeric antigen receptor (CAR)-T cells. Further investigations are necessary for a deeper understanding of the molecular mechanisms underlying HT, in order to further highlight predictive biomarkers of transformation and develop effective treatment strategies. The aim of this review is to provide a comprehensive overview of the molecular mechanisms that drive transformation to DLBCL of CLL, FL, MZLs, and LPL.

Keywords: chronic lymphocytic leukemia; follicular lymphoma; lymphoplasmacytic lymphoma; marginal zone lymphoma; Richter transformation; diffuse large B cell lymphoma; histological transformation

1. Introduction

Histological transformation (HT) into aggressive lymphoma is a turning point in a significant fraction of patients affected by indolent lymphoproliferative neoplasms, namely, chronic lymphocytic leukemia (CLL), follicular lymphoma (FL), marginal zone lymphomas (MZLs), and lymphoplasmacytic lymphoma (LPL) [1–4]. The most common histologic subtype resulting from HT is diffuse large B cell lymphoma (DLBCL), characterized by even worse clinical outcomes compared to its *de novo* counterpart [5]. DLBCL arisen from indolent non-Hodgkin lymphomas (NHLs) still represent an unmet clinical need, due to the high degree of resistance to available treatment approaches [5]. Although HT of lymphomas was first described more than 80 years ago by Gall and Mallory [6], the biological events that characterize transformation remain incompletely understood, and many questions are still unanswered. Here, we aim to provide a comprehensive overview of the molecular mechanisms that drive DLBCL transformation of CLL, FL, MZLs, and LPL.

1.1. Chronic Lymphocytic Leukemia

According to the fifth edition of the *WHO Classification of Haematolymphoid Tumours*, CLL is a B cell neoplasm characterized by the proliferation and accumulation of mature B lymphocytes in blood, lymphoid organs, and eventually, in peripheral tissues [7,8]. CLL clonal lymphocytes express CD19, CD20, CD5, CD23, surface immunoglobulin, CD22, CD79b, CD43, and CD200 [7,8]. CLL is the most frequent leukemia in the adult population, accounting for 4.6 new cases per 100,000 people per year in the US [9]. The slow progression of the disease, combined with advancements in targeted therapies in recent years, has led to relatively long survival rates, with 88.5% of cases achieving a 5-year relative survival [9,10]. A rare variant of the disease with predominant lymph node involvement is known as small lymphocytic lymphoma, which displays similar biological and molecular features as CLL. In the past, chemo-immunotherapy (CIT) was the only treatment option for CLL [11,12]. However, the introduction of pathway inhibitors, namely, Bruton tyrosine kinase (BTK) and B cell lymphoma 2 (BCL2) inhibitors, now enables prolonged progression-free survival (PFS), even in patients with high-risk characteristics such as *TP53* abnormalities and unmutated immunoglobulin heavy-chain variable (IGHV) gene status [13,14].

The most serious complication during the clinical course of CLL is Richter transformation (RT), which is characterized by the development of an aggressive type of lymphoma in CLL patients [1,7,8]. In the CIT era, RT occurred in 2–10% of CLL cases, with a higher development rate in patients who had received previous treatment [1,15]. Conversely, recent data by Hampel et al. suggest that the incidence of RT is significantly lower in patients treated with pathway inhibitors compared to those treated with CIT [16]. The reduced incidence of RT with pathway inhibitors may stem from their ability to minimize genotoxic stress, suppress aggressive subclones, modulate the microenvironment, and/or directly target the molecular pathways driving CLL progression and transformation [16,17]. RT is classified into two different histologic types: DLBCL-type and Hodgkin lymphoma (HL)-type [7,8]. DLBCL-type RT, characterized by confluent sheets of large neoplastic post-germinal center CD20+ B lymphocytes, accounts for 90% of RT cases and is more severe due to its high level of chemo-refractoriness [18,19]. Several studies reported that 70–80% of DLBCL-type RT are classified as non-germinal center B cell (non-GCB) subtype, as determined by the Hans algorithm [20,21]. However, unlike in *de novo* DLBCL, the cell of origin (COO) does not appear to affect clinical outcomes in DLBCL-type RT [22]. Approximately 80% of DLBCL-type RT result from the direct transformation of existing CLL clones, while 20% show different IGHV rearrangements, documenting a *de novo* DLBCL development [1,15]. Clonally unrelated cases tend to have a more favorable prognosis, emerging as secondary malignancies due to immune suppression in CLL, as demonstrated by ultra-deep next generation sequencing (NGS) [23]. Indeed, effective treatment for clonally related DLBCL-type RT remains an unmet medical need, as survival is extremely limited, with a median between 6 and 12 months [24]. Therefore, understanding the molecular mechanisms driving this aggressive transformation is crucial for future directions of tailoring treatment approach.

1.2. Follicular Lymphoma

FL is the second most common type of non-Hodgkin lymphoma (NHL) in the USA and Europe (20–25% of NHLs), with approximately 15,000 newly diagnosed cases per year [2,8]. It is considered the prototype of indolent lymphomas and has an equal distribution between males and females and a median age of 65 years at presentation [2]. The translocation t(14;18) (q32;q21), the genetic hallmark of FL, places the antiapoptotic *BCL2* oncogene under the control of the IGH enhancer and is present in the overwhelming majority of cases [25]. FL is a GCB lymphoma characterized by the expression of monoclonal immunoglobulins, CD19, CD20, CD10, and BCL-6 and is negative for CD5 and CD23 [26]. Given the indolent course of FL, 50% of patients are diagnosed with advanced stage and bone marrow involvement. In spite of the improvement in the diagnostic accuracy and the efficacy of CIT regimens, progression and HT still occur [27]. In fact, the rate of transformation to a more aggressive form of lymphoma is 2–3% per year. The most common histologic subtype during transformation is DLBCL, but other histological subtypes may also be seen [28]. The clonal evolution of DLBCL-type transformed FL (tFL) generally follows two primary models: (*i*) the linear model, where the transformed clone arises directly from the indolent lymphoma by acquiring additional genetic alterations, while retaining the genetic abnormalities of the indolent phase [29–31]; (*ii*) the divergent/branching model, in which both the indolent FL and the related transformed clones stem from a common progenitor clone that acquires distinct genetic changes independently at each stage [30,31]. Whole-exome sequencing analysis has revealed that the majority of tFL cases (~90%) exhibit a divergent evolution [31]. The introduction of rituximab, an anti-CD20 monoclonal antibody (mAb), has improved the 5-year OS post-transformation to 73% using CIT regimens [32]. Understanding the molecular pathways underlying transformation process can pave the way for the discovery of new therapeutic agents to further enhance treatment outcomes.

1.3. Marginal Zone Lymphomas

MZLs arise from mature B cells originating from the marginal zone of lymph nodes, which surrounds the mantle zone of the germinal center [8,33]. MZLs make up 5% to 10% of NHLs. MZLs are classified into three groups: splenic MZL (SMZL), nodal MZL (NMZL), both rare (less than 2% of all NHL), and MZL of mucosa-associated lymphoid tissue (MALT), which represents the most commonly occurring subtype of MZLs [34]. All subtypes of MZLs share morphologic and immunophenotypical similarities, although molecular and clinical characteristics differ for each entity [35]. MZLs generally have an indolent clinical course and a long natural history, with median survival approximating a decade. HT to an aggressive lymphoma is a rare event that can occur in any subtype [36]. The median time to HT in several series ranges from 1 to 15 years after diagnosis, with a frequency of about 2.4% per year, 5% at 5 years, and 10% at 12 years. One of the largest analyses on the frequency and clinical characteristics of transformed MZLs by Conconi et al. reported HT occurrence in 5% of SMZL, 4% of MALT lymphomas, and 3% of NMZL, which is similar to other series, suggesting that the risk of developing HT is similar across the different histological subtypes of MZL [3]. Limited data are available regarding the dysregulated biological pathways involved in the transformation of MZLs, with most studies primarily focusing on SMZL [33]. Transformed MZL is most commonly associated with DLBCL of the non-GCB type, often characterized by dual expression of c-MYC and BCL2 (double expressor phenotype) [3].

1.4. Lymphoplasmacytic Lymphoma

LPL is an indolent NHL marked by the clonal proliferation of small, mature B lymphocytes, plasma cells, and lymphocytes with plasmacytoid differentiation, primarily in the bone marrow but potentially affecting other lymphoid tissues and, rarely, extra-lymphoid tissues [7,8]. Two different subtypes of LPL are recognized: (*i*) Waldenström macroglobulinemia (WM) type, which accounts for 95% of the cases and is characterized by the association with IgM paraproteinemia; and (*ii*) non-WM type, a rare subtype representing 5% of LPL, which can be divided into cases with IgG or IgA paraproteinemia, non-secretory LPL, and IgM LPL without bone marrow involvement [8,37]. More than 90% of LPL cases harbor the hallmark activating mutation L625P of the *MYD88* gene, which sustains the survival of malignant B cells [29,38]. The LPL estimated incidence in the US is 0.63 new cases per 100,000 persons per year, but it varies by ethnicity, with northern Sweden having the highest reported incidence globally at 1.75 new cases per 100,000 persons per year [39]. The treatment of LPL has evolved over the years, with a reduced emphasis on CIT, and a greater focus on targeted therapies, particularly BTKi [10]. LPL patients can be further classified into four subgroups based on *MYD88* and *CXCR4* mutations, which predict their response to covalent BTKi: (*i*) *MYD88*mut*CXCR4*wt, the most common subtype, with favorable responses to BTKi; (*ii*) *MYD88*mut*CXCR4*mut, characterized by a longer time to major response, reduced progression-free survival (PFS), and lower response rates to BTKi; (*iii*) *MYD88*wt*CXCR4*wt, more aggressive and associated with poor overall survival (OS) and minimal response to BTKi; and (*iv*) *MYD88*wt*CXCR4*mut, that is rare and likely shares the prognosis of *MYD88*wt*CXCR4*wt [10,40].

Despite the indolent nature of LPL and a median survival exceeding 10 years from diagnosis, 2–10% of patients eventually develop DLBCL, typically exhibiting the non-GCB phenotype, associated with a poor prognosis [4,21,41]. A study on a cohort of 1466 LPL cases found that 20 patients developed DLBCL, with a median OS from transformation of 2.7 years, similar to that of non-GCB DLBCL [4,21]. Similar to CLL evolving into RT, the majority of DLBCL arising from LPL share the same IGHV rearrangement as the dominant LPL clone, but about 25% are clonally unrelated, with 40% of these still deriving from LPL subclones [41]. Due to the dismal outcomes associated with transformed LPL and the limited understanding of the molecular mechanisms of transformation, exploring these mechanisms is essential for developing more effective treatment strategies.

2. Molecular Mechanisms in CLL Transformation to RT

Numerous molecular alterations associated with the occurrence of DLBCL-type RT have been identified, while the development of HL-type RT is thought to resemble *de novo*

HL and may be linked, at least in part, to Epstein–Barr virus-mediated immunosuppression (Table 1) [42–44]. The primary drivers of DLBCL-type RT include genetic lesions, dysregulation of immune checkpoints, and hyperactivation of the B cell receptor (BCR) pathway [45].

2.1. TP53 Disruption

Disruption of *TP53*, through mutation and/or deletion, can either be already present in the CLL phase prior to treatment requirement (in 10–15% of the cases), representing an independent risk factor of DLBCL-type RT development, or can also be acquired at the time of transformation (Figure 1) [46,47]. *TP53* disruption has been observed in a significant proportion of DLBCL-type RT, occurring in 60–80% of clonally related cases and ~20% of clonally unrelated cases [46]. *TP53*, located on the short arm of chromosome 17 (17p13.1), is a tumor suppressor gene that encodes the p53 protein, often referred to as "the guardian of the genome", playing a crucial proapoptotic role in response to DNA damage [48]. Additionally, p53 contributes to cellular regulation through non-transcriptional mechanisms mediated by protein–protein interactions [49]. By inducing cell-cycle arrest through p21, p53 ensures sufficient time for DNA repair, preventing the replication of damaged DNA [50]. It also regulates essential repair pathways, such as homologous recombination and nucleotide excision repair [51]. Additionally, p53 initiates apoptosis if repair fails, by upregulating pro-apoptotic genes, including *BAX*, *NOXA*, and *PUMA*, while downregulating anti-apoptotic genes, such as *BCL-2*, thereby safeguarding against tumorigenesis [52,53]. Therefore, *TP53* disruption is a pro-tumorigenic event that not only promotes tumorigenesis but also diminishes the cellular response to DNA damage, contributing to chemorefractoriness.

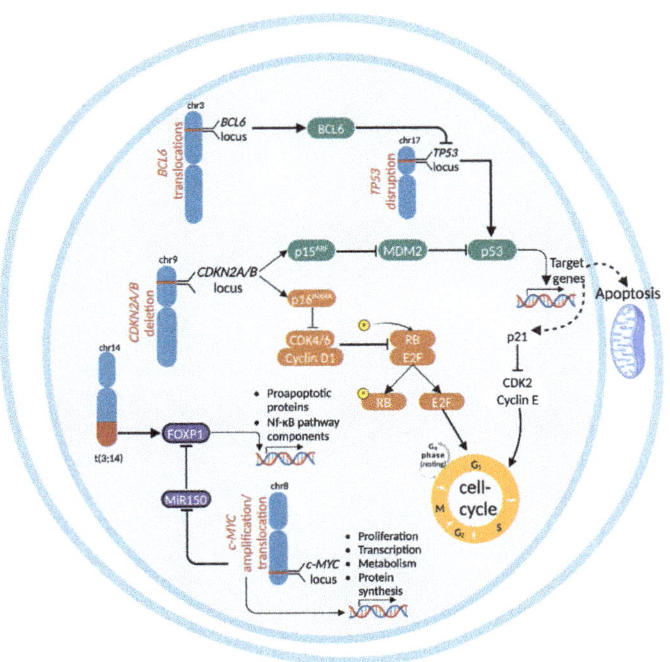

Figure 1. Main chromosomal aberrations implicated in the pathogenesis of HT. Several chromosomal aberrations drive the transformation from indolent to aggressive B cell malignancies. Deletions of the short arm of chromosome 17 disrupt *TP53*, a critical tumor suppressor responsible for DNA damage

detection and cell-cycle regulation, contributing to HT into DLBCL-type RT and FL. Similarly, deletions of the short arm of chromosome 9 result in the loss of *CDKN2A/B*, which suppresses the transcription of p16^{INK4A} and p14^{ARF}. This disruptive event enhances the G1-S phase transition and promotes MDM2-mediated inhibition of p53, leading to increased proliferation, survival, and reduced DNA damage response, which are key factors in HT across DLBCL-type RT, FL, and SMZL. Translocations involving *BCL6*, frequently observed in tFL, lead to protein overexpression that represses genes such as *TP53*, the DNA damage sensor *ATR*, and the cell-cycle-arrest regulator *CDKN1A*. The t(3;14) translocation involving the *FOXP1* and IGH genes causes FOXP1 overexpression, enhancing proliferation and survival and representing a main driver of chromosomal aberration in the HT of MALT lymphomas. Finally, translocations involving the *c-MYC* locus lead to the hyperactivation of the *c-MYC* transcriptional program, resulting in the overexpression of the c-MYC protein. This overexpression enhances proliferation and survival by promoting the expression of several target genes, including *FOXP1*, leading to DLBCL-type RT and tFL development. Created with *BioRender.com*.

2.2. *CDKN2A* Deletion

Deletion of the cyclin-dependent kinase inhibitor 2A (*CDKN2A*) gene is found in approximately 30% of DLBCL-type RT cases and is often acquired at the time of transformation, suggesting its pivotal role in driving RT in patients harboring this genetic lesion (Figure 1) [54]. *CDKN2A* encodes the cell-cycle regulators p16INK4A and p14ARF, which are responsible for the inhibition of the G1-to-S transition [55]. Under physiologic conditions, cyclin-dependent kinases 4 and 6 (CDK4 and CDK6) drive this step of the cell cycle by activating the E2F-dependent pathway, which, in turn, promotes cell entry into the S phase [56,57]. However, when hyperproliferative signals from constitutively active oncogenes exceed normal thresholds, p16INK4A is activated, inhibiting the activity of CDK4 and CDK6 [58]. Additionally, p14ARF inhibits the MDM2 E3 ubiquitin ligase (a p53 inhibitor), leading to the activation of the p53 transcriptional program, which triggers either apoptosis or cell-cycle arrest [55,59]. Consistently, *CDKN2A* deletion causes the impairment of the negative regulation of cell-cycle progression and resistance to pro-apoptotic stimuli, leading to tumorigenesis. Furthermore, the concomitant loss of *TP53* and *CDKN2A* activity disrupts the balance between positive and negative cell-cycle regulators, both of which are influenced by the BCR signaling pathway [60]. In CLL cells, where BCR signaling is hyperactivated, this imbalance leads to increased proliferation, genetic instability, and risk of HT [1].

2.3. *NOTCH1* Mutational Activation

NOTCH1 mutational activation is notably more common at the time of diagnosis of DLBCL-type RT than in CLL, with frequencies of 31% and 8.3%, respectively (Figure 2) [61]. Notch1 is a cell surface receptor activated upon the binding of a ligand of the SERRATE/JAGGED or DELTA families, located on adjacent cells [62]. This binding triggers two sequential proteolytic processes: first, an extracellular cleavage near the membrane by a disintegrin and metalloproteinase, creating a substrate for the second intramembrane cleavage by the γ-secretase complex [62,63]. This results in the intracellular release of the active Notch1 intracellular domain (NICD), which translocates to the nucleus, where it initiates the transcription of various genes that regulate cell proliferation and survival [63]. Under physiologic conditions, this signaling is terminated when specific amino acid sequences (known as degron sites) on the PEST domain are ubiquitinated, leading to the proteasome-mediated degradation of NICD [64]. The most common genetic lesions in DLBCL-type RT are disrupting mutations of the *NOTCH1* PEST domain, impairing ubiquitination and therefore prolonging Notch1 active signaling and promoting abnormal cell proliferation and tumorigenesis [65,66]. Notably, the risk of developing DLBCL-type RT is considerably higher in *NOTCH1*-mutated CLL, with a reported incidence of 45% at 15 years, compared to only 4.6% in *NOTCH1* wild-type CLL over the same period [67]. On these grounds, *NOTCH1* mutations have been identified as a significant risk factor for the development of DLBCL-type RT, with a cumulative risk of developing RT of 45% in *NOTCH1*-mutated CLL [42]. However, these data refer to the CIT era, highlighting the need for updated data in the era of targeted therapies.

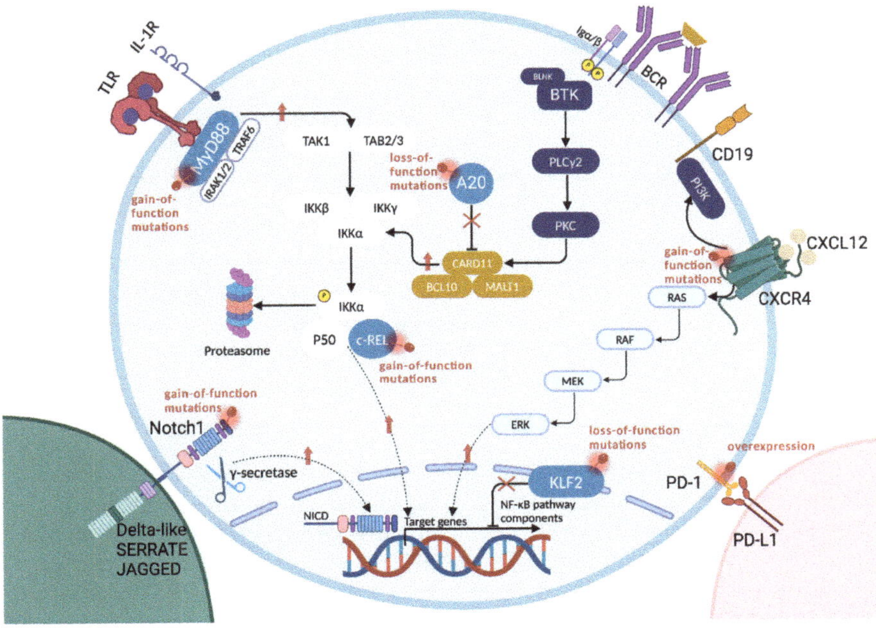

Figure 2. Genetic mutations and microenvironmental alterations implicated in HT. Several molecular pathways implicated in HT are shared among CLL, FL, MZL, and LPL. The NF-κB signaling cascade is frequently involved, driven by gain-of-function mutations in *MYD88* and *c-REL*, as well as loss-of-function mutations in *A20* and *KLF2*. Moreover, the hyperactivation of BCR signaling plays a key role in the pathogenesis of CLL and its progression to DLBCL-type RT. Also, Notch1 signaling is often dysregulated through gain-of-function mutations, leading to the upregulation of genes that promote abnormal cell proliferation and tumorigenesis. Additionally, gain-of-function mutations in *CXCR4* drive the upregulation of the MAPK1 cascade, particularly in LPL translating into a more aggressive biological behavior of the disease. In addition, microenvironmental alterations also contributes to HT. Consistently, the immune checkpoint PD-1 is frequently overexpressed in RT, which further contributes to immune evasion. Created with *BioRender.com*.

2.4. *c-MYC* Abnormalities

The *c-MYC* proto-oncogene and its encoded protein are involved in several essential cellular processes, and abnormalities in this pathway lead to aberrant cell survival, growth, metabolism, self-renewal, and genomic instability (Figure 1) [68]. The *c-MYC* transcriptional program is one of the main targets of Notch1 signaling, as it is directly activated by NCID [69]. In DLBCL-type RT, the primary genetic alterations driving *c-MYC* dysregulation include chromosomal translocations between the *c-MYC* gene, located on chromosome 8, and the IGHV regulatory regions, as well as gene amplification and mutations in the *c-MYC* promoter that enhance its activity. Another mechanism of *c-MYC* dysregulation involves loss-of-function mutations in the *MGA* gene, which normally prevents the c-MYC protein from forming a functional complex with its partner MAX. Together, genetic lesions affecting *c-MYC* and *MGA* contribute to c-MYC dysregulation in approximately 40% of DLBCL-type RT cases [54,70].

2.5. BCR Pathway Dysregulation

The hyperactivation of BCR signaling plays a key role in the pathogenesis of CLL and its progression to DLBCL-type RT, as shown in preclinical studies (Figure 2) [60,71–74]. Two potential models, which likely co-occur, have been proposed to explain this hyperactivation: antigen-dependent and antigen-independent BCR engagement [72,75]. Antigen-dependent signaling occurs when self or foreign antigens bind to the BCR, while antigen-independent signaling arises from Ig-Ig interactions within the same cell membrane, rather than from activating mutations, which are rare [76,77]. Moreover, about 30% of CLL patients share nearly identical VDJ rearrangements of the BCR Ig, forming stereotyped subsets, with subset 8 (IGHV4-39/IGHD6-13/IGHJ5) frequently linked to aggressive disease, trisomy 12, and *NOTCH1* mutations [78,79]. This subset is notably associated with the progression of CLL into DLBCL-type RT, likely due to its heightened reactivity, leading to a significant activation of CLL cells and the selection of more aggressive clones [1,80–82]. Moreover, a recent analysis of light-chain variable gene mutational status in CLL identified a significant association between the IGKV1-39/1D-39 rearrangement and *NOTCH1* mutations, trisomy 12, unmutated IGHV, and stereotyped subsets 1 and 8 [83]. These findings suggest a potential role for IGKV1-39/1D-39 in the development of DLBCL-type RT.

2.6. Dysregulation of Immune Checkpoints

PD-1 is a surface receptor on T cells that promotes apoptosis of effector T cells and supports Treg survival by interacting with its ligand, PD-L1, which is predominantly expressed on antigen-presenting cells (APCs) such as macrophages, B cells, and dendritic cells (DCs) [84]. As such, the interaction between PD1 and PD-L1 represents one major immune checkpoint inhibitor that regulates tumor escape from immune surveillance (Figure 2) [84]. DLBCL-type RT cells overexpress PD-1, while PD-L1 is primarily found on histiocytes and DCs, thus challenging the traditional PD-1/PD-L1 model of T cell suppression [85,86]. This phenomenon may be linked to the recent finding of PD-1+ regulatory B cells (Bregs) observed in various cancers [87–89]. Bregs may suppress the immune system via IL-10 production, causing T cell exhaustion, FOXP3+ Treg expansion, and recruitment of myeloid-derived suppressor cells, or by directly inhibiting T cells through a PD-L1-dependent mechanism, as in thyroid cancer [45,89]. However, the specific role of DLBCL-type RT cells as PD-1+ Bregs requires further study. Additionally, the DLBCL-type RT microenvironment exhibits elevated levels of immune checkpoints LAG3 and TIGIT [45]. LAG3, expressed on activated CD4+ and CD8+ T cells, binds MHC class II on APCs and tumor cells, including DLBCL-type RT cells [90–94]. This interaction transmits inhibitory signals, impairing T cell function and enabling tumor immune evasion [45]. TIGIT, found on natural killer (NK) and T cells, interacts with CD155 on APCs and tumor cells to suppress immune responses [95]. In DLBCL-type RT, CD226, a TIGIT competitor that activates T and NK cells via CD155, is abnormally elevated, suggesting an imbalance favoring activating signals in DLBCL-type RT tumor cells, contrasting with higher TIGIT levels in CLL cells [96–98].

The potential sensitivity of DLBCL-type RT to immune checkpoint inhibitors has prompted studies investigating PD-1/PD-L1 axis inhibition. While single-agent PD-1/PD-L1 inhibitors showed limited efficacy, promising results emerged from the phase 2 MOLTO study [45,99]. This trial evaluated a combination therapy of the PD-L1 inhibitor atezolizumab, the BCL2i venetoclax, and the anti-CD20 mAb obinutuzumab, achieving an overall response rate (ORR) of 67.9% [99]. Moreover, targeting BTK could represent a viable treatment strategy due to the overactivity of the BTK pathway [1]. Based on this rationale, the phase II RT1 trial evaluated the combination of the BTK inhibitor zanubrutinib and the PD-1 inhibitor tislelizumab in DLBCL-type RT, yielding promising results with an ORR of 58.3% and a complete response (CR) rate of 18.8% [100]. These findings suggest a potential role for PD-1 and/or BTK inhibitor-based combination therapies in the management of DLBCL-type RT, emphasizing the need for further research to optimize these approaches for improved clinical outcomes.

2.7. Novel Molecular Insights in DLBCL-Type RT

Recent studies have revealed additional genetic alterations, including somatic single nucleotide variants (SNVs) and somatic copy number alterations (CNAs), that disrupt pathways involved in DNA repair, chromatin remodeling, MAPK13 signaling, and NF-κB activity, driving DLBCL-type RT development and progression [101]. Specific DLBCL-type RT-related changes include frequent mutations in *IRF2BP2* and SNVs in genes like *CCND3*, and *SRSF1*, alongside several recurrent focal somatic CNAs [102]. Notable findings also highlight the loss of key epigenetic regulators, such as *EZH2*, *DNMT3A*, *KMT2C*, *SETD2*, *TET2*, and *ARID1A*, as well as the c-MYC inhibitor *MGA*, through either mutations or somatic CNAs [17,102]. These alterations underscore critical pathways contributing to DLBCL-type RT, including chromatin regulation, splicing, c-MYC activation, NF- κB signaling, immune evasion, inflammation, apoptosis resistance, DNA damage response, and cell-cycle dysregulation. Together, these events reflect the genetic complexity underlying DLBCL-type RT and emphasize the importance of understanding these pathways to develop targeted therapeutic strategies.

3. Molecular Mechanisms in the Transformation of FL

There is no single genetic alteration that solely drives the HT of FL [103,104]. Instead, the transformation process involves a combination of pathogenetic mechanisms affecting cell-cycle control, DNA damage response, proliferation, and microenvironmental changes that are mediated by genetic and epigenetic alterations (Table 1).

3.1. CDKN2A/B Deletion

The most frequent genomic aberration acquired during the progression to tFL is the loss of *CDKN2A/B*, occurring in approximately 45% of cases (Figure 1) [30,31,103,105,106]. The loss of *CDKN2A* can also occur in FL, although at a significantly lower frequency (10%), and is associated with poorer clinical outcomes [105]. The loss of *CDKN2A/B* disrupts cell-cycle control, driving unchecked cell proliferation and genomic instability and enhancing cellular aggressiveness [31]. Interestingly, genetic abnormalities in *TP53*, observed in around 20% of cases during tFL, are mutually exclusive with *CDKN2A/B* lesions [31].

3.2. c-MYC Abnormalities

Genetic lesions activating the *c-MYC* proto-oncogene, such as chromosomal translocations, gains and/or amplifications, and point mutations, are commonly acquired during the progression to tFL, occurring in approximately 40% of cases (Figure 1) [30,31,107]. Dysregulated c-MYC oncogenic activity can confer multiple advantages to the transformed clone through its pleiotropic roles in promoting cell growth, altering metabolism, and driving genetic instability [31]. *c-MYC* activation can also be driven by epigenetic mechanisms. Consistently, Musilova and colleagues analyzed microRNA (miRs) profiles from serial tFL biopsies and identified five miRs enriched during transformation, including miR-150 [108]. Their findings suggest the existence of a c-MYC/miR-150/FOXP1 axis, where *c-MYC* overexpression represses miR-150 levels, leading to the upregulation of the *FOXP1* gene. FOXP1, a transcription factor crucial for B cell development, is associated with the ABC subtype of DLBCL and is linked to poor clinical outcomes in both DLBCL and FL [108].

3.3. *TP53 Disruption*

TP53 disruption is uncommon in FL, occurring in approximately 5% of cases [30,109]. In contrast, *TP53* lesions are significantly more frequent in tFL, with an incidence of 25–30% (Figure 1) [107]. An early report by Lo Coco et al. demonstrated that *TP53* mutations were predominantly found in tFL samples compared to baseline FL cases, a finding subsequently validated in larger cohorts [31,107,110]. Studies have shown that *TP53* mutations typically occur without deletion of the other allele in untransformed FL, whereas biallelic alterations through deletion, loss of heterozygosity (LOH), or additional mutations are frequently observed in tFL [31,104]. However, a few studies reported inconsistent results. In a study analyzing 185 diagnostic samples of FL from the Lymphoma/Leukemia Molecular Profiling Project, *TP53* mutations were associated with shorter PFS and OS but unexpectedly showed no correlation with subsequent transformation [109]. Moreover, in that regard, Davies et al. found no direct correlation between *TP53* status and tFL, whereas their study revealed significantly higher MDM2 expression in tFL samples (72%) compared to FL samples (58%) [111]. This suggests that an alternative mechanism disrupting p53 function might be operative in tFL, potentially increasing the transformation risk [111].

3.4. *BCL-6 Genetic Lesions*

B cell lymphoma 6 (*BCL-6*) is a proto-oncogene encoding the BCL6 protein, a crucial regulator of GC development [112]. A key biological function of BCL6 in centroblasts is to promote rapid proliferation while enabling tolerance to genomic damage that occurs during clonal expansion and somatic hypermutation. BCL6 achieves this by directly repressing genes involved in DNA damage detection and checkpoint regulation, including the DNA damage sensor *ATR*, the tumor suppressor *TP53*, and the cell-cycle-arrest gene *CDKN1A* [113–115]. *BCL6* is commonly affected by mutations and/or translocations in B cell lymphomas (Figure 1) [116]. Specifically, translocations involving *BCL6* and leading to the overexpression of the protein are characteristic of "double-hit" or "triple-hit" DLBCL, which also include *BCL2* and *c-MYC* translocations [117]. These genetic events are associated with a significantly worse prognosis [118,119]. *BCL6* translocations at the time of FL diagnosis are linked to an increased risk of transformation, with a slightly higher frequency observed at HT (25% in tFL versus 10% at diagnosis) [119]. *BCL6* translocations have been detected in both GCB and ABC tFL [118], whereas in *de novo* DLBCL, they are more prevalent in ABC subtype cases [120].

3.5. *NF-κB Pathway Dysregulation*

Genetic lesions play a significant role in the pathogenesis of tFL by driving the aberrant activation of the NF-κB pathway, a critical signaling cascade that regulates DNA transcription, cytokine production, and cell survival of B lymphocytes (Figure 2) [121]. NF-κB lesions occur in one-third of tFL patients [30]. These alterations include the following: (*i*) The amplification of the proto-oncogene *c-REL* found in 11% of tFL, which encodes the c-Rel protein. c-Rel forms a dimer with p50, a component of the NF-κB complex. c-Rel-p50 dimer translocates in the nucleus and promotes the transcription of other critical components of the NF-κB pathway [30,121]. (*ii*) Loss-of-function mutations or the loss of TNF alpha-induced protein 3 (*TNFAIP3*) are identified in 15% of cases, specifically at HT. TNFAIP3 is a central regulator of inflammatory responses. It negatively modulates the TNF-induced NF-κB proinflammatory signaling pathway, supports B cell survival, and regulates TNF-mediated apoptosis [30,121,122]. (*iii*) Non-L265P *MYD88* mutations are present in both FL and tFL [30,121]. Importantly, 80% of these mutations are restricted to HT samples [30]. *MYD88* encodes the general adaptor protein MYD88, which, following ligand binding, associates to the intracellular domain of interleukin 1 and Toll-like receptors (IL1R and TLR) [123,124]. Then, it recruits the serine-threonine kinase IRAK 4, which, through the activation of IRAK1 and IRAK2 by phosphorylation, leads to the activation of TNF receptor-associated factor 6 (TRAF6) [125]. TRAF6 downstream signaling enhances cell survival by activating the BTK, canonical NF-κB, and MAPK pathways [126].

3.6. tFL Cell of Origin

When transformed cases are stratified according to their COO using the Hans algorithm, distinct mutational profiles are observed within each group [21,107,118]. Mutations in *MYD88*, *CD79B*, and *BCL10* are more frequently detected (~15–25%) in ABC-transformed cases, whereas 2p16 (*REL*) amplification is predominantly observed in GCB-transformed cases, consistent with patterns reported in DLBCL [107,121,127,128]. These findings suggest the presence of distinct subgroups of tFL, potentially arising through different transformation pathways, and associated with divergent clinical outcomes. In support of this, a prior study demonstrated increased proliferation rates at HT in a subset of tFL, as indicated by gene expression analysis [129]. This subgroup of tFL was enriched in aberrations of *TP53*, *CDKN2A/B*, and *REL*, highlighting potential mechanistic differences in transformation compared to other tFL cases.

3.7. Microenvironmental Alterations

Specific genetic mutations in FL alter the interactions between B cells and their microenvironment. In this context, the herpesvirus entry mediator (HVEM) protein, encoded by the *TNFRSF14* gene, plays a key role in regulating T-cell responses by providing either costimulatory or coinhibitory signals, depending on its interacting ligand [130,131]. The B and T-lymphocyte attenuator (BTLA) ligand, expressed on B cells, interacts with HVEM to suppress T-cell responses [131]. In FL, the *TNFRSF14* gene is frequently affected by mutations (~30–40%), deletions (~20–30%), and/or copy-neutral LOH (~10%) [30,121,132–134]. These genetic alterations reduce HVEM expression and enhance BTLA signaling [135,136]. Consequently, *TNFRSF14* aberrations disrupt the HVEM-BTLA inhibitory axis, reshaping the microenvironment of FL and promoting B cell proliferation, activation of lymphoid stroma, and an increase in follicular T-helper cells [136]. Interestingly, a recent study reported that tFL was associated with increased HVEM expression and decreased BTLA expression; however, it did not analyze the mutational status of the two genes [137]. Importantly, the transformation of FL can be further facilitated by acquired immune evasion mechanisms. In particular, these include loss-of-function mutations or deletions in HLA class I components (particularly *B2M*) and *CD58*, which encodes a protein involved in T and NK cell-mediated responses [31,121]. Both proteins are crucial for the recognition of neoplastic cells by the immune system, and their loss can be acquired upon transformation allowing the lymphoma to evade immune detection.

4. Molecular Mechanisms in the Transformation of MZLs

MZLs are rare and heterogeneous B cell malignancies, and limited data are available regarding the molecular pathways driving their transformation (Table 1) [138]. This scarcity of information poses challenges to understanding the mechanisms underlying their development and progression. In addition, conflicting data regarding the involvement of specific pathways further complicate this understanding. One such pathway in SMZL is the Notch2 signaling cascade which is affected by activating mutations in 30% of cases [139,140]. The clinical significance of these mutations remains debated. A few studies suggest that *NOTCH2* mutations correlate with better OS and PFS, while others link them to higher relapse, transformation, or mortality risks [139,141]. *NOTCH2* encodes a co-transcription factor that favors B cell differentiation into MZ B cells. Mutations typically truncate the PEST domain, preventing proteasomal degradation and leading to protein accumulation [140,142–144].

4.1. CDKN2A/B Deletion in SMZL

A recent report shows that, like other lymphomas, the deletion of 9p21 is significantly enriched in transformed SMZL (tSMZL) (Figure 1) [145]. This region contains the tumor suppressor genes *CDKN2A* and *CDKN2B*, which are involved in cell-cycle regulation. Notably, this deletion was the most frequently observed alteration in tSMZL, found in 40.6% of cases, and was absent in SMZL at diagnosis, corroborating findings from previously published studies [140,146–149].

4.2. NF-κB Pathway Dysregulation in SMZL

Regulators of the NF-κB pathway are frequently altered in tSMZL, namely *TNFAIP3* and *KLF2*, mutated in 59.4% and 31.3% of cases, respectively (Figure 2) [145]. Although these genes are frequently disrupted in SMZL, they exhibit a higher mutation frequency in tSMZL [145]. *TNFAIP3* encodes a protein that negatively regulates the NF-κB signaling pathway, through the ubiquitination of NF-κB, leading to its proteasomal degradation [150]. Previous studies have reported that *TNFAIP3* mutations occur in 7% to 15% of cases of SMZL [139,140,151,152]. Notably, Grau et al. showed that truncating loss-of-function mutations in TNFAIP3 are identified in 32% of tSMZL [145]. *KLF2* is a tumor suppressor gene that encodes a transcription factor responsible for repressing NF-κB-mediated B cell activation and their differentiation into marginal zone (MZ) B cells [153]. Loss-of-function mutations and/or the loss of *KLF2* disrupt the protein structure, preventing it from binding to DNA and carrying out its regulatory role [139,147,154,155].

4.3. Copy Number Alterations in SMZL

A recent study identified that tSMZL was characterized by a distinct set of driver CNAs, including gains in 1q, 3q, and 18q (encompassing *BCL2*), as well as losses in 1p36 (*ARID1A*), 3p21 (*SETD2*), 7q31-q32 (*KLF2, NOTCH2*), 9p21.3 (*CDKN2A/B*), and 13q14.13-q14.3 (*RB1* and *DLEU1/2*) [145]. This profile of genetic alterations largely mirrors that of *de novo* DLBCL, with the exception of 7q loss, a highly specific alteration to SMZL that is rarely observed in other small B cell neoplasms. Furthermore, the study demonstrates that, from a molecular perspective, the genome of tSMZL is significantly more complex than that of SMZL, exhibiting twice the number of alterations [145].

4.4. Epigenetic Alterations

The *KMT2D* gene has been found to be mutated in 46.9% of tSMZL [145]. KMT2D is a key epigenetic regulator involved in histone H3 lysine 4 (H3K4) methylation, which is crucial for the transcriptional activation of downstream genes by promoting an open chromatin state [156]. Similarly to *TNFAIP3* and *KLF2*, the frequency of *KMT2D* alterations in tSMZL cases is higher than that reported for SMZL [145]. Additionally, genome-wide epigenetic profiling of SMZL has revealed that aberrant promoter DNA methylation serves as a mechanism of dysregulation affecting critical pathways in the disease contributing to HT [157,158]. Elevated levels of promoter DNA methylation are found in 25% of SMZL cases and are associated with poorer OS and higher risk for HT [157]. Specifically, aberrant methylation appears to contribute to the elevated expression of (*i*) pro-survival genes, such as *TCL1B, BCL2A1,* and *FGF1*, as well as members of the BCR and NF-κB signaling pathways (including *CD79B, CARD11,* and *PIK3CB*); (*ii*) genes associated with the epigenetic regulator PRC2 complex (*EZH2, EED,* and *SUZ12*); (*iii*) genes involved in the JAK-STAT and PI3K/AKT signaling pathways, both involved in cell proliferation [157]. Therefore, aberrant methylation may also play a role in the pathogenesis and/or transformation of SMZL by the inactivation of tumor suppressors and the expression of genes sustaining tumor cell survival and proliferation.

4.5. FOXP1 Aberrations in MALT Lymphomas

A multivariate analysis in transformed MALT (tMALT) lymphomas has revealed a significant association between HT and FOXP1 expression, with FOXP1 detected in five out of eight patients with HT, four of whom exhibited the t(3;14) translocation [159]. The t(3;14) translocation has been implicated in aggressive forms of HT within MALT lymphomas (Figure 1) [159–161]. Chromosome 3 harbors the *FOXP1* gene and has been identified as a translocation partner with the *IGH* gene in both DLBCL and MALT lymphomas. Moreover, FOXP1 expression has been recognized as a poor prognostic indicator in DLBCL, particularly associated with the non-GCB subtype [162]. Sagaert et al. examined 70 distinct MALT lymphomas, correlating FOXP1 protein expression with outcome in tMALT lymphomas [159]. Immunohistochemistry and fluorescent in situ hybridization analyses detected FOXP1 expression in approximately 30% of MALT MZL cases, all of which were linked to poor clinical outcomes.

4.6. Mechanisms of HT in NMZL

The mechanisms driving the transformation of NMZL are poorly understood due to the rarity of the disease. In a study by Qian and colleagues, six cases of transformed NMZL revealed a significantly higher frequency of del(20q12) compared to non-transformed cases [163]. These transformed lymphomas also showed enrichment in extracellular matrix proteins (COL1A1 and FN1), the growth factor receptor PDGFRβ, the DNA repair protein RAD51, and the signaling molecule WNT11.

5. Molecular Mechanisms in the Transformation of LPL

The molecular alterations driving the aggressive transformation of LPL are poorly understood, with *CXCR4* mutations being the primary factors involved (Table 1) [41]. Conversely, *MYD88* L625P gain-of-function mutation, found in the vast majority of LPL, was identified only in 73% of DLBCL transformed from LPL in a recent study [29,38,41]. This finding aligns with previous studies that have identified the absence of *MYD88* mutations as a significant risk factor for the development of HT [164].

Other genetic lesions found at diagnosis or acquired at the time of transformation include mutations in *TP53*, *KMT2D*, *ARID1A*, *BTG1*, *BTG2*, *PIM1*, *CARD11*, and *CD79B*, but the full extent of their role in the transformation remains unclear [41,165,166].

CXCR4 Mutations

CXCR4 mutations are found in about 30% of LPL patients overall and in approximately 55% of transformed cases before HT (Figure 2) [41,165]. CXCR4 is a G protein coupled receptor (GPCR) activated upon CXCL12 binding under physiologic condition, functioning as a key factor in directed migration, leukocyte trafficking, and the homing of stem and progenitor cells [167]. In LPL, the most common CXCR4 alteration, occurring in 50% of cases, involves serine 338 (at position 1013), where C > G changes occur in 54% and C > A changes in 25% of cases, both producing a stop codon, while S338 frameshift mutations account for 21% of cases [165]. These mutations create premature stop codons or frameshifts, causing the production of a truncated protein which lacks its regulatory domain, thereby resulting in reduced CXCR4 desensitization and internalization, which leads to prolonged activation following CXCL12 binding [40,165]. The upregulation of the CXCR4 pathway results in abnormally high signaling mediated by diacylglycerol (DAG) and inositol trisphosphate (IP3), which lead to the aberrant activation of several intracellular pathways involved in cell proliferation, growth, and survival, including AKT signaling and therefore the MAPK 1/2 pathway [168]. Consistently, mutations of *CXCR4* translate into a more aggressive clinical behavior in LPL, eventually leading to more symptomatic and progressive disease and ultimately to HT [40].

6. Molecular Crossroads: The Shared Pathways Fueling Lymphoma Transformation

Despite the biological and clinical heterogeneity of CLL, FL, MZLs, and WM, the HT of these diseases follows converging evolutionary trajectories driven by common molecular mechanisms (Table 1). These alterations disrupt key tumor suppressor pathways, promote uncontrolled proliferation, and enhance survival signaling, ultimately leading to aggressive disease with increased genomic instability and therapy resistance [1,103,138]. The recurrence of these mechanisms across different B cell malignancies suggests a hierarchical model of transformation, where early genomic lesions, such as *TP53* disruption and *CDKN2A/B* loss, create a permissive environment for further oncogenic events, including c-*MYC* activation and NF-κB pathway dysregulation [1,103,104]. Beyond their individual roles, these alterations reflect an interplay between intrinsic tumor biology and extrinsic selective pressures. *TP53* disruption and *CDKN2A/B* loss not only drive unchecked cell-cycle progression, but also promote resistance to apoptosis, enabling the survival of genetically unstable clones that are prone to further transformation [60,103]. The acquisition of c-*MYC* abnormalities fuels metabolic reprogramming, rapid proliferation, and increased replicative stress, reinforcing tumor aggressiveness [68]. In parallel, NF-κB dysregulation sustains B cell survival through chronic inflammatory signaling and immune evasion, further shaping the tumor microenvironment to support malignant progression [121]. The convergence of these pathways across multiple lymphoma subtypes highlights key vulnerabilities that could serve as both predictive biomarkers and therapeutic targets. Identifying these genetic lesions in the pre-transformation phase could allow for early risk stratification and preemptive therapeutic intervention. Moreover, targeting these pathways through strategies aimed at restoring *TP53* function, inhibiting *c-MYC*-driven oncogenesis, or disrupting NF-κB-mediated survival signaling may hold promise in delaying, or more optimistically preventing, transformation [169–171]. Given that many of these pathways also drive resistance to conventional chemotherapy, a precision medicine approach integrating molecular profiling and targeted therapies might improve outcomes for patients at high risk of transformation, shifting the paradigm from reactive treatment to proactive disease management.

Table 1. Frequency and biological effect of genetic lesions in HT.

Disease	Gene Lesion	Biological Effect	Frequency Before HT	Frequency at HT	Reference
CLL	*TP53*	DNA damage response, cell-cycle regulation	10–15%	60–80%	[46,47]
	CDKN2A	Cell-cycle regulation	7%	30%	[54,172]
	NOTCH1	NF-κB signaling	8%	31%	[61]
	c-MYC	Proliferation and survival	-	30–35%	[54]
	MGA	c-MYC inhibition	3%	6%	[70]
	SETD2	Epigenetic regulation of gene expression	3%	30%	[173,174]

Table 1. *Cont.*

Disease	Gene Lesion	Biological Effect	Frequency Before HT	Frequency at HT	Reference
FL	*CDKN2A*	Cell-cycle regulation	5%	45%	[30,31,103,105,106]
	c-MYC	Proliferation and survival	5–10%	40%	[30,31,107]
	TP53	DNA damage response, cell-cycle regulation	5%	25%	[30,31,107,109,110]
	BCL6	B cell differentiation	10%	25%	[119]
	c-REL	NF-κB signaling	-	11%	[30]
	TNFAIP3	NF-κB signaling	5–10%	15%	[30,121,122]
	MYD88	NF-κB/BCR signaling	5%	11%	[31,121]
	CD58	Microenviroment dysregulation	-	5	[31,121]
	B2M	NF-κB signaling	-	12	[31,121]
	TNFRSF14	Microenviroment dysregulation	15–20%	30–40%	[30,121,132–134]
MZLs	*CDKN2A*	Cell-cycle regulation	5%	40%	[140,146–149]
	TNFAIP3	NF-κB signaling	15%	60%	[145]
	KLF2	NF-κB signaling and B cell differentiation	20–25%	35%	[139,145,147,154,155]
	KMT2D	Epigenetic regulation of gene expression	10–15%	45%	[157,158]
LPL	*CXCR4*	Proliferation and survival	30%	55%	[41,165]

Abbreviations: HT, histologic transformation; CLL, chronic lymphocytic leukemia; FL, follicular lymphoma; MZLs, marginal zone lymphomas; LPL, lymphoplasmacytic lymphoma.

7. Conclusions

HT is a major cause of morbidity and mortality in patients with lymphoproliferative neoplasms. Consistently, DLBCL arisen from NHLs still represent an unmet clinical need, due to the high degree of resistance to available treatment approaches. Although no unifying genetic lesion leading to HT has been identified, the central molecular pathways driving transformation include genetic and epigenetic alterations of cell-cycle control, DNA damage response, proliferation, and microenvironmental dysregulation. Importantly, several pathway alterations are shared by indolent B cell malignancies undergoing HT, including *CDKN2A* deletion, *TP53* disruption, *c-MYC* aberrations, and activating genetic lesions of NF-κB signaling. However, the clinical outcome of HT is remarkably different across distinct histologic subtypes, with DLBCL-type RT displaying poorer survival rates, implying that the prognosis relies on the initial lymphoproliferative neoplasm. Although a number of genetic alterations are associated with higher risk of HT, none of the currently investigated biomarkers is able to predict transformation with sufficient precision. In that regard, liquid biopsy, which allows to detect tumor-associated mutations in cell-free DNA in plasma, represents an innovative diagnostic approach to better mirror the spatial and/or intra-tumor heterogeneity [104,169,170]. Therefore, liquid biopsy might serve as a potential tool for transformation prediction. Most of the aforementioned data on the incidence and biology of HT derive from studies conducted in the CIT era, therefore underscoring the need for further investigations in chemo-naïve patients receiving targeted therapy, since they may exhibit distinct biological pathways. Nevertheless, at the present time, a considerable number of HT continue to involve patients with prior exposure to CIT. In light of the fact that the majority of current patients undergoing HT have received CIT, they often develop chemorefractoriness largely due to alterations of DNA damage response mechanisms, which may play a role in HT. On these grounds, targeted therapy directed towards immune checkpoints, BCL2, and BTK, as well as bispecific mAbs and chimeric antigen receptor (CAR)-T cells, have emerged as a promising treatment approach in HT [1,99,100,175,176]. In conclusion, further investigation is necessary for a deeper understanding of the molecular mechanisms underlying HT, in order to further highlight predictive biomarkers of transformation and develop effective treatment strategies.

Author Contributions: Conceptualization, N.M., S.M. and G.G.; methodology, N.M. and S.M.; writing—original draft preparation, N.M. and S.M.; S.M.; writing—review and editing, G.G., B.F.M. and A.F.A.; visualization, N.M. and S.M.; supervision, G.G.; funding acquisition, G.G. All authors have read and agreed to the published version of the manuscript.

Funding: This work was supported by Molecular bases of disease dissemination in lymphoid malignancies to optimize curative therapeutic strategies, (5 × 1000 No. 21198); the AGING Project–Department of Excellence–DIMET, Università del Piemonte Orientale, Novara, Italy; PNRR-MAD-2022-12375673 (Next Generation EU, M6/C2_CALL 2022), Italian MoH, Rome, Italy; and AIL Novara VCO ODV, Novara, Italy.

Conflicts of Interest: N.M., S.M., B.F.M. and A.F.A. declare no conflicts of interest for this specific work. G.G. declares advisory board and speaker's bureau honoraria from AbbVie, AstraZeneca, BeiGene, Hikma, Incyte, Johnson & Johnson, and Lilly.

References

1. Mouhssine, S.; Gaidano, G. Richter Syndrome: From Molecular Pathogenesis to Druggable Targets. *Cancers* **2022**, *14*, 4644. [CrossRef]
2. Cerhan, J.R. Epidemiology of Follicular Lymphoma. *Hematol. Oncol. Clin. N. Am.* **2020**, *34*, 631–646. [CrossRef] [PubMed]
3. Conconi, A.; Franceschetti, S.; Aprile von Hohenstaufen, K.; Margiotta-Casaluci, G.; Stathis, A.; Moccia, A.A.; Bertoni, F.; Ramponi, A.; Mazzucchelli, L.; Cavalli, F.; et al. Histologic transformation in marginal zone lymphomas. *Ann. Oncol.* **2015**, *26*, 2329–2335. [CrossRef]
4. Castillo, J.J.; Gustine, J.; Meid, K.; Dubeau, T.; Hunter, Z.R.; Treon, S.P. Histological transformation to diffuse large B-cell lymphoma in patients with Waldenström macroglobulinemia. *Am. J. Hematol.* **2016**, *91*, 1032–1035. [CrossRef] [PubMed]
5. Parry, E.M.; Roulland, S.; Okosun, J. DLBCL arising from indolent lymphomas: How are they different? *Semin. Hematol.* **2023**, *60*, 277–284. [CrossRef] [PubMed]
6. Gall, E.A.; Mallory, T.B. Malignant Lymphoma: A Clinico-Pathologic Survey of 618 Cases. *Am. J. Pathol.* **1942**, *18*, 381–429.
7. Swerdlow, S.H.; Campo, E.; Harris, N.L.; Jaffe, E.S.; Pileri, S.A.; Stein, H.; Thiele, J. *WHO Classification of Tumours of Haematopoietic and Lymphoid Tissues*, 4th ed.; International Agency for Research on Cancer Lyon: Lyon, France, 2017; Volume 2.
8. Alaggio, R.; Amador, C.; Anagnostopoulos, I.; Attygalle, A.D.; Araujo, I.B.O.; Berti, E.; Bhagat, G.; Borges, A.M.; Boyer, D.; Calaminici, M.; et al. The 5th edition of the World Health Organization Classification of Haematolymphoid Tumours: Lymphoid Neoplasms. *Leukemia* **2022**, *36*, 1720–1748. [CrossRef]
9. Surveillance Epidemiology and End Results Program (SEER). Cancer Stat Facts: Leukemia—Chronic Lymphocytic Leukemia (CLL). Available online: https://seer.cancer.gov/statfacts/html/clyl.html (accessed on 21 December 2024).
10. Mouhssine, S.; Maher, N.; Matti, B.F.; Alwan, A.F.; Gaidano, G. Targeting BTK in B Cell Malignancies: From Mode of Action to Resistance Mechanisms. *Int. J. Mol. Sci.* **2024**, *25*, 3234. [CrossRef]
11. Maher, N.; Mouhssine, S.; Matti, B.F.; Alwan, A.F.; Gaidano, G. Treatment Refractoriness in Chronic Lymphocytic Leukemia: Old and New Molecular Biomarkers. *Int. J. Mol. Sci.* **2023**, *24*, 10374. [CrossRef]
12. Shadman, M. Diagnosis and Treatment of Chronic Lymphocytic Leukemia: A Review. *JAMA* **2023**, *329*, 918–932. [CrossRef]
13. Mouhssine, S.; Maher, N.; Kogila, S.; Cerchione, C.; Martinelli, G.; Gaidano, G. Current Therapeutic Sequencing in Chronic Lymphocytic Leukemia. *Hematol. Rep.* **2024**, *16*, 270–282. [CrossRef] [PubMed]

14. Eichhorst, B.; Ghia, P.; Niemann, C.U.; Kater, A.P.; Gregor, M.; Hallek, M.; Jerkeman, M.; Buske, C. ESMO Clinical Practice Guideline interim update on new targeted therapies in the first line and at relapse of chronic lymphocytic leukaemia. *Ann. Oncol.* **2024**, *35*, 762–768. [CrossRef] [PubMed]
15. Rossi, D.; Spina, V.; Gaidano, G. Biology and treatment of Richter syndrome. *Blood* **2018**, *131*, 2761–2772. [CrossRef] [PubMed]
16. Hampel, P.J.; Rabe, K.G.; Wang, Y.; Hwang, S.R.; Kenderian, S.S.; Muchtar, E.; Leis, J.F.; Koehler, A.B.; Tsang, M.; Hilal, T.; et al. Incidence of Richter transformation of chronic lymphocytic leukemia/small lymphocytic lymphoma in the targeted therapy era. *Leukemia* **2024**, *39*, 503–507. [CrossRef]
17. Nadeu, F.; Royo, R.; Massoni-Badosa, R.; Playa-Albinyana, H.; Garcia-Torre, B.; Duran-Ferrer, M.; Dawson, K.J.; Kulis, M.; Diaz-Navarro, A.; Villamor, N.; et al. Detection of early seeding of Richter transformation in chronic lymphocytic leukemia. *Nat. Med.* **2022**, *28*, 1662–1671. [CrossRef]
18. Al-Sawaf, O.; Robrecht, S.; Bahlo, J.; Fink, A.M.; Cramer, P.; Tresckow, J.V.; Lange, E.; Kiehl, M.; Dreyling, M.; Ritgen, M.; et al. Richter transformation in chronic lymphocytic leukemia (CLL)—A pooled analysis of German CLL Study Group (GCLLSG) front line treatment trials. *Leukemia* **2021**, *35*, 169–176. [CrossRef]
19. Mao, Z.; Quintanilla-Martinez, L.; Raffeld, M.; Richter, M.; Krugmann, J.; Burek, C.; Hartmann, E.; Rudiger, T.; Jaffe, E.S.; Müller-Hermelink, H.K.; et al. IgVH mutational status and clonality analysis of Richter's transformation: Diffuse large B-cell lymphoma and Hodgkin lymphoma in association with B-cell chronic lymphocytic leukemia (B-CLL) represent 2 different pathways of disease evolution. *Am. J. Surg. Pathol.* **2007**, *31*, 1605–1614. [CrossRef]
20. El Hussein, S.; Medeiros, L.J.; Lyapichev, K.A.; Fang, H.; Jelloul, F.Z.; Fiskus, W.; Chen, J.; Wei, P.; Schlette, E.; Xu, J.; et al. Immunophenotypic and genomic landscape of Richter transformation diffuse large B-cell lymphoma. *Pathology* **2023**, *55*, 514–524. [CrossRef]
21. Hans, C.P.; Weisenburger, D.D.; Greiner, T.C.; Gascoyne, R.D.; Delabie, J.; Ott, G.; Müller-Hermelink, H.K.; Campo, E.; Braziel, R.M.; Jaffe, E.S.; et al. Confirmation of the molecular classification of diffuse large B-cell lymphoma by immunohistochemistry using a tissue microarray. *Blood* **2004**, *103*, 275–282. [CrossRef]
22. Wang, Y.; Tschautscher, M.A.; Rabe, K.G.; Call, T.G.; Leis, J.F.; Kenderian, S.S.; Kay, N.E.; Muchtar, E.; Van Dyke, D.L.; Koehler, A.B.; et al. Clinical characteristics and outcomes of Richter transformation: Experience of 204 patients from a single center. *Haematologica* **2020**, *105*, 765–773. [CrossRef]
23. Favini, C.; Talotta, D.; Almasri, M.; Andorno, A.; Rasi, S.; Adhinaveni, R.; Kogila, S.; Awikeh, B.; Schipani, M.; Boggione, P.; et al. Clonally unrelated Richter syndrome are truly de novo diffuse large B-cell lymphomas with a mutational profile reminiscent of clonally related Richter syndrome. *Br. J. Haematol.* **2022**, *198*, 1016–1022. [CrossRef] [PubMed]
24. Smyth, E.; Eyre, T.A.; Cheah, C.Y. Emerging Therapies for the Management of Richter Transformation. *J. Clin. Oncol.* **2023**, *41*, 395–409. [CrossRef]
25. Weiss, L.M.; Warnke, R.A.; Sklar, J.; Cleary, M.L. Molecular analysis of the t(14;18) chromosomal translocation in malignant lymphomas. *N. Engl. J. Med.* **1987**, *317*, 1185–1189. [CrossRef] [PubMed]
26. Jacobsen, E. Follicular lymphoma: 2023 update on diagnosis and management. *Am. J. Hematol.* **2022**, *97*, 1638–1651. [CrossRef] [PubMed]
27. Carbone, A.; Roulland, S.; Gloghini, A.; Younes, A.; von Keudell, G.; López-Guillermo, A.; Fitzgibbon, J. Follicular lymphoma. *Nat. Rev. Dis. Primers* **2019**, *5*, 83. [CrossRef]
28. Fischer, T.; Zing, N.P.C.; Chiattone, C.S.; Federico, M.; Luminari, S. Transformed follicular lymphoma. *Ann. Hematol.* **2018**, *97*, 17–29. [CrossRef]
29. Alcoceba, M.; García-Álvarez, M.; Medina, A.; Maldonado, R.; González-Calle, V.; Chillón, M.C.; Sarasquete, M.E.; González, M.; García-Sanz, R.; Jiménez, C. MYD88 Mutations: Transforming the Landscape of IgM Monoclonal Gammopathies. *Int. J. Mol. Sci.* **2022**, *23*, 5570. [CrossRef]
30. Okosun, J.; Bödör, C.; Wang, J.; Araf, S.; Yang, C.Y.; Pan, C.; Boller, S.; Cittaro, D.; Bozek, M.; Iqbal, S.; et al. Integrated genomic analysis identifies recurrent mutations and evolution patterns driving the initiation and progression of follicular lymphoma. *Nat. Genet.* **2014**, *46*, 176–181. [CrossRef]
31. Pasqualucci, L.; Khiabanian, H.; Fangazio, M.; Vasishtha, M.; Messina, M.; Holmes, A.B.; Ouillette, P.; Trifonov, V.; Rossi, D.; Tabbò, F.; et al. Genetics of follicular lymphoma transformation. *Cell Rep.* **2014**, *6*, 130–140. [CrossRef]
32. Link, B.K.; Maurer, M.J.; Nowakowski, G.S.; Ansell, S.M.; Macon, W.R.; Syrbu, S.I.; Slager, S.L.; Thompson, C.A.; Inwards, D.J.; Johnston, P.B.; et al. Rates and outcomes of follicular lymphoma transformation in the immunochemotherapy era: A report from the University of Iowa/MayoClinic Specialized Program of Research Excellence Molecular Epidemiology Resource. *J. Clin. Oncol.* **2013**, *31*, 3272–3278. [CrossRef]
33. Zucca, E.; Rossi, D.; Bertoni, F. Marginal zone lymphomas. *Hematol. Oncol.* **2023**, *41* (Suppl. S1), 88–91. [CrossRef] [PubMed]
34. Cerhan, J.R.; Habermann, T.M. Epidemiology of Marginal Zone Lymphoma. *Ann. Lymphoma* **2021**, *5*, 1. [CrossRef] [PubMed]
35. Cheah, C.Y.; Seymour, J.F. Marginal zone lymphoma: 2023 update on diagnosis and management. *Am. J. Hematol.* **2023**, *98*, 1645–1657. [CrossRef] [PubMed]

36. Cheah, C.Y.; Zucca, E.; Rossi, D.; Habermann, T.M. Marginal zone lymphoma: Present status and future perspectives. *Haematologica* **2022**, *107*, 35–43. [CrossRef]
37. Varettoni, M.; Boveri, E.; Zibellini, S.; Tedeschi, A.; Candido, C.; Ferretti, V.V.; Rizzo, E.; Doni, E.; Merli, M.; Farina, L.; et al. Clinical and molecular characteristics of lymphoplasmacytic lymphoma not associated with an IgM monoclonal protein: A multicentric study of the Rete Ematologica Lombarda (REL) network. *Am. J. Hematol.* **2019**, *94*, 1193–1199. [CrossRef]
38. Minderman, M.; Lantermans, H.; van der Zwaan, C.; Hoogendijk, A.J.; van den Biggelaar, M.; Kersten, M.J.; Spaargaren, M.; Pals, S.T. The oncogenic human B-cell lymphoma MYD88 L265P mutation genocopies activation by phosphorylation at the Toll/interleukin-1 receptor (TIR) domain. *Blood Cancer J.* **2023**, *13*, 125. [CrossRef]
39. McMaster, M.L. The epidemiology of Waldenström macroglobulinemia. *Semin. Hematol.* **2023**, *60*, 65–72. [CrossRef]
40. Kaiser, L.M.; Hunter, Z.R.; Treon, S.P.; Buske, C. CXCR4 in Waldenström's Macroglobulinema: Chances and challenges. *Leukemia* **2021**, *35*, 333–345. [CrossRef]
41. Berendsen, M.R.; van Bladel, D.A.G.; Hesius, E.; Berganza Irusquieta, C.; Rijntjes, J.; van Spriel, A.B.; van der Spek, E.; Pruijt, J.F.M.; Kroeze, L.I.; Hebeda, K.M.; et al. Clonal Relationship and Mutation Analysis in Lymphoplasmacytic Lymphoma/Waldenström Macroglobulinemia Associated With Diffuse Large B-cell Lymphoma. *HemaSphere* **2023**, *7*, e976. [CrossRef]
42. Condoluci, A.; Rossi, D. Biology and Treatment of Richter Transformation. *Front. Oncol.* **2022**, *12*, 829983. [CrossRef]
43. Kaźmierczak, M.; Kroll-Balcerzak, R.; Balcerzak, A.; Czechowska, E.; Gil, L.; Sawiński, K.; Szczepaniak, A.; Komarnicki, M. Hodgkin lymphoma transformation of chronic lymphocytic leukemia: Cases report and discussion. *Med. Oncol.* **2014**, *31*, 800. [CrossRef] [PubMed]
44. Tsimberidou, A.M.; Keating, M.J. Richter syndrome: Biology, incidence, and therapeutic strategies. *Cancer* **2005**, *103*, 216–228. [CrossRef] [PubMed]
45. Mahmoud, A.M.; Gaidano, G.; Mouhssine, S. Immunological Aspects of Richter Syndrome: From Immune Dysfunction to Immunotherapy. *Cancers* **2023**, *15*, 1015. [CrossRef] [PubMed]
46. Rossi, D.; Spina, V.; Deambrogi, C.; Rasi, S.; Laurenti, L.; Stamatopoulos, K.; Arcaini, L.; Lucioni, M.; Rocque, G.B.; Xu-Monette, Z.Y.; et al. The genetics of Richter syndrome reveals disease heterogeneity and predicts survival after transformation. *Blood* **2011**, *117*, 3391–3401. [CrossRef]
47. Bertossi, C.; Robrecht, S.; Ligtvoet, R.; Schneider, C.; Yosifov, D.; Zenz, T.; Jebaray, B.; Riecke, A.; Goede, V.; Fürstenau, M.; et al. S101—The Landscape of TP53 Mutations and Their Prognostic Impact in Chronic Lymphocytic Leukemia—EHA2024 Hybrid Congress. *HemaSphere* **2024**, *8*, e104. [CrossRef]
48. Duffy, M.J.; Synnott, N.C.; Crown, J. Mutant p53 as a target for cancer treatment. *Eur. J. Cancer* **2017**, *83*, 258–265. [CrossRef]
49. Zhang, H.; Xu, J.; Long, Y.; Maimaitijiang, A.; Su, Z.; Li, W.; Li, J. Unraveling the Guardian: p53's Multifaceted Role in the DNA Damage Response and Tumor Treatment Strategies. *Int. J. Mol. Sci.* **2024**, *25*, 12928. [CrossRef]
50. Engeland, K. Cell cycle regulation: p53-p21-RB signaling. *Cell Death Differ.* **2022**, *29*, 946–960. [CrossRef]
51. Gatz, S.A.; Wiesmüller, L. p53 in recombination and repair. *Cell Death Differ.* **2006**, *13*, 1003–1016. [CrossRef]
52. Chipuk, J.E.; Kuwana, T.; Bouchier-Hayes, L.; Droin, N.M.; Newmeyer, D.D.; Schuler, M.; Green, D.R. Direct activation of Bax by p53 mediates mitochondrial membrane permeabilization and apoptosis. *Science* **2004**, *303*, 1010–1014. [CrossRef]
53. Wang, J.; Thomas, H.R.; Li, Z.; Yeo, N.C.F.; Scott, H.E.; Dang, N.; Hossain, M.I.; Andrabi, S.A.; Parant, J.M. Puma, noxa, p53, and p63 differentially mediate stress pathway induced apoptosis. *Cell Death Dis.* **2021**, *12*, 659. [CrossRef] [PubMed]
54. Fabbri, G.; Khiabanian, H.; Holmes, A.B.; Wang, J.; Messina, M.; Mullighan, C.G.; Pasqualucci, L.; Rabadan, R.; Dalla-Favera, R. Genetic lesions associated with chronic lymphocytic leukemia transformation to Richter syndrome. *J. Exp. Med.* **2013**, *210*, 2273–2288. [CrossRef] [PubMed]
55. Sherr, C.J. Ink4-Arf locus in cancer and aging. *Wiley Interdiscip. Rev. Dev. Biol.* **2012**, *1*, 731–741. [CrossRef] [PubMed]
56. Sherr, C.J. The INK4a/ARF network in tumour suppression. *Nat. Rev. Mol. Cell Biol.* **2001**, *2*, 731–737. [CrossRef]
57. Burkhart, D.L.; Sage, J. Cellular mechanisms of tumour suppression by the retinoblastoma gene. *Nat. Rev. Cancer* **2008**, *8*, 671–682. [CrossRef]
58. Pellarin, I.; Dall'Acqua, A.; Favero, A.; Segatto, I.; Rossi, V.; Crestan, N.; Karimbayli, J.; Belletti, B.; Baldassarre, G. Cyclin-dependent protein kinases and cell cycle regulation in biology and disease. *Signal Transduct. Target. Ther.* **2025**, *10*, 11. [CrossRef]
59. Hernández Borrero, L.J.; El-Deiry, W.S. Tumor suppressor p53: Biology, signaling pathways, and therapeutic targeting. *Biochim. Biophys. Acta Rev. Cancer* **2021**, *1876*, 188556. [CrossRef]
60. Chakraborty, S.; Martines, C.; Porro, F.; Fortunati, I.; Bonato, A.; Dimishkovska, M.; Piazza, S.; Yadav, B.S.; Innocenti, I.; Fazio, R.; et al. B-cell receptor signaling and genetic lesions in TP53 and CDKN2A/CDKN2B cooperate in Richter transformation. *Blood* **2021**, *138*, 1053–1066. [CrossRef]
61. Fabbri, G.; Rasi, S.; Rossi, D.; Trifonov, V.; Khiabanian, H.; Ma, J.; Grunn, A.; Fangazio, M.; Capello, D.; Monti, S.; et al. Analysis of the chronic lymphocytic leukemia coding genome: Role of NOTCH1 mutational activation. *J. Exp. Med.* **2011**, *208*, 1389–1401. [CrossRef]

62. Andersson, E.R.; Sandberg, R.; Lendahl, U. Notch signaling: Simplicity in design, versatility in function. *Development* **2011**, *138*, 3593–3612. [CrossRef]
63. Rosati, E.; Baldoni, S.; De Falco, F.; Del Papa, B.; Dorillo, E.; Rompietti, C.; Albi, E.; Falzetti, F.; Di Ianni, M.; Sportoletti, P. NOTCH1 Aberrations in Chronic Lymphocytic Leukemia. *Front. Oncol.* **2018**, *8*, 229. [CrossRef] [PubMed]
64. Zhou, B.; Lin, W.; Long, Y.; Yang, Y.; Zhang, H.; Wu, K.; Chu, Q. Notch signaling pathway: Architecture, disease, and therapeutics. *Signal Transduct. Target. Ther.* **2022**, *7*, 95. [CrossRef]
65. Di Ianni, M.; Baldoni, S.; Rosati, E.; Ciurnelli, R.; Cavalli, L.; Martelli, M.F.; Marconi, P.; Screpanti, I.; Falzetti, F. A new genetic lesion in B-CLL: A NOTCH1 PEST domain mutation. *Br. J. Haematol.* **2009**, *146*, 689–691. [CrossRef] [PubMed]
66. Arruga, F.; Gizdic, B.; Bologna, C.; Cignetto, S.; Buonincontri, R.; Serra, S.; Vaisitti, T.; Gizzi, K.; Vitale, N.; Garaffo, G.; et al. Mutations in NOTCH1 PEST domain orchestrate CCL19-driven homing of chronic lymphocytic leukemia cells by modulating the tumor suppressor gene DUSP22. *Leukemia* **2017**, *31*, 1882–1893. [CrossRef] [PubMed]
67. Rossi, D.; Rasi, S.; Spina, V.; Fangazio, M.; Monti, S.; Greco, M.; Ciardullo, C.; Famà, R.; Cresta, S.; Bruscaggin, A.; et al. Different impact of NOTCH1 and SF3B1 mutations on the risk of chronic lymphocytic leukemia transformation to Richter syndrome. *Br. J. Haematol.* **2012**, *158*, 426–429. [CrossRef]
68. Dhanasekaran, R.; Deutzmann, A.; Mahauad-Fernandez, W.D.; Hansen, A.S.; Gouw, A.M.; Felsher, D.W. The MYC oncogene—The grand orchestrator of cancer growth and immune evasion. *Nat. Rev. Clin. Oncol.* **2022**, *19*, 23–36. [CrossRef]
69. Palomero, T.; Lim, W.K.; Odom, D.T.; Sulis, M.L.; Real, P.J.; Margolin, A.; Barnes, K.C.; O'Neil, J.; Neuberg, D.; Weng, A.P.; et al. NOTCH1 directly regulates c-MYC and activates a feed-forward-loop transcriptional network promoting leukemic cell growth. *Proc. Natl. Acad. Sci. USA* **2006**, *103*, 18261–18266. [CrossRef]
70. De Paoli, L.; Cerri, M.; Monti, S.; Rasi, S.; Spina, V.; Bruscaggin, A.; Greco, M.; Ciardullo, C.; Famà, R.; Cresta, S.; et al. MGA, a suppressor of MYC, is recurrently inactivated in high risk chronic lymphocytic leukemia. *Leuk. Lymphoma* **2013**, *54*, 1087–1090. [CrossRef]
71. Martines, C.; Chakraborty, S.; Vujovikj, M.; Gobessi, S.; Vaisitti, T.; Deaglio, S.; Laurenti, L.; Dimovski, A.J.; Efremov, D.G. Macrophage- and BCR- but not TLR-derived signals support the growth of CLL and Richter Syndrome murine models in vivo. *Blood* **2022**, *140*, 2335–2347. [CrossRef]
72. Herishanu, Y.; Pérez-Galán, P.; Liu, D.; Biancotto, A.; Pittaluga, S.; Vire, B.; Gibellini, F.; Njuguna, N.; Lee, E.; Stennett, L.; et al. The lymph node microenvironment promotes B-cell receptor signaling, NF-κB activation, and tumor proliferation in chronic lymphocytic leukemia. *Blood* **2011**, *117*, 563–574. [CrossRef]
73. Ringshausen, I.; Schneller, F.; Bogner, C.; Hipp, S.; Duyster, J.; Peschel, C.; Decker, T. Constitutively activated phosphatidylinositol-3 kinase (PI-3K) is involved in the defect of apoptosis in B-CLL: Association with protein kinase Cdelta. *Blood* **2002**, *100*, 3741–3748. [CrossRef] [PubMed]
74. Kohlhaas, V.; Blakemore, S.J.; Al-Maarri, M.; Nickel, N.; Pal, M.; Roth, A.; Hövelmeyer, N.; Schäfer, S.C.; Knittel, G.; Lohneis, P.; et al. Active Akt signaling triggers CLL toward Richter transformation via overactivation of Notch1. *Blood* **2021**, *137*, 646–660. [CrossRef] [PubMed]
75. Dühren-von Minden, M.; Übelhart, R.; Schneider, D.; Wossning, T.; Bach, M.P.; Buchner, M.; Hofmann, D.; Surova, E.; Follo, M.; Köhler, F.; et al. Chronic lymphocytic leukaemia is driven by antigen-independent cell-autonomous signalling. *Nature* **2012**, *489*, 309–312. [CrossRef] [PubMed]
76. Burger, J.A.; Chiorazzi, N. B cell receptor signaling in chronic lymphocytic leukemia. *Trends Immunol.* **2013**, *34*, 592–601. [CrossRef]
77. Stevenson, F.K.; Forconi, F.; Kipps, T.J. Exploring the pathways to chronic lymphocytic leukemia. *Blood* **2021**, *138*, 827–835. [CrossRef]
78. Stamatopoulos, K.; Belessi, C.; Moreno, C.; Boudjograh, M.; Guida, G.; Smilevska, T.; Belhoul, L.; Stella, S.; Stavroyianni, N.; Crespo, M.; et al. Over 20% of patients with chronic lymphocytic leukemia carry stereotyped receptors: Pathogenetic implications and clinical correlations. *Blood* **2007**, *109*, 259–270. [CrossRef]
79. Gerousi, M.; Laidou, S.; Gemenetzi, K.; Stamatopoulos, K.; Chatzidimitriou, A. Distinctive Signaling Profiles With Distinct Biological and Clinical Implications in Aggressive CLL Subsets With Stereotyped B-Cell Receptor Immunoglobulin. *Front. Oncol.* **2021**, *11*, 771454. [CrossRef]
80. Rossi, D.; Spina, V.; Cerri, M.; Rasi, S.; Deambrogi, C.; De Paoli, L.; Laurenti, L.; Maffei, R.; Forconi, F.; Bertoni, F.; et al. Stereotyped B-cell receptor is an independent risk factor of chronic lymphocytic leukemia transformation to Richter syndrome. *Clin. Cancer Res.* **2009**, *15*, 4415–4422. [CrossRef]
81. Rossi, D.; Spina, V.; Bomben, R.; Rasi, S.; Dal-Bo, M.; Bruscaggin, A.; Rossi, F.M.; Monti, S.; Degan, M.; Ciardullo, C.; et al. Association between molecular lesions and specific B-cell receptor subsets in chronic lymphocytic leukemia. *Blood* **2013**, *121*, 4902–4905. [CrossRef]
82. Gounari, M.; Ntoufa, S.; Apollonio, B.; Papakonstantinou, N.; Ponzoni, M.; Chu, C.C.; Rossi, D.; Gaidano, G.; Chiorazzi, N.; Stamatopoulos, K.; et al. Excessive antigen reactivity may underlie the clinical aggressiveness of chronic lymphocytic leukemia stereotyped subset #8. *Blood* **2015**, *125*, 3580–3587. [CrossRef]

83. Nabki, J.; Al Deeban, B.; Sium, A.M.; Cosentino, C.; Almasri, M.; Awikeh, B.; Maher, N.; Bellia, M.; Dondolin, R.; Mouhssine, S.; et al. Immunoglobulin light chain mutational status refines IGHV prognostic value in identifying chronic lymphocytic leukemia patients with early treatment requirement. *Leukemia* **2024**, *39*, 643–649. [CrossRef] [PubMed]
84. Salmaninejad, A.; Valilou, S.F.; Shabgah, A.G.; Aslani, S.; Alimardani, M.; Pasdar, A.; Sahebkar, A. PD-1/PD-L1 pathway: Basic biology and role in cancer immunotherapy. *J. Cell Physiol.* **2019**, *234*, 16824–16837. [CrossRef] [PubMed]
85. Wang, Y.; Sinha, S.; Wellik, L.E.; Secreto, C.R.; Rech, K.L.; Call, T.G.; Parikh, S.A.; Kenderian, S.S.; Muchtar, E.; Hayman, S.R.; et al. Distinct immune signatures in chronic lymphocytic leukemia and Richter syndrome. *Blood Cancer J.* **2021**, *11*, 86. [CrossRef] [PubMed]
86. Behdad, A.; Griffin, B.; Chen, Y.-H.; Ma, S.; Kelemen, K.; Lu, X.; Chen, Q.C. PD-1 is highly expressed by neoplastic B-cells in Richter transformation. *Br. J. Haematol.* **2019**, *185*, 370–373. [CrossRef]
87. Li, X.; Du, H.; Zhan, S.; Liu, W.; Wang, Z.; Lan, J.; PuYang, L.; Wan, Y.; Qu, Q.; Wang, S.; et al. The interaction between the soluble programmed death ligand-1 (sPD-L1) and PD-1(+) regulator B cells mediates immunosuppression in triple-negative breast cancer. *Front. Immunol.* **2022**, *13*, 830606. [CrossRef]
88. Xiao, X.; Lao, X.M.; Chen, M.M.; Liu, R.X.; Wei, Y.; Ouyang, F.Z.; Chen, D.P.; Zhao, X.Y.; Zhao, Q.; Li, X.F.; et al. PD-1hi Identifies a Novel Regulatory B-cell Population in Human Hepatoma That Promotes Disease Progression. *Cancer Discov.* **2016**, *6*, 546–559. [CrossRef]
89. Wang, X.; Wang, G.; Wang, Z.; Liu, B.; Han, N.; Li, J.; Lu, C.; Liu, X.; Zhang, Q.; Yang, Q.; et al. PD-1-expressing B cells suppress CD4+ and CD8+ T cells via PD-1/PD-L1-dependent pathway. *Mol. Immunol.* **2019**, *109*, 20–26. [CrossRef]
90. Augé, H.; Notarantonio, A.-B.; Morizot, R.; Quinquenel, A.; Fornecker, L.-M.; Hergalant, S.; Feugier, P.; Broséus, J. Microenvironment Remodeling and Subsequent Clinical Implications in Diffuse Large B-Cell Histologic Variant of Richter Syndrome. *Front. Immunol.* **2020**, *11*, 594841. [CrossRef]
91. Triebel, F.; Jitsukawa, S.; Baixeras, E.; Roman-Roman, S.; Genevee, C.; Viegas-Pequignot, E.; Hercend, T. LAG-3, a novel lymphocyte activation gene closely related to CD4. *Journal of Experimental Medicine* **1990**, *171*, 1393–1405. [CrossRef]
92. Hannier, S.; Triebel, F. The MHC class II ligand lymphocyte activation gene-3 is co-distributed with CD8 and CD3–TCR molecules after their engagement by mAb or peptide–MHC class I complexes. *Int. Immunol.* **1999**, *11*, 1745–1752. [CrossRef]
93. Ruffo, E.; Wu, R.C.; Bruno, T.C.; Workman, C.J.; Vignali, D.A.A. Lymphocyte-activation gene 3 (LAG3): The next immune checkpoint receptor. *Semin. Immunol.* **2019**, *42*, 101305. [CrossRef] [PubMed]
94. Hemon, P.; Jean-Louis, F.; Ramgolam, K.; Brignone, C.; Viguier, M.; Bachelez, H.; Triebel, F.; Charron, D.; Aoudjit, F.; Al-Daccak, R.; et al. MHC Class II Engagement by Its Ligand LAG-3 (CD223) Contributes to Melanoma Resistance to Apoptosis. *J. Immunol.* **2011**, *186*, 5173–5183. [CrossRef] [PubMed]
95. Harjunpää, H.; Guillerey, C. TIGIT as an emerging immune checkpoint. *Clin. Exp. Immunol.* **2020**, *200*, 108–119. [CrossRef] [PubMed]
96. Huang, Z.; Qi, G.; Miller, J.S.; Zheng, S.G. CD226: An Emerging Role in Immunologic Diseases. *Front. Cell Dev. Biol.* **2020**, *8*, 564. [CrossRef]
97. Arruga, F.; Iannello, A.; Ioannou, N.; Todesco, A.M.; Coscia, M.; Moia, R.; Gaidano, G.; Allan, J.N.; Furman, R.; Vaisitti, T.; et al. The Tigit/CD226/CD155 Immunomodulatory Axis Is Deregulated in CLL and Contributes to B-Cell Anergy. *Blood* **2021**, *138*, 3718. [CrossRef]
98. Arruga, F.; Gyau, B.B.; Iannello, A.; Vitale, N.; Vaisitti, T.; Deaglio, S. Immune Response Dysfunction in Chronic Lymphocytic Leukemia: Dissecting Molecular Mechanisms and Microenvironmental Conditions. *Int. J. Mol. Sci.* **2020**, *21*, 1825. [CrossRef]
99. Tedeschi, A.; Frustaci, A.M.; Condoluci, A.; Coscia, M.; Chiarle, R.; Zinzani, P.L.; Motta, M.; Gaidano, G.; Quaresmini, G.; Scarfò, L.; et al. Atezolizumab, venetoclax, and obinutuzumab combination in Richter transformation diffuse large B-cell lymphoma (MOLTO): A multicentre, single-arm, phase 2 trial. *Lancet Oncol.* **2024**, *25*, 1298–1309. [CrossRef]
100. Al-Sawaf, O.; Ligtvoet, R.; Robrecht, S.; Stumpf, J.; Fink, A.M.; Tausch, E.; Schneider, C.; Boettcher, S.; Mikusko, M.; Ritgen, M.; et al. Tislelizumab plus zanubrutinib for Richter transformation: The phase 2 RT1 trial. *Nat. Med.* **2024**, *30*, 240–248. [CrossRef]
101. Parry, E.M.; Ten Hacken, E.; Wu, C.J. Richter syndrome: Novel insights into the biology of transformation. *Blood* **2023**, *142*, 11–22. [CrossRef]
102. Parry, E.M.; Leshchiner, I.; Guièze, R.; Johnson, C.; Tausch, E.; Parikh, S.A.; Lemvigh, C.; Broséus, J.; Hergalant, S.; Messer, C.; et al. Evolutionary history of transformation from chronic lymphocytic leukemia to Richter syndrome. *Nat. Med.* **2023**, *29*, 158–169. [CrossRef]
103. Rossi, D. XIII. Molecular pathogenesis of transformed lymphomas. *Hematol. Oncol.* **2015**, *33* (Suppl. S1), 70–74. [CrossRef] [PubMed]
104. Alcoceba, M.; García-Álvarez, M.; Okosun, J.; Ferrero, S.; Ladetto, M.; Fitzgibbon, J.; García-Sanz, R. Genetics of Transformed Follicular Lymphoma. *Hemato* **2022**, *3*, 615–633. [CrossRef]

105. Alhejaily, A.; Day, A.G.; Feilotter, H.E.; Baetz, T.; Lebrun, D.P. Inactivation of the CDKN2A tumor-suppressor gene by deletion or methylation is common at diagnosis in follicular lymphoma and associated with poor clinical outcome. *Clin. Cancer Res.* **2014**, *20*, 1676–1686. [CrossRef] [PubMed]
106. Elenitoba-Johnson, K.S.; Gascoyne, R.D.; Lim, M.S.; Chhanabai, M.; Jaffe, E.S.; Raffeld, M. Homozygous deletions at chromosome 9p21 involving p16 and p15 are associated with histologic progression in follicle center lymphoma. *Blood* **1998**, *91*, 4677–4685. [CrossRef]
107. Kridel, R.; Chan, F.C.; Mottok, A.; Boyle, M.; Farinha, P.; Tan, K.; Meissner, B.; Bashashati, A.; McPherson, A.; Roth, A.; et al. Histological Transformation and Progression in Follicular Lymphoma: A Clonal Evolution Study. *PLoS Med.* **2016**, *13*, e1002197. [CrossRef]
108. Musilova, K.; Devan, J.; Cerna, K.; Seda, V.; Pavlasova, G.; Sharma, S.; Oppelt, J.; Pytlik, R.; Prochazka, V.; Prouzova, Z.; et al. miR-150 downregulation contributes to the high-grade transformation of follicular lymphoma by upregulating FOXP1 levels. *Blood* **2018**, *132*, 2389–2400. [CrossRef]
109. O'Shea, D.; O'Riain, C.; Taylor, C.; Waters, R.; Carlotti, E.; Macdougall, F.; Gribben, J.; Rosenwald, A.; Ott, G.; Rimsza, L.M.; et al. The presence of TP53 mutation at diagnosis of follicular lymphoma identifies a high-risk group of patients with shortened time to disease progression and poorer overall survival. *Blood* **2008**, *112*, 3126–3129. [CrossRef]
110. Lo Coco, F.; Gaidano, G.; Louie, D.C.; Offit, K.; Chaganti, R.S.; Dalla-Favera, R. p53 mutations are associated with histologic transformation of follicular lymphoma. *Blood* **1993**, *82*, 2289–2295. [CrossRef]
111. Davies, A.J.; Lee, A.M.; Taylor, C.; Clear, A.J.; Goff, L.K.; Iqbal, S.; Cuthbert-Heavens, D.; Calaminici, M.; Norton, A.J.; Lister, T.A.; et al. A limited role for TP53 mutation in the transformation of follicular lymphoma to diffuse large B-cell lymphoma. *Leukemia* **2005**, *19*, 1459–1465. [CrossRef]
112. Yang, H.; Green, M.R. Epigenetic Programing of B-Cell Lymphoma by BCL6 and Its Genetic Deregulation. *Front. Cell Dev. Biol.* **2019**, *7*, 272. [CrossRef]
113. Phan, R.T.; Dalla-Favera, R. The BCL6 proto-oncogene suppresses p53 expression in germinal-centre B cells. *Nature* **2004**, *432*, 635–639. [CrossRef] [PubMed]
114. Ranuncolo, S.M.; Polo, J.M.; Dierov, J.; Singer, M.; Kuo, T.; Greally, J.; Green, R.; Carroll, M.; Melnick, A. Bcl-6 mediates the germinal center B cell phenotype and lymphomagenesis through transcriptional repression of the DNA-damage sensor ATR. *Nat. Immunol.* **2007**, *8*, 705–714. [CrossRef] [PubMed]
115. Phan, R.T.; Saito, M.; Basso, K.; Niu, H.; Dalla-Favera, R. BCL6 interacts with the transcription factor Miz-1 to suppress the cyclin-dependent kinase inhibitor p21 and cell cycle arrest in germinal center B cells. *Nat. Immunol.* **2005**, *6*, 1054–1060. [CrossRef] [PubMed]
116. Ye, B.H.; Lista, F.; Lo Coco, F.; Knowles, D.M.; Offit, K.; Chaganti, R.S.; Dalla-Favera, R. Alterations of a zinc finger-encoding gene, BCL-6, in diffuse large-cell lymphoma. *Science* **1993**, *262*, 747–750. [CrossRef]
117. Kim, H.; Kim, H.J.; Kim, S.H. Diagnostic Approach for Double-Hit and Triple-Hit Lymphoma Based on Immunophenotypic and Cytogenetic Characteristics of Bone Marrow Specimens. *Ann. Lab. Med.* **2020**, *40*, 361–369. [CrossRef]
118. Kridel, R.; Mottok, A.; Farinha, P.; Ben-Neriah, S.; Ennishi, D.; Zheng, Y.; Chavez, E.A.; Shulha, H.P.; Tan, K.; Chan, F.C.; et al. Cell of origin of transformed follicular lymphoma. *Blood* **2015**, *126*, 2118–2127. [CrossRef]
119. Akasaka, T.; Lossos, I.S.; Levy, R. BCL6 gene translocation in follicular lymphoma: A harbinger of eventual transformation to diffuse aggressive lymphoma. *Blood* **2003**, *102*, 1443–1448. [CrossRef]
120. Iqbal, J.; Greiner, T.C.; Patel, K.; Dave, B.J.; Smith, L.; Ji, J.; Wright, G.; Sanger, W.G.; Pickering, D.L.; Jain, S.; et al. Distinctive patterns of BCL6 molecular alterations and their functional consequences in different subgroups of diffuse large B-cell lymphoma. *Leukemia* **2007**, *21*, 2332–2343. [CrossRef]
121. Bouska, A.; McKeithan, T.W.; Deffenbacher, K.E.; Lachel, C.; Wright, G.W.; Iqbal, J.; Smith, L.M.; Zhang, W.; Kucuk, C.; Rinaldi, A.; et al. Genome-wide copy-number analyses reveal genomic abnormalities involved in transformation of follicular lymphoma. *Blood* **2014**, *123*, 1681–1690. [CrossRef]
122. Liu, T.; Zhang, L.; Joo, D.; Sun, S.C. NF-κB signaling in inflammation. *Signal Transduct. Target. Ther.* **2017**, *2*, 17023. [CrossRef]
123. Burns, K.; Martinon, F.; Esslinger, C.; Pahl, H.; Schneider, P.; Bodmer, J.L.; Di Marco, F.; French, L.; Tschopp, J. MyD88, an adapter protein involved in interleukin-1 signaling. *J. Biol. Chem.* **1998**, *273*, 12203–12209. [CrossRef] [PubMed]
124. Dunne, A.; Ejdeback, M.; Ludidi, P.L.; O'Neill, L.A.; Gay, N.J. Structural complementarity of Toll/interleukin-1 receptor domains in Toll-like receptors and the adaptors Mal and MyD88. *J. Biol. Chem.* **2003**, *278*, 41443–41451. [CrossRef] [PubMed]
125. Qian, Y.; Commane, M.; Ninomiya-Tsuji, J.; Matsumoto, K.; Li, X. IRAK-mediated translocation of TRAF6 and TAB2 in the interleukin-1-induced activation of NFkappa B. *J. Biol. Chem.* **2001**, *276*, 41661–41667. [CrossRef] [PubMed]
126. Deguine, J.; Barton, G.M. MyD88: A central player in innate immune signaling. *F1000Prime Rep.* **2014**, *6*, 97. [CrossRef]
127. Bouska, A.; Zhang, W.; Gong, Q.; Iqbal, J.; Scuto, A.; Vose, J.; Ludvigsen, M.; Fu, K.; Weisenburger, D.D.; Greiner, T.C.; et al. Combined copy number and mutation analysis identifies oncogenic pathways associated with transformation of follicular lymphoma. *Leukemia* **2017**, *31*, 83–91. [CrossRef]

128. Kwiecinska, A.; Ichimura, K.; Berglund, M.; Dinets, A.; Sulaiman, L.; Collins, V.P.; Larsson, C.; Porwit, A.; Lagercrantz, S.B. Amplification of 2p as a genomic marker for transformation in lymphoma. *Genes Chromosomes Cancer* **2014**, *53*, 750–768. [CrossRef]
129. Davies, A.J.; Rosenwald, A.; Wright, G.; Lee, A.; Last, K.W.; Weisenburger, D.D.; Chan, W.C.; Delabie, J.; Braziel, R.M.; Campo, E.; et al. Transformation of follicular lymphoma to diffuse large B-cell lymphoma proceeds by distinct oncogenic mechanisms. *Br. J. Haematol.* **2007**, *136*, 286–293. [CrossRef]
130. Cai, G.; Freeman, G.J. The CD160, BTLA, LIGHT/HVEM pathway: A bidirectional switch regulating T-cell activation. *Immunol. Rev.* **2009**, *229*, 244–258. [CrossRef]
131. Murphy, K.M.; Nelson, C.A.; Sedý, J.R. Balancing co-stimulation and inhibition with BTLA and HVEM. *Nat. Rev. Immunol.* **2006**, *6*, 671–681. [CrossRef]
132. O'Shea, D.; O'Riain, C.; Gupta, M.; Waters, R.; Yang, Y.; Wrench, D.; Gribben, J.; Rosenwald, A.; Ott, G.; Rimsza, L.M.; et al. Regions of acquired uniparental disomy at diagnosis of follicular lymphoma are associated with both overall survival and risk of transformation. *Blood* **2009**, *113*, 2298–2301. [CrossRef]
133. García-Álvarez, M.; Alonso-Álvarez, S.; Prieto-Conde, I.; Jiménez, C.; Sarasquete, M.E.; Chillón, M.C.; Medina, A.; Balanzategui, A.; Maldonado, R.; Antón, A.; et al. Genetic complexity impacts the clinical outcome of follicular lymphoma patients. *Blood Cancer J.* **2021**, *11*, 11. [CrossRef] [PubMed]
134. Cheung, K.J.; Shah, S.P.; Steidl, C.; Johnson, N.; Relander, T.; Telenius, A.; Lai, B.; Murphy, K.P.; Lam, W.; Al-Tourah, A.J.; et al. Genome-wide profiling of follicular lymphoma by array comparative genomic hybridization reveals prognostically significant DNA copy number imbalances. *Blood* **2009**, *113*, 137–148. [CrossRef] [PubMed]
135. Kotsiou, E.; Okosun, J.; Besley, C.; Iqbal, S.; Matthews, J.; Fitzgibbon, J.; Gribben, J.G.; Davies, J.K. TNFRSF14 aberrations in follicular lymphoma increase clinically significant allogeneic T-cell responses. *Blood* **2016**, *128*, 72–81. [CrossRef] [PubMed]
136. Boice, M.; Salloum, D.; Mourcin, F.; Sanghvi, V.; Amin, R.; Oricchio, E.; Jiang, M.; Mottok, A.; Denis-Lagache, N.; Ciriello, G.; et al. Loss of the HVEM Tumor Suppressor in Lymphoma and Restoration by Modified CAR-T Cells. *Cell* **2016**, *167*, 405–418.e413. [CrossRef]
137. Carreras, J.; Lopez-Guillermo, A.; Kikuti, Y.Y.; Itoh, J.; Masashi, M.; Ikoma, H.; Tomita, S.; Hiraiwa, S.; Hamoudi, R.; Rosenwald, A.; et al. High TNFRSF14 and low BTLA are associated with poor prognosis in Follicular Lymphoma and in Diffuse Large B-cell Lymphoma transformation. *J. Clin. Exp. Hematop.* **2019**, *59*, 1–16. [CrossRef]
138. Casulo, C.; Friedberg, J. Transformation of marginal zone lymphoma (and association with other lymphomas). *Best. Pr. Res. Clin. Haematol.* **2017**, *30*, 131–138. [CrossRef]
139. Parry, M.; Rose-Zerilli, M.J.; Ljungström, V.; Gibson, J.; Wang, J.; Walewska, R.; Parker, H.; Parker, A.; Davis, Z.; Gardiner, A.; et al. Genetics and Prognostication in Splenic Marginal Zone Lymphoma: Revelations from Deep Sequencing. *Clin. Cancer Res.* **2015**, *21*, 4174–4183. [CrossRef]
140. Rossi, D.; Trifonov, V.; Fangazio, M.; Bruscaggin, A.; Rasi, S.; Spina, V.; Monti, S.; Vaisitti, T.; Arruga, F.; Famà, R.; et al. The coding genome of splenic marginal zone lymphoma: Activation of NOTCH2 and other pathways regulating marginal zone development. *J. Exp. Med.* **2012**, *209*, 1537–1551. [CrossRef]
141. Kiel, M.J.; Velusamy, T.; Betz, B.L.; Zhao, L.; Weigelin, H.G.; Chiang, M.Y.; Huebner-Chan, D.R.; Bailey, N.G.; Yang, D.T.; Bhagat, G.; et al. Whole-genome sequencing identifies recurrent somatic NOTCH2 mutations in splenic marginal zone lymphoma. *J. Exp. Med.* **2012**, *209*, 1553–1565. [CrossRef]
142. Bray, S.J. Notch signalling: A simple pathway becomes complex. *Nat. Rev. Mol. Cell Biol.* **2006**, *7*, 678–689. [CrossRef]
143. Sanchez-Irizarry, C.; Carpenter, A.C.; Weng, A.P.; Pear, W.S.; Aster, J.C.; Blacklow, S.C. Notch subunit heterodimerization and prevention of ligand-independent proteolytic activation depend, respectively, on a novel domain and the LNR repeats. *Mol. Cell Biol.* **2004**, *24*, 9265–9273. [CrossRef] [PubMed]
144. Trøen, G.; Wlodarska, I.; Warsame, A.; Hernández Llodrà, S.; De Wolf-Peeters, C.; Delabie, J. NOTCH2 mutations in marginal zone lymphoma. *Haematologica* **2008**, *93*, 1107–1109. [CrossRef] [PubMed]
145. Grau, M.; López, C.; Navarro, A.; Frigola, G.; Nadeu, F.; Clot, G.; Bastidas-Mora, G.; Alcoceba, M.; Baptista, M.J.; Blanes, M.; et al. Unraveling the genetics of transformed splenic marginal zone lymphoma. *Blood Adv.* **2023**, *7*, 3695–3709. [CrossRef] [PubMed]
146. Salido, M.; Baró, C.; Oscier, D.; Stamatopoulos, K.; Dierlamm, J.; Matutes, E.; Traverse-Glehen, A.; Berger, F.; Felman, P.; Thieblemont, C.; et al. Cytogenetic aberrations and their prognostic value in a series of 330 splenic marginal zone B-cell lymphomas: A multicenter study of the Splenic B-Cell Lymphoma Group. *Blood* **2010**, *116*, 1479–1488. [CrossRef]
147. Rinaldi, A.; Mian, M.; Chigrinova, E.; Arcaini, L.; Bhagat, G.; Novak, U.; Rancoita, P.M.; De Campos, C.P.; Forconi, F.; Gascoyne, R.D.; et al. Genome-wide DNA profiling of marginal zone lymphomas identifies subtype-specific lesions with an impact on the clinical outcome. *Blood* **2011**, *117*, 1595–1604. [CrossRef]
148. Martínez, N.; Almaraz, C.; Vaqué, J.P.; Varela, I.; Derdak, S.; Beltran, S.; Mollejo, M.; Campos-Martin, Y.; Agueda, L.; Rinaldi, A.; et al. Whole-exome sequencing in splenic marginal zone lymphoma reveals mutations in genes involved in marginal zone differentiation. *Leukemia* **2014**, *28*, 1334–1340. [CrossRef]

149. Fresquet, V.; Robles, E.F.; Parker, A.; Martinez-Useros, J.; Mena, M.; Malumbres, R.; Agirre, X.; Catarino, S.; Arteta, D.; Osaba, L.; et al. High-throughput sequencing analysis of the chromosome 7q32 deletion reveals IRF5 as a potential tumour suppressor in splenic marginal-zone lymphoma. *Br. J. Haematol.* **2012**, *158*, 712–726. [CrossRef]
150. Pujari, R.; Hunte, R.; Khan, W.N.; Shembade, N. A20-mediated negative regulation of canonical NF-κB signaling pathway. *Immunol. Res.* **2013**, *57*, 166–171. [CrossRef]
151. Jaramillo Oquendo, C.; Parker, H.; Oscier, D.; Ennis, S.; Gibson, J.; Strefford, J.C. Systematic Review of Somatic Mutations in Splenic Marginal Zone Lymphoma. *Sci. Rep.* **2019**, *9*, 10444. [CrossRef]
152. Bonfiglio, F.; Bruscaggin, A.; Guidetti, F.; Terzi di Bergamo, L.; Faderl, M.; Spina, V.; Condoluci, A.; Bonomini, L.; Forestieri, G.; Koch, R.; et al. Genetic and phenotypic attributes of splenic marginal zone lymphoma. *Blood* **2022**, *139*, 732–747. [CrossRef]
153. Wittner, J.; Schuh, W. Krüppel-like factor 2: A central regulator of B cell differentiation and plasma cell homing. *Front. Immunol.* **2023**, *14*, 1172641. [CrossRef]
154. Spina, V.; Khiabanian, H.; Messina, M.; Monti, S.; Cascione, L.; Bruscaggin, A.; Spaccarotella, E.; Holmes, A.B.; Arcaini, L.; Lucioni, M.; et al. The genetics of nodal marginal zone lymphoma. *Blood* **2016**, *128*, 1362–1373. [CrossRef] [PubMed]
155. Clipson, A.; Wang, M.; de Leval, L.; Ashton-Key, M.; Wotherspoon, A.; Vassiliou, G.; Bolli, N.; Grove, C.; Moody, S.; Escudero-Ibarz, L.; et al. KLF2 mutation is the most frequent somatic change in splenic marginal zone lymphoma and identifies a subset with distinct genotype. *Leukemia* **2015**, *29*, 1177–1185. [CrossRef] [PubMed]
156. Froimchuk, E.; Jang, Y.; Ge, K. Histone H3 lysine 4 methyltransferase KMT2D. *Gene* **2017**, *627*, 337–342. [CrossRef] [PubMed]
157. Arribas, A.J.; Rinaldi, A.; Mensah, A.A.; Kwee, I.; Cascione, L.; Robles, E.F.; Martinez-Climent, J.A.; Oscier, D.; Arcaini, L.; Baldini, L.; et al. DNA methylation profiling identifies two splenic marginal zone lymphoma subgroups with different clinical and genetic features. *Blood* **2015**, *125*, 1922–1931. [CrossRef]
158. Arribas, A.J.; Bertoni, F. Methylation patterns in marginal zone lymphoma. *Best. Pr. Res. Clin. Haematol.* **2017**, *30*, 24–31. [CrossRef]
159. Sagaert, X.; de Paepe, P.; Libbrecht, L.; Vanhentenrijk, V.; Verhoef, G.; Thomas, J.; Wlodarska, I.; De Wolf-Peeters, C. Forkhead box protein P1 expression in mucosa-associated lymphoid tissue lymphomas predicts poor prognosis and transformation to diffuse large B-cell lymphoma. *J. Clin. Oncol.* **2006**, *24*, 2490–2497. [CrossRef]
160. Streubel, B.; Vinatzer, U.; Lamprecht, A.; Raderer, M.; Chott, A. T(3;14)(p14.1;q32) involving IGH and FOXP1 is a novel recurrent chromosomal aberration in MALT lymphoma. *Leukemia* **2005**, *19*, 652–658. [CrossRef]
161. Haralambieva, E.; Adam, P.; Ventura, R.; Katzenberger, T.; Kalla, J.; Höller, S.; Hartmann, M.; Rosenwald, A.; Greiner, A.; Muller-Hermelink, H.K.; et al. Genetic rearrangement of FOXP1 is predominantly detected in a subset of diffuse large B-cell lymphomas with extranodal presentation. *Leukemia* **2006**, *20*, 1300–1303. [CrossRef]
162. Barrans, S.L.; Fenton, J.A.; Banham, A.; Owen, R.G.; Jack, A.S. Strong expression of FOXP1 identifies a distinct subset of diffuse large B-cell lymphoma (DLBCL) patients with poor outcome. *Blood* **2004**, *104*, 2933–2935. [CrossRef]
163. Qian, L.; Soderquist, C.; Schrank-Hacker, A.; Strauser, H.; Dupoux, V.; Tang, C.N.; Smith, J.R.; Sun, A.; Majumdar, S.; Nguyen, T.; et al. Deletion 20q12 is associated with histological transformation of nodal marginal zone lymphoma to diffuse large B-cell lymphoma. *Am. J. Hematol.* **2020**, *95*, 238–244. [CrossRef] [PubMed]
164. Treon, S.P.; Gustine, J.; Xu, L.; Manning, R.J.; Tsakmaklis, N.; Demos, M.; Meid, K.; Guerrera, M.L.; Munshi, M.; Chan, G.; et al. MYD88 wild-type Waldenstrom Macroglobulinaemia: Differential diagnosis, risk of histological transformation, and overall survival. *Br. J. Haematol.* **2018**, *180*, 374–380. [CrossRef] [PubMed]
165. Hunter, Z.R.; Xu, L.; Yang, G.; Zhou, Y.; Liu, X.; Cao, Y.; Manning, R.J.; Tripsas, C.; Patterson, C.J.; Sheehy, P.; et al. The genomic landscape of Waldenstrom macroglobulinemia is characterized by highly recurring MYD88 and WHIM-like CXCR4 mutations, and small somatic deletions associated with B-cell lymphomagenesis. *Blood* **2014**, *123*, 1637–1646. [CrossRef] [PubMed]
166. Varettoni, M.; Zibellini, S.; Defrancesco, I.; Ferretti, V.V.; Rizzo, E.; Malcovati, L.; Gallì, A.; Della Porta, M.G.; Boveri, E.; Arcaini, L.; et al. Pattern of somatic mutations in patients with Waldenström macroglobulinemia or IgM monoclonal gammopathy of undetermined significance. *Haematologica* **2017**, *102*, 2077–2085. [CrossRef]
167. Rossi, D.; Zlotnik, A. The biology of chemokines and their receptors. *Annu. Rev. Immunol.* **2000**, *18*, 217–242. [CrossRef]
168. Cao, Y.; Hunter, Z.R.; Liu, X.; Xu, L.; Yang, G.; Chen, J.; Patterson, C.J.; Tsakmaklis, N.; Kanan, S.; Rodig, S.; et al. The WHIM-like CXCR4(S338X) somatic mutation activates AKT and ERK, and promotes resistance to ibrutinib and other agents used in the treatment of Waldenstrom's Macroglobulinemia. *Leukemia* **2015**, *29*, 169–176. [CrossRef]
169. Wang, H.; Guo, M.; Wei, H.; Chen, Y. Targeting p53 pathways: Mechanisms, structures and advances in therapy. *Signal Transduct. Target. Ther.* **2023**, *8*, 92. [CrossRef]
170. Weber, L.I.; Hartl, M. Strategies to target the cancer driver MYC in tumor cells. *Front. Oncol.* **2023**, *13*, 1142111. [CrossRef]
171. Yu, H.; Lin, L.; Zhang, Z.; Zhang, H.; Hu, H. Targeting NF-κB pathway for the therapy of diseases: Mechanism and clinical study. *Signal Transduct. Target. Ther.* **2020**, *5*, 209. [CrossRef]
172. Teierle, S.M.; Huang, Y.; Kittai, A.S.; Bhat, S.A.; Grever, M.R.; Rogers, K.A.; Zhao, W.; Jones, D.; Byrd, J.C.; Avenarius, M.R.; et al. FISH Analysis of CDKN2A/B Deletion in Chronic Lymphocytic Leukemia. *Blood* **2022**, *140*, 7016–7017. [CrossRef]

173. Klintman, J.; Appleby, N.; Stamatopoulos, B.; Ridout, K.; Eyre, T.A.; Robbe, P.; Pascua, L.L.; Knight, S.J.L.; Dreau, H.; Cabes, M.; et al. Genomic and transcriptomic correlates of Richter transformation in chronic lymphocytic leukemia. *Blood* **2021**, *137*, 2800–2816. [CrossRef]
174. Parker, H.; Rose-Zerilli, M.J.J.; Larrayoz, M.; Clifford, R.; Edelmann, J.; Blakemore, S.; Gibson, J.; Wang, J.; Ljungström, V.; Wojdacz, T.K.; et al. Genomic disruption of the histone methyltransferase SETD2 in chronic lymphocytic leukaemia. *Leukemia* **2016**, *30*, 2179–2186. [CrossRef]
175. Mouhssine, S.; Maher, N.; Gaidano, G. A STEP ahead for CAR-T cell therapy of large B cell lymphoma: Understanding the molecular determinants of resistance. *Transl. Cancer Res.* **2023**, *12*, 2970–2975. [CrossRef]
176. Sworder, B.J.; Kurtz, D.M.; Alig, S.K.; Frank, M.J.; Shukla, N.; Garofalo, A.; Macaulay, C.W.; Shahrokh Esfahani, M.; Olsen, M.N.; Hamilton, J.; et al. Determinants of resistance to engineered T cell therapies targeting CD19 in large B cell lymphomas. *Cancer Cell* **2023**, *41*, 210–225.e215. [CrossRef]

cancers

MDPI

Review

Pathobiological Features and Therapeutic Opportunities Linked to TNF Family Member Expression in Classic Hodgkin Lymphoma

Mohamed N. Alibrahim [1,*], Annunziata Gloghini [2] and Antonino Carbone [3,*]

1 Faculty of Medicine, Zagazig University, Zagazig 44511, Egypt
2 Department of Avanced Pathology, Fondazione IRCCS, Istituto Nazionale dei Tumori, 20133 Milano, Italy; annunziata.gloghini@istitutotumori.mi.it
3 Centro di Riferimento Oncologico, Istituto di Ricovero e Cura a Carattere Scientifico, National Cancer Institute, 33081 Aviano, Italy
* Correspondence: 20512020101186@medicine.zu.edu.eg (M.N.A.); acarbone@cro.it (A.C.); Tel.: +39-0434-659085 (A.C.)

Simple Summary: The tumor necrosis factor (TNF) family plays a crucial role in the pathogenesis of Hodgkin lymphoma. Hodgkin Reed–Sternberg cells, the diagnostic hallmark of the disease, exhibit overexpression of TNF receptor family members such as CD30 and CD40. These molecules have a critical roles in the survival and proliferation of Hodgkin Reed–Sternberg cells. Therefore, targeting these TNF receptors represents a promising therapeutic strategy. Therapies that target CD30 have already shown efficacy in clinical settings. Checkpoint inhibitors such as nivolumab and pembrolizumab have demonstrated high response rates in patients with classic Hodgkin lymphoma, particularly in those who have not responded to conventional therapies. By understanding how TNF signaling interacts with immune checkpoints, researchers can design more effective treatment regimens that simultaneously target multiple pathways.

Abstract: The tumor necrosis factor (TNF) family, which includes 19 ligands and 29 receptors, influences cellular proliferation, differentiation, and apoptosis. The TNF family plays a crucial role in the pathogenesis of Hodgkin lymphoma (HL), particularly through its influence on the tumor microenvironment (TME). Hodgkin Reed–Sternberg (HRS) cells, the hallmark of classic HL (cHL), exhibit overexpression of TNF receptor family members such as CD30 and CD40. Given the critical roles of CD30 and CD40 in the survival and proliferation of HRS cells within the TME, targeting these TNF receptors represents a promising therapeutic strategy; therapies that target CD30 have already shown efficacy in clinical settings. The programmed death-1 (PD-1)/programmed death-ligand 1 (PD-L1) axis plays a crucial role in immune evasion by HRS cells, which express PD-L1 that interacts with PD-1 on T cells, leading to T cell exhaustion and a diminished immune response against the tumor. By blocking this interaction, checkpoint inhibitors such as nivolumab and pembrolizumab have demonstrated high response rates in patients with cHL, particularly in those who have not responded to conventional therapies. The integration of immune checkpoint inhibitors (ICIs) with standard chemotherapy regimens has improved outcomes for patients with advanced-stage cHL. By understanding how TNF signaling interacts with immune checkpoints, researchers can design more effective treatment regimens that simultaneously target multiple pathways. Combining TNF inhibitors with checkpoint blockade therapies may enhance the overall anti-tumor response by addressing both direct tumor signaling and the immune evasion mechanisms employed by tumor cells.

Keywords: TNF family; CD40; CD30; classic Hodgkin lymphoma; tumor microenvironment; pathobiology; new therapies

1. Introduction

This review adopts a narrative approach to synthesize existing evidence on the role of the tumor necrosis factor (TNF) family in the pathogenesis and treatment of Hodgkin lymphoma (HL). The literature search was conducted in November 2024 using major academic databases, including PubMed and Google Scholar. Articles were included based on their relevance to the biological functions of TNF family members, their interactions with Hodgkin Reed–Sternberg cells (HRS), and their therapeutic implications in HL. This narrative review aims to provide a comprehensive discussion of these topics to guide future research and clinical applications.

The TNF family comprises a complex group of cytokines and their receptors that play critical roles in the regulation of immune responses, inflammation, and cell survival. This family includes 19 ligands and 29 receptors, which interact through a complex signaling network that influences various cellular activities, including proliferation, differentiation, and apoptosis [1–5] (Table 1).

Table 1. Key members of the TNF ligand and receptor families [1,3].

Ligand	Ligand Symbol	Receptor(s)	Receptor Symbol(s)
tumor necrosis factor-alpha	TNF-α (TNFSF2) *	TNFR1, TNFR2	TNFRSF1A, TNFRSF1B
lymphotoxin-alpha	LT-α (TNFSF1)	TNFR1	TNFRSF1A
lymphotoxin-beta	LT-β (TNFSF3)	LTβR	TNFRSF3
OX40 ligand	OX40L (TNFSF4)	OX40	TNFRSF4
CD40 ligand *	CD40L (TNFSF5)	CD40 *	TNFRSF5
fas ligand	FasL (TNFSF6)	Fas	TNFRSF6
CD27 ligand	CD27L (TNFSF7)	CD27	TNFRSF7
CD30 ligand *	CD30L (TNFSF8)	CD30 *	TNFRSF8
4-1BB ligand	4-1BBL (TNFSF9)	4-1BB	TNFRSF9
TNF-related apoptosis-inducing ligand	TRAIL (TNFSF10)	TRAIL-R1, TRAIL-R2, TRAIL-R3, TRAIL-R4	TNFRSF10A, TNFRSF10B, TNFRSF10C, TNFRSF10D
receptor activator of nuclear factor kappa-B ligand	RANKL (TNFSF11)	RANK	TNFRSF11A
TNF-related weak inducer of apoptosis	TWEAK (TNFSF12)	Fn14	TNFRSF12A
a proliferation-inducing ligand *	APRIL (TNFSF13) *	BCMA, TACI	TNFRSF13A, TNFRSF13B
B-cell-activating factor *	BAFF (TNFSF13B)	BAFF-R *, BCMA, TACI	TNFRSF13C, TNFRSF13A, TNFRSF13B
LIGHT	LIGHT (TNFSF14)	LTβR, HVEM, DcR3	TNFRSF3, TNFRSF14, TNFRSF6B
TL1A	TL1A (TNFSF15)	DR3, DcR3	TNFRSF25, TNFRSF6B
glucocorticoid-induced TNFR ligand	GITRL (TNFSF18)	GITR	TNFRSF18
ectodysplasin A1	EDA-A1 (TNFSF14)	EDAR	TNFRSF27
ectodysplasin A2	EDA-A2	XEDAR	TNFRSF27

This table outlines key members of the TNF ligand and receptor families, which play essential roles in immune regulation, cell survival, and apoptosis. Receptors or families marked with an asterisk (*) are specifically discussed in this study, highlighting their roles in classic Hodgkin lymphoma and their therapeutic potential.

One of the most well-studied members of the TNF family is TNF-α, which is known for its pro-inflammatory properties [6,7]; TNF-α can exert both tumor-suppressive and tumor-promoting effects depending on the context, highlighting its dual role in cancer biology, which has led to the exploration of TNF-α as a target for immunotherapeutic strategies [8]. For instance, TNF-α binding to Tumor Necrosis Factor Receptor 1 (TNFR1) typically induces apoptosis in tumor cells and inflammation, while TNFR2 signaling can promote cell survival and proliferation, thus contributing to tumor progression [8,9].

The TNF family also includes several other important members such as BAFF (B-cell-activating factor) and APRIL (a proliferation-inducing ligand), which are crucial for B cell survival and differentiation [10,11]. Dysregulated BAFF expression has been shown in murine models to lead to autoimmune-like conditions, including an SLE-like phenotype, through its impact on B cell homeostasis and autoantibody production [12]. TNF family members are essential for the development and activation of both innate and adaptive immune responses [13]. They facilitate T cell co-stimulation and B cell activation, thereby enhancing the immune response against pathogens [6,10,13]. The signaling pathways activated by TNF receptors, such as nuclear factor-kappa B (NF-κB), are critical for orchestrating these immune processes [4]. Additionally, the TNF family has been implicated in therapeutic strategies for cancer and inflammatory diseases, highlighting their significance in both health and disease [14,15].

The TNF family plays a crucial role in the pathogenesis of HL, particularly through its influence on the TME. HRS cells, the hallmark of cHL, exhibit overexpression of TNF receptor family members such as CD30 and CD40 [16–20], which are integral to their survival and proliferation within the TME. This overexpression is linked to the constitutive activation of NF-κB and alterations in the Janus kinase/signal transducer and activation of transcription (Jak/STAT) signaling pathway, contributing to immune evasion and tumor growth [21,22]. Furthermore, the TME in classic HL is characterized by a diverse group of inflammatory and immune cells, which interact with HRS cells and can modulate responses to therapies, including checkpoint blockade [23]. The unique characteristics of the TME in HL, influenced by TNF family signaling, underscore the complexity of therapeutic strategies aimed at this malignancy [23,24].

Starting from 1985, significant advancements in the understanding of the pathobiology of HL have been achieved [19]. Since 1995, research has focused on the characterization of HL, particularly on the expression of functional CD40 on HRS cells. CD40 was found to play a crucial role in tumor growth and signaling pathways [25]. This work laid the groundwork for understanding the TME in HL. By 2010, the emphasis had shifted towards the complex interactions within the TME, highlighting how immune cells contribute to tumor growth and immune evasion in classic HL (cHL) [21].

The role of cytokines and the JAK-STAT signaling pathway became central to understanding the immune landscape surrounding HL, which was critical for developing targeted therapies.

Subsequently, the focus has expanded to include the implications of checkpoint blockade therapies in HL, revealing how the TME influences treatment outcomes. Studies indicate that the effectiveness of Programmed Cell Death Ligand 1 (PD-1/PD-L1) blockade varies significantly among different lymphomas, underscoring the importance of the TME in therapeutic responses [23,26–28].

This evolution illustrates a growing appreciation of the interplay between tumor biology and immune modulation over nearly three decades. The aim of this study is to examine the role of TNF family members in the pathogenesis and treatment of cHL. It focuses on how TNF family signaling, particularly through CD30, CD40, and TNF-α, influences the survival and proliferation of HRS cells within the TME. Additionally, the study explores therapeutic opportunities, including immunotherapies and checkpoint inhibitors, that target these pathways, offering potential advancements in the treatment of cHL.

2. Early Research on TNF Family Members in cHL

CD40, a member of the TNF receptor superfamily, is expressed on HRS cells and has been identified as a functional receptor capable of transducing growth signals. Ref. [25] demonstrated that CD40 is expressed in 100% of HL cases examined, highlighting its widespread presence in HRS cells [25]. This expression is significant because it suggests that CD40 may play a pivotal role in the biology of HL, particularly in how these malignant cells evade apoptosis and sustain their proliferation. The interaction of CD40 with its ligand, CD40L, which is expressed on activated T cells, can lead to various downstream signaling cascades that promote cell survival and proliferation, including the activation of NF-κB pathways [25,29].

The survival advantage provided by CD40 signaling in HRS cells is multifaceted. Activation of CD40 can lead to the upregulation of anti-apoptotic proteins, thereby inhibiting programmed cell death. This is particularly relevant in the context of HL, where HRS cells often exhibit resistance to apoptosis, a hallmark of cancer cell survival. Furthermore, CD40 signaling can enhance the production of cytokines and chemokines that create a supportive TME, facilitating the further growth and survival of HRS cells [25].

The ability of CD40 to modulate immune responses also plays a role in the evasion of immune surveillance, allowing HRS cells to persist in the host despite the presence of an immune response.

In addition to CD40, CD30 is another TNF receptor that is characteristically expressed in HRS cells. CD30 signaling has been implicated in promoting cell proliferation and survival through similar mechanisms, including the activation of NF-κB and other survival pathways. The co-expression of CD30 and CD40 in HRS cells suggests a synergistic effect where both receptors contribute to the malignant phenotype of these cells. The interplay between these receptors and their respective ligands can create a complex signaling network that supports the aggressive nature of HL [25].

Given the critical roles of CD30 and CD40 in the survival and proliferation of HRS cells, targeting these TNF receptors represents a promising therapeutic strategy. Inhibiting CD40 signaling could disrupt the survival signals that HRS cells rely on, potentially leading to increased apoptosis and reduced tumor growth. Similarly, therapies that target CD30 have already shown efficacy in clinical settings, as evidenced by the success of brentuximab vedotin, an antibody–drug conjugate targeting CD30, which has demonstrated significant responses in patients with relapsed or refractory HL [25,30].

The therapeutic implications of targeting TNF receptors, particularly CD30 and CD40, may enhance treatment outcomes for patients with HL. The expression of these TNF family receptors plays a crucial role in the biology of HRS cells by contributing to their survival and proliferation, underscoring CD40 as a potential therapeutic target. As research continues to clarify the signaling networks involved in HL, targeting these TNF receptors could offer new treatment strategies aimed at improving patient outcomes [25].

3. The Role of the Tumor Microenvironment

The TME plays a pivotal role in the progression and survival of tumors, particularly in cHL, where the interplay between malignant HRS cells and the surrounding non-malignant cells is critical. The TME is composed of various cell types, including immune cells, fibroblasts, and endothelial cells, which collectively influence tumor behavior and immune evasion mechanisms [31]. Understanding the interactions between HRS cells and the TME, especially through the TNF family receptors such as CD30 and CD40, is essential for developing targeted therapies that can modulate these interactions to improve patient outcomes. The CD30 receptor, a member of the TNF receptor superfamily, is expressed on HRS cells and has been shown to mediate various signaling pathways that promote cell survival and proliferation [21]. When CD30 interacts with its ligand, it activates downstream signaling cascades that can inhibit apoptosis and enhance the survival of HRS cells within the TME. This interaction not only supports the malignant cells but also alters the behavior of surrounding immune cells, leading to an immunosuppressive TME that favors tumor growth and immune evasion [21].

The presence of CD30 in the TME is thus a double-edged sword; while it promotes the survival of HRS cells, it also contributes to the recruitment and activation of regulatory T cells and other immune suppressive elements that further protect the tumor from immune surveillance. Similarly, CD40, another TNF family receptor, plays a significant role in the interactions between HRS cells and the immune components of the TME. CD40 is typically expressed on antigen-presenting cells and is crucial for T cell activation and the generation of an effective immune response. However, in the context of cHL, the engagement of CD40 on HRS cells can lead to the secretion of various cytokines that modulate the immune response, often skewing it towards an anti-inflammatory profile. This results in a microenvironment that not only supports tumor growth but also inhibits effective anti-tumor immunity [21].

The ability of HRS cells to manipulate CD40 signaling underscores the complexity of the TME and highlights the potential for therapeutic interventions that target these pathways. The interplay between HRS cells and the TME is further complicated by the presence of various cytokines and chemokines that are produced by both malignant and non-malignant cells. These soluble factors can create a feedback loop that enhances tumor survival and promotes immune evasion. For instance, the secretion of interleukin-10 (IL-10), which has a potent anti-inflammatory effect, by HRS cells can inhibit the function of effector T cells and promote the differentiation of regulatory T cells, thereby dampening the overall immune response against the tumor, Furthermore, IL-10 functions as a growth and differentiation factor for B cells [5,21].

This dynamic interaction between cytokines and the cellular components of the TME is a critical area of research, as it offers insights into potential therapeutic targets that could disrupt these pro-tumorigenic signals. Therapeutic strategies aimed at modulating the TME in cHL are increasingly focused on disrupting the interactions mediated by TNF family receptors like CD30 and CD40. One promising approach involves the use of monoclonal antibodies that specifically target these receptors, thereby blocking their signaling pathways and potentially restoring effective immune responses against the tumor. For example, anti-CD30 therapies have shown efficacy in reducing tumor burden and improving patient outcomes by directly targeting HRS cells and altering the TME [21]. Additionally, combining these targeted therapies with immune checkpoint inhibitors (ICIs) may enhance the overall anti-tumor response by restoring the function of exhausted T cells and promoting a more favorable immune environment.

In conclusion, the role of the TME in classic HL is multifaceted, with TNF family receptors such as CD30 and CD40 serving as critical mediators of the interactions between HRS cells and the surrounding non-malignant cells. These interactions not only promote tumor survival but also facilitate immune evasion, making them attractive targets for therapeutic intervention. Continued research into the mechanisms by which HRS cells manipulate the TME will be essential for the development of novel therapies aimed at improving outcomes for patients with cHL.

4. Pro-Inflammatory and Anti-Inflammatory Roles of TNF Family Members

Members of the TNF family are integral to the pathophysiology of cHL, exerting both pro-inflammatory and anti-inflammatory effects that shape the tumor microenvironment. TNFR family members, such as CD30 and CD40, are overexpressed on HRS cells, mediating critical survival signals via pathways like NF-κB and JAK-STAT, which drive cell proliferation, inhibit apoptosis, and promote immune evasion. TNF-α, along with cytokines such as IL-6 and IL-1β, contributes to a pro-inflammatory microenvironment by recruiting eosinophils, mast cells, and Th2-polarized T cells, which enhance tumor growth through angiogenesis, stromal remodeling, and suppression of cytotoxic immune responses. These cytokines also induce chemokines such as CCL5, CCL17, and CCL22, which further attract immune-suppressive regulatory T cells (Tregs), reinforcing an immune-evasive niche [5,21,32,33].

Concurrently, anti-inflammatory mediators like IL-10 and TGF-β play key roles in dampening effective anti-tumor immune responses. IL-10, secreted by both HRS cells and tumor-associated macrophages (TAMs), suppresses Th1-mediated immunity and cytotoxic T cell activation, while TGF-β inhibits T cell proliferation and promotes fibrosis, which is a hallmark of the nodular sclerosis subtype of cHL. The immunosuppressive effects of these cytokines are compounded by the polarization of macrophages towards an M2 phenotype, characterized by secretion of anti-inflammatory factors that foster immune escape and further support HRS cell survival [5,21,34].

Interestingly, this cytokine imbalance is influenced by Epstein–Barr virus (EBV) infection in approximately 40% of cHL cases. EBV contributes to the dysregulated cytokine milieu by inducing the expression of viral proteins that mimic B cell survival signals, driving IL-10 production, and enhancing the recruitment of immune-suppressive cells. This dual role of TNF family members highlights their importance as both promoters of tumor growth and immune escape mechanisms. Consequently, therapeutic strategies targeting these pathways, such as inhibitors of CD30 and immune checkpoint blockade with PD-1/PD-L1 inhibitors, are being explored to disrupt these interactions, offering promising avenues to improve outcomes for cHL patients [21,33,34].

5. Immune Evasion and Tumor Survival

cHL utilizes a range of immune evasion mechanisms that are pivotal for tumor survival and progression. Central to these strategies are the HRS cells, which, despite being a minority in the tumor, heavily influence the surrounding immune microenvironment. HRS cells secrete chemokines such as CCL5, CCL17, and CCL20 to recruit Th2 and Tregs, fostering a supportive environment that inhibits cytotoxic T cell and natural killer (NK) cell activity [24,35–37]. Furthermore, HRS cells often downregulate or mutate antigen-presenting machinery, including major histocompatibility complex (MHC) class I and II molecules, impairing their recognition by immune effector cells. This is compounded by the expression of HLA-G, which protects HRS cells from NK cell-mediated lysis, and genetic alterations such as β2-microglobulin mutations that disrupt MHC-I assembly [36–39].

Additionally, HRS cells leverage immune checkpoint molecules like PD-L1 and PD-L2 to deactivate PD-1-expressing T cells in the microenvironment, thus suppressing anti-tumor immunity. Amplification of the 9p24.1 locus, which includes the PD-L1, PD-L2, and JAK2 genes, further upregulates these pathways, strengthening immune evasion [39]. HRS cells also modulate their interactions with $CD4^+$ T cells by expressing surface molecules such as CD80 and CD40, which engage with immune cells to provide survival signals while simultaneously shielding the tumor from effective immune attack [35]. The loss of co-stimulatory molecules like CD58 and the epigenetic silencing of CIITA, a key regulator of MHC-II, further impair immune recognition [36,39].

The tumor microenvironment in cHL is also characterized by a significant infiltration of immunosuppressive cells, including macrophages, which are often polarized toward a tumor-supportive phenotype. HRS cells induce these macrophages to express PD-L1, creating localized immunosuppressive niches. This interplay not only protects the tumor from immune-mediated destruction but also promotes its growth and survival [39]. The complexity of these immune evasion strategies underscores the potential for targeted therapies, such as immune checkpoint inhibitors, to disrupt the tumor's immune shield and restore anti-tumor immunity [35,36,38,39].

6. Checkpoint Blockade and Advanced Immunotherapies

The blockade of the PD-1/PD-L1 pathways in HRS cells represents a significant advancement in the therapeutic landscape of cHL [40–44]. This shift is primarily attributed to the unique TME associated with cHL, which is rich in inflammatory and immune cells that facilitate tumor immune evasion. The introduction of checkpoint inhibitors, such as nivolumab and pembrolizumab, has revolutionized treatment options for patients with relapsed or refractory cHL, showcasing the potential of harnessing the immune system to combat malignancies [23,44].

The PD-1/PD-L1 axis plays a crucial role in immune evasion by HRS cells, which express PD-L1 that interacts with PD-1 on T cells, leading to T cell exhaustion and a diminished immune response against the tumor. By blocking this interaction, checkpoint inhibitors can reactivate T cells, enhancing the immune response against HRS cells. Clinical studies have shown that the presence of PD-L1 in the TME is a key factor in the effectiveness of these therapies, correlating with the level of immune suppression and tumor progression in cHL [23,45].

Nivolumab and pembrolizumab have shown remarkable success in clinical trials, significantly improving outcomes for patients with relapsed or refractory cHL. For instance, nivolumab has been associated with an overall response rate of approximately 65% in patients who have undergone multiple lines of prior therapy, with some achieving complete remission. Similarly, pembrolizumab has demonstrated comparable efficacy, further solidifying the role of PD-1 blockade in the treatment paradigm of cHL. These results underscore the importance of the PD-1/PD-L1 pathway as a therapeutic target in a disease marked by an immunosuppressive microenvironment. The therapeutic shift toward checkpoint inhibitors is supported by an evolving understanding of the TME in cHL. In cHL, the TME includes various immune cells, such as T cells, B cells, and macrophages, that can influence the effectiveness of PD-1/PD-L1 blockade [23,28,46].

The identification of biomarkers predicting response to checkpoint inhibitors is a significant area of ongoing research, as understanding the molecular characteristics of HRS cells and the TME may enable stratification of patients most likely to benefit from PD-1/PD-L1 blockade. High levels of PD-L1 expression on HRS cells, for example, have been associated with improved responses to nivolumab and pembrolizumab, suggesting that patient selection based on TME characteristics could enhance treatment outcomes.

In conclusion, the blockade of the PD-1/PD-L1 pathways in HRS cells represents a transformative advancement in the management of cHL. The introduction of checkpoint inhibitors like nivolumab and pembrolizumab has not only improved response rates in relapsed or refractory cases but has also reshaped the understanding of the TME's role in cancer immunotherapy [23,46].

7. Resistance to Chemotherapy and Immunotherapy in HL

HL demonstrates resistance to chemotherapy and immunotherapy through complex mechanisms involving the TME and immune evasion. Resistance to chemotherapy arises as HRS cells exploit protective signals within the TME. Soluble factors such as cytokines (e.g., IL-10) and chemokines (e.g., CCL5, CCL17) shield tumor cells from chemotherapy-induced cytotoxicity by reducing apoptosis and enhancing survival pathways. Direct interactions with immune-suppressive cells, including regulatory Tregs and M2-polarized macrophages, further protect HRS cells and promote drug resistance, Additionally, genetic factors such as the multidrug resistance 1 (MDR1) C3435T polymorphism have been implicated in HL susceptibility, although its role in mediating chemotherapy response is not well-established. Drug resistance is also driven by mechanisms like increased drug efflux through P-glycoprotein overexpression, underscoring the multifaceted nature of treatment resistance [36,47–50].

In immunotherapy, resistance is driven by overexpression of PD-L1 on HRS cells, mediated by 9p24.1 amplification and EBV-induced signaling pathways, which suppress T cell activity via the PD-1/PD-L1 axis. Furthermore, downregulation of MHC class I and II molecules, as well as mutations in components of the antigen presentation machinery, impair tumor recognition by cytotoxic T cells. The immune-suppressive environment, supported by myeloid-derived suppressor cells (MDSCs) and cytokines such as IL-10, further limits the efficacy of immune checkpoint inhibitors, sustaining tumor survival [48,49,51]. Resistance to BV in HL arises primarily from mechanisms such as resistance to Monomethyl Auristatin E (MMAE), the cytotoxic component, and overexpression of the MDR1 gene, which encodes P-glycoprotein, actively pumping MMAE out of cells. Unlike other cancers, CD30 expression is typically retained, suggesting resistance is mediated through intracellular drug handling rather than target loss. Additional factors contributing to resistance include the TME, which promotes resistance through metabolic reprogramming, extracellular matrix remodeling, immune suppression, and angiogenesis. Epigenetic modifications also play a role by altering the expression of key regulatory genes. Furthermore,

CD30 ectodomain shedding mediated by a disintegrin and metalloproteinase (ADAM) proteins reduces effective drug docking, and defective linker-payload processing limits MMAE delivery. Strategies to address resistance include modifying payload–linker combinations, utilizing nanoparticle-based targeting, combining BV with immune checkpoint inhibitors, and exploring Clustered Regularly Interspaced Short Palindromic Repeats and CRISPR-associated protein 9 (CRISPR-Cas9) to counteract MDR1 overexpression or CD30 shedding [52,53].

8. The Potential of TNF Family Members as Biomarkers for HL Prognosis or Treatment Response

The TNF family and its related pathways offer potential as biomarkers in assessing HL prognosis and treatment responses. Research underscores that members of the TNF family, such as TNF-α, TRAIL, BAFF, and APRIL, and their receptors (e.g., CD30, CD40, BCMA, TACI) [1–5], are intricately involved in HL pathogenesis, primarily through their roles in the tumor microenvironment and their interactions with HRS cells. These interactions regulate cell survival, apoptosis, and immune evasion [33,54–57].

Elevated serum levels of TNF-α have been observed in HL patients and correlate with disease severity and poorer outcomes in some studies [58,59]. However, the role of soluble TNF receptors, though indicative of disease activity, remains complex, as these molecules may originate from both tumor and immune cells. TRAIL, another TNF superfamily member, shows potential for inducing apoptosis in HL cells when anti-apoptotic pathways are inhibited, emphasizing its utility in therapeutic strategies. Similarly, BAFF and APRIL, which support the survival of B cells, including malignant HRS cells, have been associated with lower failure-free survival in HL, highlighting their prognostic significance [33,57].

The expression of TNF-related receptors such as CD30 and CD40 on HRS cells is central to their malignant behavior. These receptors mediate interactions with immune cells, fostering an immunosuppressive microenvironment and enhancing tumor progression. CD30, in particular, has emerged as a therapeutic target, with drugs like brentuximab vedotin demonstrating efficacy in refractory HL cases. Beyond therapeutic targeting, the overexpression of TNF ligands and receptors within the tumor microenvironment further supports their utility as biomarkers [54,56].

In summary, TNF family members hold promise as biomarkers for predicting HL outcomes and guiding treatment decisions. Their dual roles in the TME—facilitating tumor growth and serving as therapeutic targets—underscore the importance of integrating these pathways into biomarker panels for a more personalized approach to HL management.

9. New Therapeutic Opportunities

ICIs, such as nivolumab and pembrolizumab, have demonstrated high response rates in patients with cHL, particularly in those who have not responded to conventional therapies. For instance, nivolumab has been shown to achieve an objective response rate of approximately 70% in relapsed or refractory cHL, with many patients experiencing durable remissions [60–62].

The mechanism of action of these inhibitors involves blocking the interaction between PD-1 on T cells and PD-L1 on tumor cells, thereby reinvigorating the immune response against the malignancy. This is particularly relevant in cHL, where HRS cells often express PD-L1, facilitating immune evasion [60]. The integration of ICIs with chemotherapy regimens has been a focal point of research. Studies have indicated that combining nivolumab with standard chemotherapy regimens, such as ABVD (doxorubicin, bleomycin, vinblastine, and dacarbazine), can improve outcomes for patients with advanced-stage cHL [63].

The rationale behind this combination is that while chemotherapy targets rapidly dividing cells, ICIs can enhance the immune system's ability to recognize and attack residual cancer cells post-chemotherapy. This synergistic approach aims to reduce the risk of relapse and improve long-term survival rates [63]. However, the use of ICIs is associated with a unique spectrum of immune-related adverse events (irAEs), which can complicate treatment. These irAEs can affect various organ systems, including the skin, gastrointestinal tract, liver, and endocrine organs, and may require prompt recognition and management [63].

The management of these toxicities is critical, as they can lead to treatment discontinuation or dose modification, potentially impacting overall efficacy. Guidelines suggest a multidisciplinary approach to managing irAEs, which may include the use of corticosteroids or other immunosuppressive agents to mitigate severe reactions while maintaining the effectiveness of the ICI therapy [63]. In addition to managing irAEs, ongoing research is focused on identifying biomarkers that can predict responses to ICI therapy. The expression of PD-L1 on tumor cells has been associated with better responses to PD-1 inhibitors, although this relationship is not entirely consistent across all patients [62,64].

Understanding the TME and the presence of immune suppressive cells, such as regulatory T cells and myeloid-derived suppressor cells, may also provide insights into resistance mechanisms and help tailor combination therapies for individual patients [62].

The potential for combining ICIs with other therapeutic modalities, such as targeted therapies and novel agents like brentuximab vedotin, is also being explored [65]. Brentuximab vedotin, an antibody–drug conjugate targeting CD30, has shown efficacy in patients with relapsed or refractory cHL and is being studied in combination with ICIs to enhance therapeutic outcomes [64,66]. This combination approach aims to leverage the strengths of both immunotherapy and targeted therapy, potentially leading to improved response rates and reduced relapse rates. Moreover, the role of radiotherapy in conjunction with ICIs is being investigated. While traditional radiotherapy has been a cornerstone in the treatment of localized cHL, its integration with ICIs could enhance the immune response by inducing immunogenic cell death and promoting the release of tumor antigens [63].

This strategy may help to overcome some of the limitations associated with conventional therapies and improve overall treatment efficacy. In summary, the landscape of cHL treatment is rapidly evolving with the incorporation of ICIs into standard therapeutic regimens. The combination of ICIs with chemotherapy, targeted therapies, and radiotherapy presents a promising avenue for enhancing treatment efficacy while addressing the management of immune-related adverse events [67–70] (Table 2). Ongoing research into biomarkers and resistance mechanisms will be crucial in optimizing these combination strategies and personalizing treatment for patients with cHL.

Table 2. Selected studies using a combination of immune checkpoint inhibitors (ICIs) with chemotherapy, targeted therapy, and radiotherapy.

Combination Therapy	Possible Benefits	References
ICIs with chemotherapy (e.g., Nivolumab + AVD)	N+AVD showed longer progression-free survival and a more favorable side-effect profile compared to BV+AVD in adolescents and adults with advanced-stage (III or IV) classic Hodgkin lymphoma	https://doi.org/10.1056/NEJMoa2405888 Herrera et al., 2024 [69]
ICIs with targeted therapy (Brentuximab Vedotin)	BV+Nivo as a first salvage therapy demonstrated high response rates and durable progression-free survival with manageable side effects in relapsed/refractory classic Hodgkin lymphoma, especially benefiting those proceeding directly to autologous stem cell transplant	https://doi.org/10.1182/blood.2020009178 Advani et al., 2021 [67]
ICIs (nivolumab, pembrolizumab) with radiotherapy	ICI-RT showed high response rates and durable complete remission in relapsed/refractory Hodgkin lymphoma, serving as an effective bridge to stem cell transplant with no major safety issues	https://doi.org/10.4081/hr.2021.9080 Lucchini et al., 2021 [70]

Abbreviations: AVD: adriamycin (Doxorubicin), vinblastine, dacarbazine, BV: brentuximab vedotin, ICIs: immune checkpoint inhibitors, Nivo: nivolumab, RT: radiotherapy.

10. Comparative Analysis of Pathobiological Features and Therapeutic Opportunities Linked to TNF Family Member Expression in cHL Across Time Periods

In the early stages of research on TNF family members, the focus was predominantly on their direct signaling capabilities. For instance, the role of TNF-alpha in inducing apoptosis and its involvement in inflammatory responses were well documented, leading to the development of therapeutic agents targeting TNF receptors directly. However, as our understanding of the immune system evolved, it became clear that the interactions between TNF family members and various immune cells within the TME are far more complex than previously thought.

The cHL microenvironment exemplifies this complexity, where HRS cells interact with a diverse array of immune cells, including T and B cells, macrophages, and eosinophils, creating a dynamic milieu that influences tumor growth and immune evasion [21].

The constitutive activation of NF-κB and alterations in the JAK-STAT signaling pathway in HRS cells contribute to the overexpression of TNF receptor family members, such as CD30 and CD40, highlighting the importance of these interactions in the context of tumor biology [21].

As the understanding of the TME deepened, researchers began to appreciate how the immune context can modulate the effects of TNF family signaling. For example, the presence of immune checkpoint molecules such as PD-1 and PD-L1 in the TME can significantly alter the response to TNF family signaling, leading to immune evasion and tumor progression. This realization has prompted a shift in treatment strategies from merely targeting TNF family receptors to leveraging immune checkpoint inhibitors that can enhance anti-tumor immunity.

The lessons learned from cHL, where the TME is rich in inflammatory cells, have been particularly instructive in this regard. Studies have shown that checkpoint blockade therapies can be more effective in tumors with a robust immune infiltrate, suggesting that the TME's composition is a critical determinant of therapeutic outcomes [23].

The evolution of treatment strategies reflects this growing understanding of the TME's role in modulating TNF family signaling. Early therapeutic approaches focused on directly inhibiting TNF-alpha or its receptors, which provided some clinical benefit but often fell short in terms of long-term efficacy. The emergence of immune checkpoint inhibitors, such as those targeting PD-1 and PD-L1, represents a paradigm shift in how we approach cancer treatment. These agents work by reinvigorating exhausted T cells within the TME, thereby enhancing the immune response against tumors that may have previously escaped immune surveillance due to the immunosuppressive effects of the TME [23]. This shift underscores the importance of considering the TME not just as a passive backdrop for tumor growth, but as an active participant in the regulation of immune responses. Furthermore, the interplay between TNF family members and the immune microenvironment has implications for the development of combination therapies. By understanding how TNF signaling interacts with immune checkpoints, researchers can design more effective treatment regimens that simultaneously target multiple pathways [56] (Table 3).

Table 3. Molecular targets of classic Hodgkin lymphoma and therapeutic agents.

Microenvironmental Cell Markers	Therapeutic Agent
PD-1	Checkpoint inhibitors (Nivolumab, Pembrolizumab)
PD-L1	Checkpoint inhibitors
Deregulated pathways	
NFκB	NFκB inhibitors (Bortezomib)
AKT/MAPK	MAPK inhibitors (Sorafenib)
	HDAC inhibitors
PI3K/AKT	Idelalisib
mTOR	mTOR inhibitor (Everolimus)

Table 3. *Cont.*

Microenvironmental Cell Markers	Therapeutic Agent
HRS markers	
CD30	Monoclonal antibodies anti-CD30 (Brentuximab Vedotin)
CD40	Monoclonal antibodies anti-CD40
MUM1/IRF4	Immunomodulator (Lenalidomide)
Notch1	Notch inhibitor
CD20	Monoclonal antibodies anti-CD20 (Rituximab)
CD80	Monoclonal antibodies anti-CD80 (Galiximab)
HDACs	HDAC inhibitors
EBV	Patient-derived cytotoxic T cells

Abbreviations: AKT/MAPK: AKT/Mitogen-activated protein kinase, CD20: cluster of differentiation 20, CD30: cluster of differentiation 30, CD40: cluster of differentiation 40, CD80: cluster of differentiation 80, EBV: Epstein–Barr Virus, HDACs: histone deacetylases, mTOR: mammalian target of rapamycin, MUM1/IRF4: multiple myeloma oncogene 1/Interferon regulatory factor 4, NFκB: nuclear factor kappa B, Notch1: notch homolog 1, PD-1: programmed cell death protein 1, PD-L1: programmed death-ligand 1, PI3K/AKT: phosphoinositide 3-kinase/AKT.

11. Conclusions

The study of TNF family involvement in HL spans nearly four decades, with initial findings in 1985 and 1995 identifying CD30 and CD40, respectively, on HRS cells and recognizing that these TNF family members play a key role in HRS cell growth and interaction with the immune system [25]. Subsequent research furthered our understanding of HL's TME and how it supports immune escape, revealing that cytokine networks around malignant cells significantly contribute to tumor proliferation and immune suppression [21]. Recent studies have shown that targeting the PD-1/PD-L1 pathway in the TME enhances immunotherapy, marking a promising advancement in immune-targeting therapies [23].

Prospectively, combining TNF inhibitors with checkpoint blockade therapies may enhance the overall anti-tumor response by addressing both direct tumor signaling and the immune evasion mechanisms employed by tumors [71–73] (Figure 1). This integrated approach, which has already demonstrated effectiveness in solid tumors, is particularly relevant in the context of cHL, where the unique characteristics of the TME can be exploited to improve therapeutic outcomes [23]. Further insights into microenvironment biology and clinical trials testing predictive biomarkers and novel compounds (Table 4) will likely lead to additional improvements in outcomes.

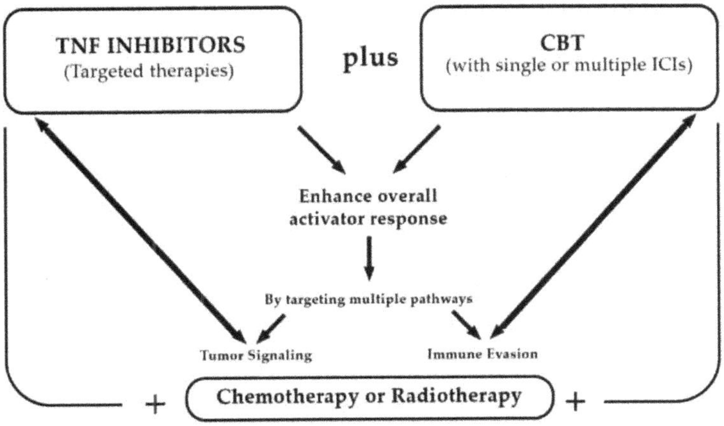

Figure 1. New strategy using integrated approach and combination therapies in classic Hodgkin lymphoma.

Table 4. Selection of ongoing phase I/II trials evaluating novel therapeutic combination in classic Hodgkin lymphoma (cHL).

Target	Agent/Intervention	Phase	Clinical Setting	Trial Identifier
CD30	Autologous CAR.CD30 EBV specific-CTLs	1	Relapsed CD30+ Hodgkin lymphoma and CD30+ non-Hodgkin lymphoma	NCT01192464
CD30	CD30 CAR T cell	1/2	Post-AutoHSCT for poor-risk Hodgkin lymphoma	NCT06617286
CD30	LCAR-HL30 cells	1	Relapsed/refractory Hodgkin lymphoma and anaplastic large cell lymphoma	NCT06494371
CD30	Modified BV-AVD-R regimen	2	Children with intermediate/high-risk cHL	NCT06201507
CD30	ATLCAR.CD30.CCR4 cells	1/2	Relapsed/refractory CD30+ Hodgkin lymphoma	NCT06090864
CD30	GEN3017	1/2	Relapsed/refractory CD30+ Hodgkin and non-Hodgkin lymphoma	NCT06018129
CD30	AFM13 + AB-101	2	Recurrent/refractory Hodgkin lymphoma and CD30+ peripheral T cell lymphoma	NCT05883449
CD30	Brentuximab Vedotin + CHP	2	Newly diagnosed CD30+ peripheral T cell lymphoma (PTCL)	NCT05673785
CD30	Brentuximab Vedotin + AVD	Observational	Pediatric patients with newly diagnosed CD30+ Hodgkin lymphoma	NCT05481437
CD30	CD30-directed CAR-T cells in combination with Nivolumab	1	R\R cHL post-frontline therapy failure	NCT05352828
CD30	Anti-CD30 CAR-T Cell Injection	1	R\R CD30+ lymphoma	NCT05208853
CD30	BV	Observational	R\R CD30+ lymphoma	NCT04998331
CD30	HSP-CAR30 T cells	1/2	R\R Hodgkin and T cell lymphoma	NCT04653649
LAG-3, PD-1, CD47	AK129 with or without AK117	1/2	R\R classic Hodgkin lymphoma, PD-1/L1 inhibitor treatment failure	NCT06642792
PD-1	Pembrolizumab	2	First-line treatment, advanced-stage cHL	NCT06045195
PD-1	Tislelizumab	2	Frontline, de novo Hodgkin lymphoma unsuitable for standard chemotherapy	NCT05977673
PD-1	Zimberelimab + AVD	2	First-line, early-stage Hodgkin lymphoma	NCT05900765
PD-1	Prolgolimab ± Bendamustine	2	Second-line, relapsed R\R cHL	NCT05757466
CD30 + PD-1	Autologous CD30.CAR-T + Nivolumab	1	R\R cHL after frontline therapy	NCT05352828
PD-1	PD-1 inhibitor + Decitabine + ASCT	2	Second-line for R\R cHL	NCT05137886
PD-1	PD-1 inhibitor ± GVD regimen	2	R\R cHL	NCT04624984
JAK	Ruxolitinib	2	R\R cHL	NCT02164500
JAK	INCB047986	1	Advanced malignancies, including Hodgkin and non-Hodgkin lymphoma	NCT01929941
CD30	CD30 CAR T cell	1/2	Poor-risk cHL post-AutoHSCT	NCT06617286
CD30	HSP-CAR30	1/2	R\R Hodgkin and T cell lymphoma	NCT04653649
PD-1	PD-1 inhibitor	1	R\R Hodgkin lymphoma	NCT04134325
BTK	Ibrutinib	2	R\R cHL	NCT02824029
CD30	CD30.CAR-EBVST cells	1	R\R CD30-positive lymphomas	NCT04288726
CD30, PD-1	Chidamide + Decitabine + Anti-PD-1 antibody and Brentuximab Vedotin + Bendamustine + Anti-PD-1 antibody	2	R/R cHL, transplant-ineligible, or refused a transplant	NCT06563778
PD-1	Pembrolizumab coformulated with Hyaluronidase	2	R/R cHL or R/R PMBCL	NCT06504394

Table 4. *Cont.*

Target	Agent/Intervention	Phase	Clinical Setting	Trial Identifier
PD-1	Chidamide + Decitabine + Anti-PD-1 antibody (experimental), Brentuximab Vedotin + Anti-PD-1 antibody (comparator)	2	R/R cHL for ineligible patients or those refusing a transplant	NCT06393361
PD-1	Azacitidine + PD-1 therapy	2	Relapsed/refractory classic Hodgkin lymphoma	NCT06190067

Abbreviations: AK117: experimental drug targeting CD47, AK129: experimental drug targeting PD-1, ASCT: autologous stem cell transplant, AutoHSCT: autologous hematopoietic stem cell transplant, AVD: Adriamycin (doxorubicin), Vinblastine, and Dacarbazine, BTK: Bruton's Tyrosine Kinase, CAR-T: Chimeric Antigen Receptor T cells, cHL: classic Hodgkin lymphoma, EBVST cells: Epstein–Barr virus-specific T cells, GVD: Gemcitabine, Vinorelbine, and Doxorubicin, HSP-CAR30 T cells: Heat Shock Protein–Chimeric Antigen Receptor targeting CD30 T cells, INCB047986: experimental JAK inhibitor, JAK: Janus Kinase, LAG-3: Lymphocyte Activation Gene-3, PD-1: Programmed Death-1, PMBCL: primary mediastinal B cell lymphoma, R/R: relapsed/refractory.

Author Contributions: Conceptualization, M.N.A., A.G. and A.C.; Writing—original draft preparation, M.N.A.; Writing—review and editing, A.C. All authors have read and agreed to the published version of the manuscript.

Funding: This research received no external funding.

Conflicts of Interest: The authors declare no conflict of interest.

References

1. Dostert, C.; Grusdat, M.; Letellier, E.; Brenner, D. The TNF Family of Ligands and Receptors: Communication Modules in the Immune System and Beyond. *Physiol. Rev.* **2019**, *99*, 115–160. [CrossRef] [PubMed]
2. Li, H.; Liu, S.; Li, C.; Xiao, Z.; Hu, J.; Zhao, C. TNF Family-Based Signature Predicts Prognosis, Tumor Microenvironment, and Molecular Subtypes in Bladder Carcinoma. *Front. Cell Dev. Biol.* **2021**, *9*, 800967. [CrossRef]
3. Orlinick, J.R.; Chao, M.V. TNF-related ligands and their receptors. *Cell. Signal.* **1998**, *10*, 543–551. [CrossRef]
4. Zhang, C.; Zhang, G.; Sun, N.; Zhang, Z.; Zhang, Z.; Luo, Y.; Che, Y.; Xue, Q.; He, J. Comprehensive molecular analyses of a TNF family-based signature with regard to prognosis, immune features, and biomarkers for immunotherapy in lung adenocarcinoma. *EBioMedicine* **2020**, *59*, 102959. [CrossRef]
5. Skinnider, B.F.; Mak, T.W. The role of cytokines in classical Hodgkin lymphoma. *Blood* **2002**, *99*, 4283–4297. [CrossRef]
6. Wang, Q.W.; Lin, W.W.; Zhu, Y.J. Comprehensive analysis of a TNF family based-signature in diffuse gliomas with regard to prognosis and immune significance. *Cell Commun. Signal.* **2022**, *20*, 6. [CrossRef]
7. Groves, R.W.; Allen, M.H.; Ross, E.L.; Barker, J.N.; MacDonald, D.M. Tumour necrosis factor alpha is pro-inflammatory in normal human skin and modulates cutaneous adhesion molecule expression. *Br. J. Dermatol.* **1995**, *132*, 345–352. [CrossRef]
8. Zhu, Z.; Tang, C.; Xu, T.; Zhao, Z. Molecular Analysis of Prognosis and Immune Pathways of Pancreatic Cancer Based on TNF Family Members. *J. Oncol.* **2021**, *2021*, 2676996. [CrossRef]
9. Parameswaran, N.; Patial, S. Tumor necrosis factor-alpha signaling in macrophages. *Crit. Rev. Eukaryot. Gene Expr.* **2010**, *20*, 87–103. [CrossRef]
10. Tafalla, C.; Granja, A.G. Novel Insights on the Regulation of B Cell Functionality by Members of the Tumor Necrosis Factor Superfamily in Jawed Fish. *Front. Immunol.* **2018**, *9*, 1285. [CrossRef]
11. Chiu, A.; Qiao, X.; He, B.; Hyjjek, E.; Lee, J.; Cesarman, E.; Chadburn, A.; Knowles, D.M.; Cerutti, A. The TNF family members BAFF and APRIL play an important role in Hodgkin lymphoma. *Blood* **2005**, *106*, 22. [CrossRef]
12. Wang, J.; Lo, J.C.; Foster, A.; Yu, P.; Chen, H.M.; Wang, Y.; Tamada, K.; Chen, L.; Fu, Y.X. The regulation of T cell homeostasis and autoimmunity by T cell-derived LIGHT. *J. Clin. Investig.* **2001**, *108*, 1771–1780. [CrossRef]
13. Šedý, J.; Bekiaris, V.; Ware, C.F. Tumor necrosis factor superfamily in innate immunity and inflammation. *Cold Spring Harbor Perspect. Biol.* **2015**, *7*, a016279. [CrossRef]
14. Diaz Arguello, O.A.; Haisma, H.J. Apoptosis-Inducing TNF Superfamily Ligands for Cancer Therapy. *Cancers* **2021**, *13*, 1543. [CrossRef]
15. Younes, A.; Kadin, M.E. Emerging applications of the tumor necrosis factor family of ligands and receptors in cancer therapy. *J. Clin. Oncol.* **2003**, *21*, 3526–3534. [CrossRef]
16. Banchereau, J.; de Paoli, P.; Valle, A.; Garcia, E.; Rousset, F. Long-term human B cell lines dependent on interleukin-4 and antibody to CD40. *Science* **1991**, *251*, 70–72. [CrossRef] [PubMed]
17. Durkop, H.; Latza, U.; Hummel, M.; Eitelbach, F.; Seed, B.; Stein, H. Molecular cloning and expression of a new member of the nerve growth factor receptor family that is characteristic for Hodgkin's disease. *Cell* **1992**, *68*, 421–427. [CrossRef]
18. O'Grady, J.T.; Stewart, S.; Lowrey, J.; Howie, S.E.; Krajewski, A.S. CD40 expression in Hodgkin's disease. *Am. J. Pathol.* **1994**, *144*, 21–26. [PubMed]

19. Stein, H.; Mason, D.Y.; Gerdes, J.; O'Connor, N.; Wainscoat, J.; Pallesen, G.; Gatter, K.; Falini, B.; Delsol, G.; Lemke, H.; et al. The expression of the Hodgkin's disease associated antigen Ki-1 in reactive and neoplastic lymphoid tissue: Evidence that Reed-Sternberg cells and histiocytic malignancies are derived from activated lymphoid cells. *Blood* **1985**, *66*, 848–858. [CrossRef] [PubMed]
20. Kim, L.H.; Eow, G.I.; Peh, S.C.; Poppema, S. The role of CD30, CD40 and CD95 in the regulation of proliferation and apoptosis in classical Hodgkin's lymphoma. *Pathology* **2003**, *35*, 428–435. [CrossRef]
21. Aldinucci, D.; Gloghini, A.; Pinto, A.; De Filippi, R.; Carbone, A. The classical Hodgkin's lymphoma microenvironment and its role in promoting tumour growth and immune escape. *J. Pathol.* **2010**, *221*, 248–263. [CrossRef]
22. Jardin, F. NFkB Pathway and Hodgkin Lymphoma. *Biomedicines* **2022**, *10*, 2153. [CrossRef]
23. Carbone, A.; Gloghini, A.; Carlo-Stella, C. Tumor microenvironment contribution to checkpoint blockade therapy: Lessons learned from Hodgkin lymphoma. *Blood* **2023**, *141*, 2187–2193. [CrossRef]
24. Poppema, S. Immunobiology and pathophysiology of Hodgkin lymphomas. *Hematol. Am. Soc. Hematol. Educ. Program.* **2005**, *2005*, 231–238. [CrossRef]
25. Carbone, A.; Gloghini, A.; Gattei, V.; Aldinucci, D.; Degan, M.; De Paoli, P.; Zagonel, V.; Pinto, A. Expression of functional CD40 antigen on Reed-Sternberg cells and Hodgkin's disease cell lines. *Blood* **1995**, *85*, 780–789. [CrossRef]
26. Armand, P. Immune checkpoint blockade in hematologic malignancies. *Blood* **2015**, *125*, 3393–3400. [CrossRef]
27. Mina, A.A.; Vakkalagadda, C.; Pro, B. Novel Therapies and Approaches to Relapsed/Refractory HL Beyond Chemotherapy. *Cancers* **2019**, *11*, 421. [CrossRef] [PubMed]
28. Lin, A.Y.; Schnitter, J.M.; Gordon, L.I. Immune Checkpoint Blockade for the Treatment of Hodgkin Lymphoma. *Immunotargets Ther.* **2022**, *11*, 1–10. [CrossRef]
29. Foyil, K.V.; Bartlett, N.L. Anti-CD30 Antibodies for Hodgkin lymphoma. *Curr. Hematol. Malig. Rep.* **2010**, *5*, 140–147. [CrossRef] [PubMed]
30. Song, Y.; Guo, Y.; Huang, H.; Li, W.; Ke, X.; Feng, J.; Xu, W.; Miao, H.; Kinley, J.; Song, G.; et al. Phase II single-arm study of brentuximab vedotin in Chinese patients with relapsed/refractory classical Hodgkin lymphoma or systemic anaplastic large cell lymphoma. *Expert. Rev. Hematol.* **2021**, *14*, 867–875. [CrossRef] [PubMed]
31. Steidl, C.; Connors, J.M.; Gascoyne, R.D. Molecular pathogenesis of Hodgkin's lymphoma: Increasing evidence of the importance of the microenvironment. *J. Clin. Oncol.* **2011**, *29*, 1812–1826. [CrossRef] [PubMed]
32. Maggio, E.; van den Berg, A.; Diepstra, A.; Kluiver, J.; Visser, L.; Poppema, S. Chemokines, cytokines and their receptors in Hodgkin's lymphoma cell lines and tissues. *Ann. Oncol.* **2002**, *13* (Suppl. S1), 52–56. [CrossRef] [PubMed]
33. Hamed Anber, N.; El-Sebaie, A.H.; Darwish, N.H.E.; Mousa, S.A.; Shamaa, S.S. Prognostic value of some inflammatory markers in patients with lymphoma. *Biosci. Rep.* **2019**, *39*, BSR20182174. [CrossRef] [PubMed]
34. Calabretta, E.; d'Amore, F.; Carlo-Stella, C. Immune and Inflammatory Cells of the Tumor Microenvironment Represent Novel Therapeutic Targets in Classical Hodgkin Lymphoma. *Int. J. Mol. Sci.* **2019**, *20*, 5503. [CrossRef]
35. Wein, F.; Weniger, M.A.; Hoing, B.; Arnolds, J.; Huttmann, A.; Hansmann, M.L.; Hartmann, S.; Kuppers, R. Complex Immune Evasion Strategies in Classical Hodgkin Lymphoma. *Cancer Immunol. Res.* **2017**, *5*, 1122–1132. [CrossRef]
36. Upadhyay, R.; Hammerich, L.; Peng, P.; Brown, B.; Merad, M.; Brody, J.D. Lymphoma: Immune evasion strategies. *Cancers* **2015**, *7*, 736–762. [CrossRef]
37. de la Cruz-Merino, L.; Lejeune, M.; Nogales Fernandez, E.; Henao Carrasco, F.; Grueso Lopez, A.; Illescas Vacas, A.; Pulla, M.P.; Callau, C.; Alvaro, T. Role of immune escape mechanisms in Hodgkin's lymphoma development and progression: A whole new world with therapeutic implications. *Clin. Dev. Immunol.* **2012**, *2012*, 756353. [CrossRef]
38. de Charette, M.; Houot, R. Hide or defend, the two strategies of lymphoma immune evasion: Potential implications for immunotherapy. *Haematologica* **2018**, *103*, 1256–1268. [CrossRef]
39. Menter, T.; Tzankov, A. Mechanisms of Immune Evasion and Immune Modulation by Lymphoma Cells. *Front. Oncol.* **2018**, *8*, 54. [CrossRef]
40. Armand, P.; Zinzani, P.L.; Lee, H.J.; Johnson, N.A.; Brice, P.; Radford, J.; Ribrag, V.; Molin, D.; Vassilakopoulos, T.P.; Tomita, A.; et al. Five-year follow-up of KEYNOTE-087: Pembrolizumab monotherapy for relapsed/refractory classical Hodgkin lymphoma. *Blood* **2023**, *142*, 878–886. [CrossRef]
41. Domingo-Domenech, E.; Sureda, A. Treatment of Hodgkin Lymphoma Relapsed after Autologous Stem Cell Transplantation. *J. Clin. Med.* **2020**, *9*, 1384. [CrossRef] [PubMed]
42. Pophali, P.; Varela, J.C.; Rosenblatt, J. Immune checkpoint blockade in hematological malignancies: Current state and future potential. *Front. Oncol.* **2024**, *14*, 1323914. [CrossRef] [PubMed]
43. Xie, W.; Medeiros, L.J.; Li, S.; Yin, C.C.; Khoury, J.D.; Xu, J. PD-1/PD-L1 Pathway and Its Blockade in Patients with Classic Hodgkin Lymphoma and Non-Hodgkin Large-Cell Lymphomas. *Curr. Hematol. Malig. Rep.* **2020**, *15*, 372–381. [CrossRef] [PubMed]
44. Ansell, S.M. PD-1 Blockade in Classic Hodgkin Lymphoma. *JCO Oncol. Pract.* **2021**, *17*, 72–73. [CrossRef] [PubMed]
45. Vari, F.; Arpon, D.; Keane, C.; Hertzberg, M.S.; Talaulikar, D.; Jain, S.; Cui, Q.; Han, E.; Tobin, J.; Bird, R.; et al. Immune evasion via PD-1/PD-L1 on NK cells and monocyte/macrophages is more prominent in Hodgkin lymphoma than DLBCL. *Blood* **2018**, *131*, 1809–1819. [CrossRef]

46. Ansell, S.M.; Brockelmann, P.J.; von Keudell, G.; Lee, H.J.; Santoro, A.; Zinzani, P.L.; Collins, G.P.; Cohen, J.B.; de Boer, J.P.; Kuruvilla, J.; et al. Nivolumab for relapsed/refractory classical Hodgkin lymphoma: 5-year survival from the pivotal phase 2 CheckMate 205 study. *Blood Adv.* **2023**, *7*, 6266–6274. [CrossRef]
47. Aldinucci, D.; Celegato, M.; Casagrande, N. Microenvironmental interactions in classical Hodgkin lymphoma and their role in promoting tumor growth, immune escape and drug resistance. *Cancer Lett.* **2016**, *380*, 243–252. [CrossRef]
48. Brockelmann, P.J.; Borchmann, P.; Engert, A. Current and future immunotherapeutic approaches in Hodgkin lymphoma. *Leuk. Lymphoma* **2016**, *57*, 2014–2024. [CrossRef]
49. Veldman, J.; Visser, L.; Berg, A.V.D.; Diepstra, A. Primary and acquired resistance mechanisms to immune checkpoint inhibition in Hodgkin lymphoma. *Cancer Treat. Rev.* **2020**, *82*, 101931. [CrossRef]
50. Mhaidat, N.M.; Alshogran, O.Y.; Khabour, O.F.; Alzoubi, K.H.; Matalka, I.I.; Haddadin, W.J.; Mahasneh, I.O.; Aldaher, A.N. Multi-drug resistance 1 genetic polymorphism and prediction of chemotherapy response in Hodgkin's Lymphoma. *J. Exp. Clin. Cancer Res.* **2011**, *30*, 68. [CrossRef]
51. Michot, J.M.; Lazarovici, J.; Ghez, D.; Danu, A.; Ferme, C.; Bigorgne, A.; Ribrag, V.; Marabelle, A.; Aspeslagh, S. Challenges and perspectives in the immunotherapy of Hodgkin lymphoma. *Eur. J. Cancer* **2017**, *85*, 67–77. [CrossRef] [PubMed]
52. Chen, R.; Hou, J.; Newman, E.; Kim, Y.; Donohue, C.; Liu, X.; Thomas, S.H.; Forman, S.J.; Kane, S.E. CD30 Downregulation, MMAE Resistance, and MDR1 Upregulation Are All Associated with Resistance to Brentuximab Vedotin. *Mol. Cancer Ther.* **2015**, *14*, 1376–1384. [CrossRef]
53. Bera, D.; Roy, D. Brentuximab vedotin resistance in classic Hodgkin's lymphoma and its therapeutic strategies: A review. *Future J. Pharm. Sci.* **2024**, *10*, 15. [CrossRef]
54. Carlo-Stella, C.; Santoro, A. Microenvironment-related biomarkers and novel targets in classical Hodgkin's lymphoma. *Biomark. Med.* **2015**, *9*, 807–817. [CrossRef] [PubMed]
55. Aoki, T.; Steidl, C. Novel Biomarker Approaches in Classic Hodgkin Lymphoma. *Cancer J.* **2018**, *24*, 206–214. [CrossRef] [PubMed]
56. Mottok, A.; Steidl, C. Biology of classical Hodgkin lymphoma: Implications for prognosis and novel therapies. *Blood* **2018**, *131*, 1654–1665. [CrossRef]
57. Staege, M.S.; Kewitz, S.; Bernig, T.; Kuhnol, C.; Mauz-Korholz, C. Prognostic Biomarkers for Hodgkin Lymphoma. *Pediatr. Hematol. Oncol.* **2015**, *32*, 433–454. [CrossRef]
58. Warzocha, K.; Bienvenu, J.; Ribeiro, P.; Moullet, I.; Dumontet, C.; Neidhardt-Berard, E.M.; Coiffier, B.; Salles, G. Plasma levels of tumour necrosis factor and its soluble receptors correlate with clinical features and outcome of Hodgkin's disease patients. *Br. J. Cancer* **1998**, *77*, 2357–2362. [CrossRef] [PubMed]
59. Cuccaro, A.; Bartolomei, F.; Cupelli, E.; Galli, E.; Giachelia, M.; Hohaus, S. Prognostic factors in hodgkin lymphoma. *Mediterr. J. Hematol. Infect. Dis.* **2014**, *6*, e2014053. [CrossRef]
60. Bröckelmann, P.J. Unfolding the potential of anti-programmed cell death protein 1 blockade in Hodgkin lymphoma-combination and personalisation? *Br. J. Haematol.* **2022**, *196*, 10–12. [CrossRef]
61. Morad, G.; Helmink, B.A.; Sharma, P.; Wargo, J.A. Hallmarks of response, resistance, and toxicity to immune checkpoint blockade. *Cell* **2021**, *184*, 5309–5337. [CrossRef] [PubMed]
62. Tarekegn, K.; Colon Ramos, A.; Singh, B.; Sequeira Gross, H.G.; Gupta, S. Checkpoint Inhibitors in Relapsed/Refractory Classical Hodgkin Lymphoma. *World J. Oncol.* **2021**, *12*, 81–84. [CrossRef]
63. Kuczmarski, T.M.; Lynch, R.C. Managing common toxicities associated with checkpoint inhibitor and chemotherapy combinations for untreated classic Hodgkin lymphoma. *Br. J. Haematol.* **2024**, *205*, 100–108. [CrossRef] [PubMed]
64. Zhang, X.Y.; Collins, G.P. Checkpoint Inhibitors and the Changing Face of the Relapsed/Refractory Classical Hodgkin Lymphoma Pathway. *Curr. Oncol. Rep.* **2022**, *24*, 1477–1488. [CrossRef]
65. Herrera, A.F.; LeBlanc, M.L.; Castellino, S.M.; Li, H.; Rutherford, S.C.; Evens, A.M.; Davison, K.; Punnett, A.; Hodgson, D.C.; Parsons, S.K. SWOG S1826, a randomized study of nivolumab (N)-AVD versus brentuximab vedotin (BV)-AVD in advanced stage (AS) classic Hodgkin lymphoma (HL). *J. Clin. Oncol.* **2023**, *41*. [CrossRef]
66. Lai, C.; Kandahari, A.M.; Ujjani, C. The Evolving Role of Brentuximab Vedotin in Classical Hodgkin Lymphoma. *Blood Lymphat. Cancer* **2019**, *9*, 63–71. [CrossRef]
67. Advani, R.H.; Moskowitz, A.J.; Bartlett, N.L.; Vose, J.M.; Ramchandren, R.; Feldman, T.A.; LaCasce, A.S.; Christian, B.A.; Ansell, S.M.; Moskowitz, C.H.; et al. Brentuximab vedotin in combination with nivolumab in relapsed or refractory Hodgkin lymphoma: 3-year study results. *Blood* **2021**, *138*, 427–438. [CrossRef]
68. Burlile, J.F.; Frechette, K.M.; Breen, W.G.; Hwang, S.R.; Higgins, A.S.; Nedved, A.N.; Harmsen, W.S.; Pulsipher, S.D.; Witzig, T.E.; Micallef, I.N.; et al. Patterns of progression after immune checkpoint inhibitors for Hodgkin lymphoma: Implications for radiation therapy. *Blood Adv.* **2024**, *8*, 1250–1257. [CrossRef]
69. Herrera, A.F.; LeBlanc, M.; Castellino, S.M.; Li, H.; Rutherford, S.C.; Evens, A.M.; Davison, K.; Punnett, A.; Parsons, S.K.; Ahmed, S.; et al. Nivolumab+AVD in Advanced-Stage Classic Hodgkin's Lymphoma. *N. Engl. J. Med.* **2024**, *391*, 1379–1389. [CrossRef]
70. Lucchini, E.; Rusconi, C.; Levis, M.; Ricci, F.; Santoro, A.; Ricardi, U.; Volpetti, S.; Matrone, F.; di Russo, A.; Caizzi, M.; et al. Immune checkpoint inhibitors in combination with radiotherapy as salvage treatment for relapsed/refractory classical Hodgkin lymphoma: A retrospective analysis in 12 patients. *Hematol. Rep.* **2021**, *13*, 9080. [CrossRef]

71. Salomon, R.; Dahan, R. Next Generation CD40 Agonistic Antibodies for Cancer Immunotherapy. *Front. Immunol.* **2022**, *13*, 940674. [CrossRef] [PubMed]
72. van Hooren, L.; Vaccaro, A.; Ramachandran, M.; Vazaios, K.; Libard, S.; van de Walle, T.; Georganaki, M.; Huang, H.; Pietila, I.; Lau, J.; et al. Agonistic CD40 therapy induces tertiary lymphoid structures but impairs responses to checkpoint blockade in glioma. *Nat. Commun.* **2021**, *12*, 4127. [CrossRef] [PubMed]
73. Vonderheide, R.H. CD40 Agonist Antibodies in Cancer Immunotherapy. *Annu. Rev. Med.* **2020**, *71*, 47–58. [CrossRef] [PubMed]

cancers

MDPI

Review

Increased c-MYC Expression Associated with Active IGH Locus Rearrangement: An Emerging Role for c-MYC in Chronic Lymphocytic Leukemia

Kenza Guiyedi [1,†], Milène Parquet [1,†], Said Aoufouchi [2], Jasmine Chauzeix [1,3], David Rizzo [1,3], Israa Al Jamal [1,4], Jean Feuillard [1,3], Nathalie Gachard [1,3] and Sophie Peron [1,*]

1 Centre National de la Recherche Scientifique (CNRS), Unité Mixte de Recherche (UMR) 7276/INSERM U1262, Université de Limoges, 87000 Limoges, France
2 Gustave Roussy, B-Cell and Genome Plasticity Team, CNRS UMR9019, Villejuif, France and Université Paris-Saclay, 91400 Orsay, France
3 Laboratoire d'Hématologie Biologique, Centre Hospitalier Universitaire de Limoges, 87000 Limoges, France
4 Faculty of Sciences, GSBT Genomic Surveillance and Biotherapy Team, Mont Michel Campus, Lebanese University, Tripoli 1300, Lebanon
* Correspondence: sophie.peron@unilim.fr; Tel.: +33-519-564-210
† These authors contributed equally to this work.

Simple Summary: This review explores the role of c-MYC in Chronic Lymphocytic Leukemia (CLL) and its impact on genetic instability and disease progression. As a key oncogene, c-MYC encodes a transcription factor involved in regulating cell cycle, growth, and apoptosis. We previously described CLL cases enriched with unmutated IGHV genes, *MYC* overexpression and with active rearrangement of the *IGH* immunoglobulin heavy chain (IGH) locus. The *MYC* overexpression seems to promote increased DNA damage, including double-strand breaks (DSBs), chromosomal translocations linked to repair errors during DNA repair. Highlighting c-MYC's dual role, this review try to show how *MYC* overexpression it not only driving cell proliferation but also contributes to genomic instability.

Abstract: This review examines the pivotal role of c-MYC in Chronic Lymphocytic Leukemia (CLL), focusing on how its overexpression leads to increased genetic instability, thereby accelerating disease progression. *MYC*, a major oncogene, encodes a transcription factor that regulates essential cellular processes, including cell cycle control, proliferation, and apoptosis. In CLL cases enriched with unmutated immunoglobulin heavy chain variable (IGHV) genes, *MYC* is significantly overexpressed and associated with active rearrangements in the *IGH* immunoglobulin heavy chain locus. This overexpression results in substantial DNA damage, including double-strand breaks, chromosomal translocations, and an increase in abnormal repair events. Consequently, c-MYC plays a dual role in CLL: it promotes aggressive cell proliferation while concurrently driving genomic instability through its involvement in genetic recombination. This dynamic contributes not only to CLL progression but also to the overall aggressiveness of the disease. Additionally, the review suggests that c-MYC's influence on genetic rearrangements makes it an attractive target for therapeutic strategies aimed at mitigating CLL malignancy. These findings underscore c-MYC's critical importance in advancing CLL progression, highlighting the need for further research to explore its potential as a target in future treatment approaches.

Keywords: MYC; CLL; recombination; DNA repair; B cell

1. Introduction

Chronic Lymphocytic Leukemia (CLL) is a highly frequent B-cell cancer, mostly affecting the elderly. With a very variable clinical presentation, the overall survival of CLL patients ranges from years to decades and remains incurable. CLL is an indolent lymphoma characterized by the accumulation of small monomorphic round CD19+ CD23+ CD5+ B lymphocytes (B-cells). CLL diagnosis is based on the detection of monoclonal CLL cells in the blood with lymphocytosis exceeding 5 g/L and infiltration of lymphoid organs. There is a preneoplastic stage termed monoclonal B-cell lymphocytosis (MBL) characterized by a blood clonal B-cell count <5 g/L without tumor infiltration of tissues or any feature diagnostic of another B-cell lymphoma or disorder. When CLL cells infiltrate secondary lymphoid organs, leading to lymphadenopathy or splenomegaly and lymphocytosis is <5 g/L in the blood, the diagnosis leads to small lymphocytic lymphoma (SLL) [1]. After CLL diagnosis, patient care and clinical management are decided based on Rai or Binet staging classifications [2,3]. Both classifications, easy to use for physicians, reliably predict disease progression. Moreover, karyotype and genetic heterogeneity influence overall survival and therapeutic response. The CLL International Prognostic Index (CLL-IPI) classifies patients by combining Rai and Binet scores with genetic parameters and provides treatment recommendations [4]. Several chromosomal abnormalities are recurrent in CLL and are predictive of CLL progression and/or resistance to therapy. Among these recurrent abnormalities, the most frequent and/or important in the decision tree for treatment include deletion of the long arm of chromosome 13 (del (13q)), trisomy 12, and other markers of poor prognosis, which are deletion of the long arm of chromosome 11 (del (11q)) and deletion of the short arm of chromosome 17 (del (17p)). Other poor prognosis genetic abnormalities involve the Notch pathway (*NOTCH1* mutations), NF-kappa B activation (*BIRC3* or *MYD88* mutations), splicing (*SF3B1* mutations), the DNA lesion sensor *ATM* or the *TP53* anti-oncogene (mutations and/or deletion) [4]. In addition, since the publication of papers by Hamblin and Damned [5,6], it is globally accepted that the mutational status of genes coding for the variable (*IGHV*) part of the immunoglobulin (Ig) heavy-chain (IGH) locus is an indicator of CLL prognosis. CLL patients with an unmutated variable region (umCLLs) have a poor prognosis, while patients whose IGHV gene has been targeted by mutations (mCLL) have a higher survival rate. Recently, we described CLLs with poor prognoses characterized by high c-MYC expression associated with active *IGH* locus recombination [7]. An increase in *IGH* recombination induced by MYC overexpression was observed in the CH12F3 B lymphocyte cell line, confirming that c-MYC potentiates the rearrangement of the IGH locus. Nevertheless, the precise mechanism remains questionable, and in this review, we recapitulate data from the literature concerning c-MYC to discuss the potential modus operandi and impact of c-MYC-induced *IGH* recombination on tumoral CLL B-cells.

2. c-MYC, a Master Orchestrator of B Cell Fate

The human MYC proto-oncogenes encode a family of three transcription factors: c-MYC, N-MYC, and L-MYC, encoded by genes located on chromosomes 8, 2, and 1, respectively. These MYC paralogs have similar structural architectures and functions but differential patterns of expression with distinct times and locations of expression during cell differentiation. The *MYC* oncogene was discovered because of its homology with the v-MYC oncogene in avian acute leukemia virus (MC29) [8]. More precisely, the c-MYC oncogene is located at 8q24 [9] and is composed of three exons (Figure 1): the non-coding exon 1 and exons 2 and 3 encoding the c-MYC transcription factor. In humans, four *MYC* promoters have been identified: P0 (human-specific and representing 5% of transcripts) [10], P1 (10 to 25% of transcripts), P2 (75 to 90% of transcripts) [11], and P3 (5% of transcripts) [12].

c-MYC protein is composed of 439 amino acids and is subdivided into three domains (Figure 1). The N-terminal transactivation domain (TAD) (residue 1–143), necessary for the biological activity of c-MYC and c-MYC mediated transcriptional activation, contains three short conserved segments, termed MYC boxes (MB0, MBI, and MBII). MBI, by recruiting cyclin CDK complexes, promotes transcriptional elongation by RNA polymerase II (RNAPII) phosphorylation [13]. MBII is required for c-MYC targeting of transcriptional activation and repression and controls c-MYC turnover. The central domain includes an acidic PEST domain (residues 226–229) adjacent to MBIII and MBIV conserved boxes. MBIII can recruit the histone deacetylase 3 (HDAC3) and is engaged in transcriptional repression.

MBIV contains the first nuclear localization site (NLS1, aa 320–328) and is required for c-MYC pro-apoptotic function. The C-terminal domain consists of a basic region that contains the second nuclear localization site (NLS2, aa 364–374), a helix-loop-helix (HLH) domain, and a leucine zipper. The bHLH-Zip domain forms stable heterodimer c-MYC/Max that directly binds DNA sequences at the canonical Enhancer box (E_box, 5′CACGTG-3′) to stimulate transcription [14].

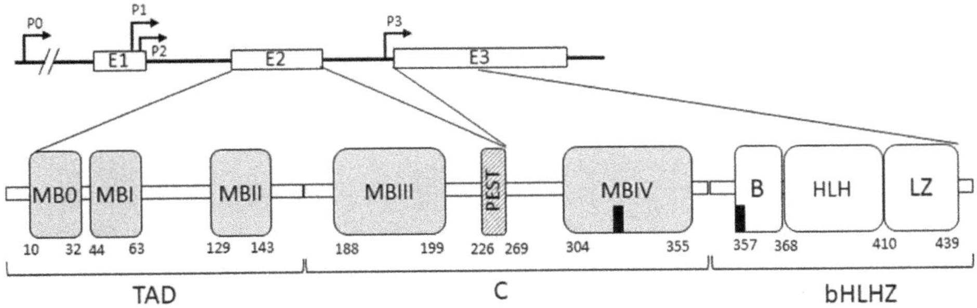

Figure 1. *MYC* gene and protein: the gene contains 4 promoters (P) and 3 exons (upper diagram). The protein contains 3 domains (lower diagram): the N-terminal transactivation domain (TAD) includes 3 conserved MYC boxes: MB0, MBI, and MBII. The Central domain (C) includes an acidic PEST domain, a MYC box IV, and a nuclear Localization site, NLS (black square). The C-terminal b-HLHZip domain includes a basic region (B), which contains a second NLS, a helix-loop-helix (HLH), and a leucine zipper (LZ). Not drawn to scale (from [15–18]).

The oncoprotein c-MYC is a ubiquitous transcription factor that regulates 10–15% of all human genes [19]. The ability of this protein to bind the consensus sequence (CACGTG) of gene E-boxes [20] results in a broad field of action and a key role in cell homeostasis, participating in the regulation of cell cycle, cell growth, survival, cellular metabolism, etc. [21,22], and also suggests a strict transcriptional regulation [23]. As MYC factors exert fundamental roles in cell functions in all tissues, they play a central role in B-cells, starting at B lymphoid lineage establishment to terminal differentiation into memory B-cells and plasma cells.

During early B-cell development, c-MYC regulates B-cell proliferation and survival depending on the developmental stage (Table 1). The crucial role of c-MYC in B-cell progenitors has been demonstrated using a mouse model of c-MYC conditional inactivation in B-cells, which showed a significant increase in apoptosis at all stages of early B development as well as the requirement for c-MYC in cell proliferation with a specificity depending on each developmental stage [24,25]. The expansion of pro-B cells has been shown to depend on the opposing effects of EBF1 and PAX5 on *MYC* gene regulation [26]. Presumably, c-MYC regulation relies on the balance between these two transcription factors, switching from negative to positive regulation in the same network, and also depends on the response to cytokines, including IL-7 [24]. In pre-B cells, c-MYC inactivation arrests cell development and blocks differentiation in immature B-cells. Specifically, during the maturation and expansion of large pre-B cells into small pre-B cells, c-MYC and *N-MYC* are expressed. In immature B cells, only c-MYC is expressed and then, in mature B cells, c-MYC is expressed only in mature B cells after their activation [27].

Following bone marrow exit, naïve mature B lymphocytes express a complete BCR and enter the peripheral immune system. In primary B cells, c-*MYC* expression induces an increase in intracellular calcium, which is itself necessary for stimulation of B cell proliferation and differentiation by c-MYC, as shown by Habib et al., 2007. c-MYC reduces PMCA4, which promotes proliferation, differentiation, and immunogenic response of B cells dependent on NFAT [27]. Furthermore, constitutive c-MYC signaling alone or supported by calcium/NFAT signals promotes B cell activation/differentiation or tumorigenesis. Finally, these authors also reported that *MYC* expression inducing high calcium levels increased NFAT translocation in the absence of upstream signaling from pre-BCR/BCR and increased the short isoform level of NFAT (NFATc1A), which does not promote apoptosis in lymphocytes compared to other NFATc1 isoforms of. Therefore c-MYC inducing high levels of calcium in B cells influences transcriptional and biological responses in a major fashion. In peripheral B cells, Habib and al. report that deletion of c-MYC and N-MYC together decreases peritoneal B cells, which express $CD5^{low}$, IgM^{high} (B1 B cells), and conventional B cells ($CD5^{neg}$, IgM^{high}, B2 B cells) suggesting that c-MYC is essential for the development of both populations. Antigen (Ag) binding by the BCR provides signals to B-cells to relocalize in the secondary lymphoid organs. Depending on the Ag nature, B-cells undergo T-cell independent or dependent activation. Follicular B-cells confronted with co-activation by CD4+ T lymphocytes and antigen-presenting cells will be activated. This process triggers class switch recombination (CSR) and germinal center (GC) formation, allowing the generation and selection of B-cells able to produce high-affinity antibodies. GC is a dynamic structure defined by dark (DZ) and light (LZ) zones, which are both characterized by differential marker expression at the surface of B-cells and stromal cells and different cellular compositions of the microenvironment. In the GC, B cells proliferate strongly: at this stage, B-cells are called centroblasts and undergo somatic hypermutation (SHM) of their immunoglobulin genes in the DZ. Conversely, in the LZ, B-cells, the centrocytes, reduce their proliferation and undergo cell selection dependent on Ag-BCR affinity. Selected cells generate plasma cells after terminal differentiation or memory B cells. B-cells with low BCR affinity for Ag enter either apoptosis or a state of energy. These cells can also be redirected to the DZ for a new cycle of selection. In order to control and regulate the GC reaction, different molecular pathways are involved. c-MYC represses the cell cycle inhibitor p-27 and, to a lesser extent, p-21 and is essential for the initiation and maintenance of the GC process [28]. In the DZ, where B-cell proliferation occurs at a high rate, BCL6 is expressed in order to protect the cell from apoptosis and to induce proliferation. c-MYC determines the number of cell divisions that occur in the DZ and is essential in GC initiation, maintenance, and positive selection [29–35]. In the LZ, CD40 and BCR activation signals induce increased PI3K/AKT pathway signaling, which in turn induces *MYC* expression [29,36]. This is also supported by the NF-κB and mTORC1 pathways. It has been shown that *MYC* expression is quantitatively proportional to the strength of T cell help in the LZ [30]. Cells that have actively received help from T cells promote the development of plasma cells and limit differentiation into memory cells [32,37,38]. The transcriptional program switch from B-cells to plasma cells is governed by BLIMP1, which suppresses *MYC* expression [28,31]. During the transit between the LZ and DZ, *MYC* and *BCL6* are both expressed and will be regulated depending on other zone-specific regulatory factors.

Table 1. c-MYC plays a crucial role in regulating B-cell proliferation and survival, but its effects vary depending on the B-cell's developmental stage. In pro-B cells, c-MYC regulates the cell cycle, driving pro-B cell proliferation and promoting V(D)J recombinations at the *IGH* locus. In pre-B cells, c-MYC promotes pre-B cell expansion, allowing the cells to divide before Ig light chain V(D)J recombination. As B cells mature and leave the bone marrow, *MYC* expression turns off. Once B cells recognize antigens, *MYC* expression is induced after BCR engagement to drive the rapid proliferation necessary for an effective immune response, and in the germinal center (GC), *MYC* is induced in centroblasts in the dark zone of the GC, where proliferation and somatic hypermutation occur to produce high-affinity antibodies.

Compartment	Bone Marrow		Peripheral Blood		Secondary Lymphoïd Organs			Peripheral Blood	Bone Marrow
					Extrafollicular Zone	Germinal Center			
						Dark Zone	Light Zone		
B cell Developmental Stage	Pro-B	Pre-B	Transitional B Cell	Mature B Cell	Activated B Cell	Centroblast	Centrocyte	Memory B Cell	Plasma Cell
MYC Expression	+	+	-	-	+	+	-	-	-

As previously mentioned, c-MYC is an important player in B lymphocyte development. Its expression is induced in waves depending on the stage, and its activity has been particularly implicated in the proliferation and apoptosis of B cells, modulated by other factors regulating these mechanisms.

3. c-MYC Implication in B Cell Cancers

Cancer Genome Atlas reports that c-MYC and its paralogs are involved in 28% of cancers [39]. c-MYC deregulation is more prevalent than N-MYC and L-MYC in human cancer types for both hematological and solid tumors [40]. Different mechanisms drive abnormal *MYC* activation in human cancers. This can be achieved through genetic alteration, gene expression deregulation, or protein modification (Table 2). The first implication of c-MYC in human cancer arose with the identification of the *MYC* gene translocation to the *IGH* locus in human Burkitt lymphoma (BL) [11]. c-MYC plays a role in several types of B-cell acute leukemia: B lymphoblastic leukemia (B-ALL) with t(9; 22) and *BCR::ABL1* rearrangement, B-ALL with t(v; 11) *KMT2A* rearrangement, B-ALL with the t(12; 21) and *ETV6::RUNX1* rearrangement and B-ALL with other chromosomal rearrangements as *TCF3::PBX1* in leukemia derived from pre-B lymphocytes [24]. In the last WHO revision of the classification of tumors of hematopoietic and lymphoid tissues, there is a subtype with *MYC* rearrangement distinct from other B-ALL types and Burkitt lymphoma (ref OMS 2022). c-MYC is also involved in many germinal center-derived lymphomas and the histological transformation of indolent mature B-cell malignancies (follicular lymphoma, CLL, MALT, DLBCL, BL, and PBL) [24,31].

In precursor B cells, BCL2 cooperates with c-MYC to promote their proliferation, but this cooperation can allow these cells to become tumorigenic, as shown by Strasser et al., 1990 when using double transgenic Eμ-Bcl2/myc mice, which developed more tumors compared to Eμ-myc only mice [41]. Bissonnette et al., 1992 and Fanidi et al. 1992 also demonstrated the inhibition of apoptosis mediated by MYC of BCL2 in some cells but not the proliferative function of c-myc [42,43]. Other proteins cooperate with c-MYC in tumorigenesis such as RAS, MYB, PIM1, RAF1 [44].

Its role in B lymphomagenesis begins in hematopoietic malignancies, where genomic abnormalities involving the *MYC* gene are almost always found. In Burkitt lymphoma, *MYC* rearrangement with one of the *IG* gene loci is a genetic hallmark of this lymphoma. In 80% of cases, *MYC* translocation will be with the *IGH* locus in 14q32 or with *IGK* or *IGL* genes at 2p12 or 22q11 in 10% of cases each [24,45]. In DLBCL, the c-MYC protein is detected in approximately 40% of patients at diagnosis without gene alterations and its rearrangement is found in approximately 10% of DLBCL cases [24,31,33,45]. As for Burkitt's lymphoma, Ig genes are the most frequent *MYC* partners, but they can also translocate in 35–50% with *BCL6*, *BCL2*, *PAX5*, or even *IKZK1* [24]. In follicular lymphoma, up to 75% of histological transformations to a high-grade lymphoma (t-FL) show *MYC* mutations/translocations/amplification or gains associated with *MYC* overexpression and other alterations. In PBL, recurrent somatic mutations in the gene encoding *BLIMP1* (*PRDM1*) occur in 50% of cases, and in 80% of diagnoses, c-MYC and BLIMP1 proteins are co-expressed. As in BL and DLBCL, *MYC* rearrangements are also characteristic of an aggressive subset in PBL. Along with *MYC* gains, they are found in ~50% of cases as partners and in ~85% of cases, the *IG* genes with t(8; 14) *MYC::IGH* as common fusion product. c-MYC pathway deregulation is also one of the major events in MALT transformation to DLBCL in 40–80% of cases. R.L. Brinster's group was the first to highlight *MYC* implication in tumoral transformation using a mouse model expressing *MYC* under the dependence of Eμ or Ig enhancers [46]. This work showed that abnormal *MYC* expression induces B-cell lymphomagenesis resulting in aggressive B-cell lymphomas. Other transgenic or knock-in mouse models, which expressed the *MYC* oncogene under Eμ/3′RR transcriptional control provided a better understanding of c-MYC-dependent B-cell malignant transformation mechanisms [47].

Benhamou et al. (2018) described an axis composed of c-MYC, a miRNA cluster (miR17-92), and the signaling molecule PTEN, which is critical for balancing PI3K activity in its control of B cell-fate decisions for positive and negative selection of immature B cells [48]. In their study, the authors pointed out that CD19 is a major B Cell Receptor (BCR)-independent regulator of c-MYC levels in B cell neoplasms by activating the PI3K/AKT/GSK3 axis to promote B cell transformation and lymphoma progression [49,50]. The same team also demonstrated the role of the c-MYC/miR17-92/PTEN axis in Recombination Activating Genes (RAG) expression in early B cell development. Indeed, they used different genetically modified mouse models and showed that altered function of the c-MYC/miR17-92/PTEN axis alters the PI3K/AKT/FOXO1 pathway, leading to deregulation of RAG expression and blockage of pre-BCR assembly [48].

Table 2. Molecular mechanisms associated with abnormal MYC activation in human cancers.

Type of Abnormalities	Observed Defects	References
Genetic aberrations	Insertional mutagenesis Chromosomal translocations	(Dudley et al. 2002) [51] (Erikson et al. 1983) (Meyer et Penn 2008) [52,53]
Gene expression	Gene amplification Deregulation of pathways that influence *MYC* oncogene transcription	(Collins et Groudine 1982) [54] (Alitalo et al. 1983) [55] (Nau et al. 1985) [56] (Mariani-Costantini et al. 1988) [57] (Münzel et al. 1991) [58] (Augenlicht et al. 1997) (Kalkat et al. 2017) [59,60] (He et al. 1998) (Yagi et al. 2002) [61,62]
Protein modifications	Post-translational modifications of MYC proteins	(Allen-Petersen et Sears 2019) [63] (Sears et al. 2000) [64]

4. c-MYC in Chronic Lymphocytic Leukemia

In CLL, *MYC* expression is detectable in proliferating pseudo-follicle CLL cells (>25% of c-MYC-positive cells). Moreover, deregulation of the c-MYC pathway and its gain in activity are frequently observed (>70% of cases). Simultaneous aberrations in *MYC* and *TP53* are very common in Richter's transformation of CLL. In most cases, the *MYC* gene remains intact [24,31].

Causes of *MYC* over-activation, in addition to its genetic change, are diverse: gain of the miR-17-92 cluster regulating *MYC*, mutations/deletions of *MYC* regulator *MGA*, mutations in *MYC*-transcriptional activator *NOTCH1*, T-cell interactions via CD40L or interactions with stromal or nurse-like cells producing BAFF and activating NF-κB and/or BCR signaling, mutation in BCR signaling related genes (*CD79* or *CARD11*) driving the expression of *MYC* in general. An interesting mechanism of BCR activation leading to *MYC* induction has also been proposed in CLL. IgM signaling promotes increased translation of *MYC* RNA and thus participates in a feedback loop that enhances c-MYC activity by eIF4. In some cases, the Ig surface of CLL cells is completely glycosylated in a constant region and contains mannosylated sites in the variable region. These glycosylated Ig interact with lectins, and dissimilar to the normal surface Ig reaction with antigen, binding to lectins does not cause surface Ig endocytosis but leads to persistent signaling and strong ascending regulation of *MYC*.

Overall, BCR signaling in *MYC*-driven malignant B cells is an advantage, but BCR-mediated signaling can be considered an important but replaceable means of ensuring c-MYC activity. In addition, higher levels of c-MYC are able to affect BCR signaling, including its binding to the promoter regions of CD79a and SYK and their overexpression in precancerous B lymphocytes in *Eμ-myc* mice. High c-MYC levels can also downregulate inhibitors of BCR signaling by regulating microRNAs or even lead to its deregulation by inducing the *miR-17-92* cluster (*miR-19*, *miR-18α*) [65–68]. In this case, genomic regions containing these miRNAs are frequently upregulated/amplified in Richter's transformation of CLL and clinically aggressive CLL. Moreover, in *t*-FL, when c-MYC binds to the region upstream of *miR-150*, its level is reduced. *Mir-150* regulates transcription factor FOXP1, and it has been shown that lower *miR-150* levels and higher FOXP1 levels are associated with shorter survival in CLL [31].

Deregulation of BCR signaling is not the only impact of c-MYC-mediated microRNA regulation. Many hallmarks of cancer are impacted such as proliferation, cell survival, angiogenesis, vasculogenesis, metabolism and apoptosis by repression of *miR-15a/16-1*, *miR-28*, *miR-29*, and *miR-34α*, upregulated of *miR-17-92*, *miR-106a/b*, *miR-20a*, *miR-93*, *miR-17*, and *miR-23a/b* or cooperation of c-MYC with *miR-548* and *miR-155* for example. It should be noted that *miR-34α* is also known to be the first link between DNA damage response and BCR signaling by limiting the propensity of the BCR pathway via repression of FOXP1 during the DNA damage response, thus ensuring interruption of this pro-proliferative signal. In addition, in B-cell malignancies, in vivo, induction of *miR-34α* and arresting BCR signaling by using Fludarabine and doxorubicin did not work in malignant cells with impaired p53 pathways. Therefore, the repression of both *miR-150* and *miR-34α* by *MYC*-activating aberrations in combination with *TP53* deletion/mutation, which also represses *miR-34α* and which are both very common in Richter's transformation of CLL, could lead to prominent BCR signaling upregulation. The down-modulation of both *miR-150* and *miR-34a* has been reported as a constant phenomenon.

5. MYC and Treatment for CLL

Up to 2020, immunochemotherapy was the backbone of first-line CLL treatment [69]. Based on fludarabine eligibility, the reference treatment was fludarabine, cyclophosphamide, and rituximab combinaison (FCR) with 6 or 4 monthly cycles depending on age. In FCR-ineligible patients, immunotherapy remained the gold standard using the combinaison of bendamustine and rituximab or chlorambucil and obinutuzumab. The presence of a significant TP53 alteration contraindicates immunochemotherapy, making search tests mandatory before any treatment. When TP53 is altered, continuous ibrutinib treatment is the gold standard. Current guidelines recommend targeted therapies as first-line treatment, mainly BTK inhibitors and venetoclax alone, in combination or combined with obinutuzumab. FCR is no longer proposed as a first-line treatment.

MYC exhibits prognostic significance in chronic lymphocytic leukemia (CLL) concerning two types of 8q24 abnormalities: translocations involving MYC and genomic gains. MYC translocations are infrequent, while gains of one or more copies of the MYC gene occur in less than 1% and 5–6% of CLL cases, respectively. The poor prognosis associated with these alterations correlates with an aggressive clinical course and reduced overall survival (OS) [70]. Additionally, MYC gain may be associated with del (17p), which further exacerbates the prognosis [71]. However, alterations in MYC do not yield specific treatment recommendations, and their impact on therapeutic outcomes remains inadequately characterized. The potential role of MYC activation in mediating drug resistance, particularly to BTK inhibitors, is currently unknown.

Recently, MYC has been identified as a potential therapeutic target in CLL. Targeting the PHB-eIF4F complex inhibits MYC translation, resulting in reduced proliferation and survival of CLL cells. These results suggest that disruption of the PHB-eIF4F interaction could represent a promising therapeutic strategy for CLL patients by down-regulating MYC activity and modifying the malignant phenotype [72]. Furthermore, MYC orchestrates metabolic reprogramming in CLL cells, and a study by Simon-Molas et al. [73] shows that it is possible to target MYC-mediated metabolic alterations to improve therapeutic outcomes in CLL.

6. c-MYC Contribution to DNA Damage and Recombination

Under normal conditions in B lymphocytes, waves of *MYC* expression are coordinated with specific development stages, and when *MYC* is expressed, its activity is principally interpreted regarding its role in regulating proliferation and apoptosis. In addition, in CLL and Richter's syndrome, the contribution of c-MYC to tumorigenesis is related to its role in cell proliferation. In our previous work [7], we observed an association between increased *MYC* expression and an active recombination process at the constant part of the IGH locus in CLL cases. This observation reinforces the concept that c-MYC promotes DNA rearrangement, which can occur through several mechanisms.

First, increased c-MYC-induced proliferation can be the source of non-programmed double-strand breaks (DSBs), leading to genetic rearrangement if misrepaired. Indeed, DNA DSBs incessantly arise in the cell genome. These non-programmed DSBs are mainly produced through the conversion of single-strand breaks (SSB). The latter are produced during replication fork collapse or stalling because of topological stress [74–79]. In certain circumstances, replication and transcription proceed on the same genes and collide (TRI for transcription–replication interaction) [80–82], resulting in DNA DSBs [83]. Other replication obstacles, such as non-B DNA structures (R loops, G quadruplexes, hairpins, cruciform) or chromatin alteration, can also lead to DNA lesions and breakage. Transcriptional stress has also been described as source a of DSBs. Altogether, with increased cell replication and transcription, it is highly conceivable that increased c-MYC activity can potentialize DNA recombination.

Furthermore, c-MYC may be involved in mechanisms that are not restricted to its transcription factor function-inducing players implicated in cell proliferation, as c-MYC can be involved directly in processes leading to genetic rearrangements. Littler et al. examined the ability of c-MYC to modulate mitosis [84]. The authors developed the Tet-on FC-*MYC* cell model with a tetracycline-sensitive *MYC* transgene and mutated endogenous *MYC* alleles. Thus, they were able to study the role of c-MYC in mitotic chromosome segregation by modulating c-MYC expression either weakly (so-called MYC-Low system) or strongly (so-called MYC-High system). What was observed in MYC-High cells was a change in mitotic timing, as well as a morphological change in the metaphase mitotic spindle. In MYC-High cells, the rupture of the nuclear envelope in metaphase was accelerated, while the transition from metaphase to anaphase was delayed (increased metaphase in MYC-High cells). In addition, automated analysis of high-throughput images revealed a change in the proportions of the mitotic spindle during metaphase in terms of length and width in MYC-High cells (increased width and reduced length). Yet, as Levine and Holland's review of work identifying the sources of mitotic errors in human tumors and their effect on cell fitness and transformation reminds us, mitosis is a delicate event that must be executed with great fidelity to ensure genomic stability [85]. Data in the literature also argue that c-MYC potentializes DNA breaks, which promote DNA recombination. For example, Vafa et al. reported that overexpression of c-MYC in vitro in resting human fibroblasts led to an increase in reactive oxygen species (ROS) correlated with DNA DSB accumulation, which was not associated with cell entry into apoptosis [86]. The same study showed that cells could enter the cycle despite the presence of DNA damage by overriding P53-mediated cell cycle arrest, leading to the proliferation of damaged cells. On the other hand, Ray et al. [87] demonstrated that c-MYC could also induce widespread DNA breaks both in vitro (in normal human foreskin fibroblasts) and in vivo (in murine lymphocytes) without ROS intervention. Furthermore, Karlsson et al. showed that, in addition to increasing DSB generation in vitro (in normal human fibroblasts inducibly overexpressing c-MYC or NHF-MYCER), c-MYC overexpression led to a global defect in the repair of these lesions by affecting the homologous recombination (HR), non-homologous end joining (NHEJ) and single-strand annealing (SSA) pathways; but not the nucleotide excision repair (NER) pathway [88]. These results were obtained by testing the efficiency of DSB repair in a hamster cell line designed to contain a GFP-coupled recombination substrate and inducible for *MYC* overexpression (DRAA8-MYCER). These same studies also highlighted the role of c-MYC in inducing genetic rearrangements. Indeed, karyotypic analysis of 100 NHF-MYCER metaphases showed that *MYC* overexpression induces chromatid breaks, deletions, and translocations. Finally, Chiang et al. pointed out that Nijmegen syndrome, a rare genetic disorder of chromosomal instability associated with cancer predisposition, radiosensitivity, and chromosomal instability, involves the NBS1 (p95 or nibrin) gene, part of the MRE11/RAD50/NBS1 complex which is an essential component of the NHEJ repair system [89]. In vitro, in lymphoblastoid cell lines overexpressing c-MYC either constitutively (CBMyc.Max) or in an inducible manner (EREB. TCMyc), as well as in a murine fibroblast cell line (NIH3T3) overexpressing c-MYC, the team's work showed that NBS1 was a direct transcriptional target of the c-MYC transcription factor. Furthermore, RNA interference experiments in this study confirmed the upregulation of *NBS1* expression by endogenous c-MYC in kidney epithelial cell lines (293T) and in the HeLa cancer line. Chiang et al. also reported in this study that regulation of NBS1 by c-MYC required an e-box site in intron 1 of the *NBS1* gene; and that overexpression of *NBS1* induced cell proliferation (in Rat1a line cells).

Overall, there are several indications of c-MYC involvement in the induction of DNA lesions, and this is likely correlated with its expression level, meaning that oncogenic transformation, which turns on new origins of replication and increases transcription, combined with *MYC* overexpression, could be synergistic and lead to genetic rearrangements.

7. Discussion: Proposed Mode of Action of c-MYC Promoting Sμ-3′RR IGH Locus Rearrangement in CLL

Genomic instability is one of the hallmarks of cancer. Reciprocal chromosome translocations are well-known in B lymphomas as a primary event in lymphomagenesis [90]. In CLL, without being frequent, chromosome translocations are not rare, occurring in 20% of cases [91], and 4 to 9% involve *IG* loci, mainly *IGH* [92]. The prognosis of CLL is then rather pejorative. It has been shown that the cytogenetic abnormalities characteristic of tumoral B cells in CLL are linked to the accumulation of DNA damage with DSBs within these cells [93]. Indeed, translocations originate from illegitimate recombination between two chromosomes as a result of DSB repair errors. Most DSBs are normally generated during physiological processes. This is the case for DSBs at the IG loci produced by RAG enzymes during V(D)J rearrangements or induced by Activation-Induced cytidine Deaminase (AID) during Ig class recombination. The DSBs can also be generated randomly as accidents of normal cellular mechanisms or following DNA damage by endogenous or exogenous agents. Unscheduled DSBs are increased by oncogenic stress due to increased rates of replication and transcription and a rise in oxidative stress leading to the production of ROS, which has been shown to be involved in the direct and indirect production of DSBs [94,95]. We observed an increase in *IGH* locus transcription in CLL cases with high *MYC* expression [7], and CLL cells have increased ROS production [96,97]. In these cases, c-MYC could participate in the genomic instability and specifically potentialize *IGH* locus recombination. Intra-*IGH* constant region recombination remains to be elucidated. Generally, in B lymphomas, translocations involving *IGH* usually result in altered expression of the partner gene, frequently an oncogene, but in CLL, the deregulation induced by these translocations is poorly understood. Thirty to forty partner loci have been listed and, in most cases, without identification of the genes involved. However, the spatial conformation of the *IGH* locus in B-cells can account for the bias of increased recombination within the *IGH* locus constant region. Recombination events depend on interactions between distantly located DNA ends that can occur because of the three-dimensional (3D) chromatin architecture. In B-cells, it has been shown the *IGH* constant region is poised in a chromatin loop holding the enhancers Eμ and 3′α, which are separated by more than 100 kb, in close proximity [98]. This structure facilitates contact between Sμ, which is proximal to Eμ, and 3′α and enables recombination between these DNA regions after DSB lesions. Whether the IGH spatial conformation is altered or not and how it favors Sμ-3′RRrec in CLL remain to be determined. Another intriguing question is the impact of such IGH recombination on the homeostasis of the tumoral CLL cell pool. By abrogating IGH expression and BCR expression at the B-cell surface, recombination of the IGH locus between the switch μ (Sμ) region and the 3′ regulatory regions (3′RR) [99–102], (Sμ-3′RRrec), is supposed to induce cell death after BCR-mediated survival signaling suppression. In our study, tumor cells from CLL Sμ-3′RRrec High cases all expressed IGM at the cell surface. The Sμ-3RRrec junctions we detected were most likely amplified from the non-functional IGH allele. Nevertheless, it can be assumed that this recombination can also target the functional allele, leading to CLL apoptosis. The study conducted was performed on biological material extracted from blood mononuclear cells, which are resting cells, and Sμ-3′RRrec detection in secondary lymphoid organs where CLL cells can be activated within proliferative centers and where a direct effect of Sμ-3′RRrec could be observed on the CLL cell pool homeostasis would be of great interest.

8. Conclusions

c-MYC oncogenic activity is well established concerning the activation of cell proliferation through transcription of target genes and alteration of cell cycle control which are among its classical functions. However, c-MYC appears to have non-classical functions, such as induction of DNA lesions and genetic recombination. Whether these functions contribute to tumor transformation and evolution is difficult to evaluate, but evidence of increased c-MYC expression associated with active IGH locus recombination in CLLs of poor prognosis suggests that these non-classical functions c-MYC promote more aggressive tumorigenesis. This can be achieved through genome instability. On the other hand, c-MYC-induced Sμ-3′RRrec on the IGH locus, inducing BCR loss at the cell surface, potentially influences the CLL B-cell fate. Therefore, the impact of non-classical c-MYC functions on the maintenance of genomic stability appears crucial in tumor genesis. Future studies could address how high levels of c-MYC modulate chromatin architecture and impact cellular transcriptional control in CLL. This modulation affects long-range genomic interactions, enabling the cell to coordinate the expression of genes located far apart on the genome, thereby facilitating changes in transcriptional programs. In the context of neoplastic transformation, this chromatin reorganization likely contributes to the deregulated gene expression that underpins cancer aggressiveness. Thus, the identification of MYC-specific alterations in gene expression in CLL could participate in the development of targeted therapies.

Author Contributions: Writing—review and editing, K.G., M.P., N.G., S.A., J.C., D.R., I.A.J., J.F. and S.P.; funding acquisition, S.P. All authors have read and agreed to the published version of the manuscript.

Funding: This research was funded by la Ligue Contre le Cancer (Comité de la Haute-Vienne, CD87 and Comité de la Creuse CD23), the Fondation ARC pour la recherche sur le cancer (ARCPJA2022060005138). M.P. is supported by Région Nouvelle Aquitaine and Université de Limoges. K.G. is supported by Région Nouvelle Aquitaine and Délégation INSERM Nouvelle Aquitaine. S.P. is a National Institute of Health and Medical Research (INSERM) investigator.

Acknowledgments: The authors would like to thank Jeanne Cook Moreau (UMR CNRS 7276 INSERM 1276) for careful English editing.

Conflicts of Interest: The authors declare no conflicts of interest.

References

1. Hallek, M.; Cheson, B.D.; Catovsky, D.; Caligaris-Cappio, F.; Dighiero, G.; Döhner, H.; Hillmen, P.; Keating, M.J.; Montserrat, E.; Rai, K.R.; et al. Guidelines for the Diagnosis and Treatment of Chronic Lymphocytic Leukemia: A Report from the International Workshop on Chronic Lymphocytic Leukemia Updating the National Cancer Institute-Working Group 1996 Guidelines. *Blood* **2008**, *111*, 5446–5456. [CrossRef] [PubMed]
2. Rai, K.R.; Sawitsky, A.; Cronkite, E.P.; Chanana, A.D.; Levy, R.N.; Pasternack, B.S. Clinical Staging of Chronic Lymphocytic Leukemia. *Blood* **1975**, *46*, 219–234. [CrossRef] [PubMed]
3. Binet, J.L.; Auquier, A.; Dighiero, G.; Chastang, C.; Piguet, H.; Goasguen, J.; Vaugier, G.; Potron, G.; Colona, P.; Oberling, F.; et al. A New Prognostic Classification of Chronic Lymphocytic Leukemia Derived from a Multivariate Survival Analysis. *Cancer* **1981**, *48*, 198–206. [CrossRef]
4. Rossi, D.; Khiabanian, H.; Spina, V.; Ciardullo, C.; Bruscaggin, A.; Famà, R.; Rasi, S.; Monti, S.; Deambrogi, C.; De Paoli, L.; et al. Clinical Impact of Small TP53 Mutated Subclones in Chronic Lymphocytic Leukemia. *Blood* **2014**, *123*, 2139–2147. [CrossRef] [PubMed]
5. Hamblin, T.J.; Davis, Z.; Gardiner, A.; Oscier, D.G.; Stevenson, F.K. Unmutated Ig V(H) Genes Are Associated with a More Aggressive Form of Chronic Lymphocytic Leukemia. *Blood* **1999**, *94*, 1848–1854. [CrossRef] [PubMed]

6. Damle, R.N.; Wasil, T.; Fais, F.; Ghiotto, F.; Valetto, A.; Allen, S.L.; Buchbinder, A.; Budman, D.; Dittmar, K.; Kolitz, J.; et al. Ig V Gene Mutation Status and CD38 Expression as Novel Prognostic Indicators in Chronic Lymphocytic Leukemia. *Blood* **1999**, *94*, 1840–1847. [CrossRef] [PubMed]
7. Al Jamal, I.; Parquet, M.; Guiyedi, K.; Aoufouchi, S.; Le Guillou, M.; Rizzo, D.; Pollet, J.; Dupont, M.; Boulin, M.; Faumont, N.; et al. IGH 3′RR Recombination Uncovers a Non-Germinal Center Imprint and c-MYC-Dependent IGH Rearrangement in Unmutated Chronic Lymphocytic Leukemia. *Haematologica* **2024**, *109*, 466–478. [CrossRef] [PubMed]
8. Dalla-Favera, R.; Bregni, M.; Erikson, J.; Patterson, D.; Gallo, R.C.; Croce, C.M. Human C-Myc Onc Gene Is Located on the Region of Chromosome 8 That Is Translocated in Burkitt Lymphoma Cells. *Proc. Natl. Acad. Sci. USA* **1982**, *79*, 7824–7827. [CrossRef]
9. Neel, B.G.; Jhanwar, S.C.; Chaganti, R.S.; Hayward, W.S. Two Human C-Onc Genes Are Located on the Long Arm of Chromosome 8. *Proc. Natl. Acad. Sci. USA* **1982**, *79*, 7842–7846. [CrossRef]
10. Bentley, D.L.; Groudine, M. A Block to Elongation Is Largely Responsible for Decreased Transcription of C-Myc in Differentiated HL60 Cells. *Nature* **1986**, *321*, 702–706. [CrossRef]
11. Taub, R.; Moulding, C.; Battey, J.; Murphy, W.; Vasicek, T.; Lenoir, G.M.; Leder, P. Activation and Somatic Mutation of the Translocated C-Myc Gene in Burkitt Lymphoma Cells. *Cell* **1984**, *36*, 339–348. [CrossRef] [PubMed]
12. Luo, Q.; Li, J.; Cenkci, B.; Kretzner, L. Autorepression of C-Myc Requires Both Initiator and E2F-Binding Site Elements and Cooperation with the P107 Gene Product. *Oncogene* **2004**, *23*, 1088–1097. [CrossRef] [PubMed]
13. Kato, G.J.; Barrett, J.; Villa-Garcia, M.; Dang, C.V. An Amino-Terminal c-Myc Domain Required for Neoplastic Transformation Activates Transcription. *Mol. Cell. Biol.* **1990**, *10*, 5914–5920. [CrossRef] [PubMed]
14. Ayer, D.E.; Kretzner, L.; Eisenman, R.N. Mad: A Heterodimeric Partner for Max That Antagonizes Myc Transcriptional Activity. *Cell* **1993**, *72*, 211–222. [CrossRef]
15. Jha, R.K.; Kouzine, F.; Levens, D. MYC Function and Regulation in Physiological Perspective. *Front. Cell Dev. Biol.* **2023**, *11*, 1268275. [CrossRef]
16. Baluapuri, A.; Wolf, E.; Eilers, M. Target-Gene Independent Functions of MYC Oncogenes. *Nat. Rev. Mol. Cell Biol.* **2020**, *21*, 255–267. [CrossRef]
17. Cencioni, C.; Scagnoli, F.; Spallotta, F.; Nasi, S.; Illi, B. The "Superoncogene" Myc at the Crossroad between Metabolism and Gene Expression in Glioblastoma Multiforme. *Int. J. Mol. Sci.* **2023**, *24*, 4217. [CrossRef] [PubMed]
18. Sakamuro, D.; Prendergast, G.C. New Myc-Interacting Proteins: A Second Myc Network Emerges. *Oncogene* **1999**, *18*, 2942–2954. [CrossRef]
19. Strobl, L.J.; Kohlhuber, F.; Mautner, J.; Polack, A.; Eick, D. Absence of a Paused Transcription Complex from the C-Myc P2 Promoter of the Translocation Chromosome in Burkitt's Lymphoma Cells: Implication for the c-Myc P1/P2 Promoter Shift. *Oncogene* **1993**, *8*, 1437–1447. [PubMed]
20. Conacci-Sorrell, M.; McFerrin, L.; Eisenman, R.N. An Overview of MYC and Its Interactome. *Cold Spring Harb. Perspect. Med.* **2014**, *4*, a014357. [CrossRef]
21. Pelengaris, S.; Khan, M. The Many Faces of C-MYC. *Arch. Biochem. Biophys.* **2003**, *416*, 129–136. [CrossRef] [PubMed]
22. Dang, C.V.; O'Donnell, K.A.; Zeller, K.I.; Nguyen, T.; Osthus, R.C.; Li, F. The C-Myc Target Gene Network. *Semin. Cancer Biol.* **2006**, *16*, 253–264. [CrossRef] [PubMed]
23. Spencer, C.A.; Groudine, M. Control of C-Myc Regulation in Normal and Neoplastic Cells. *Adv. Cancer Res.* **1991**, *56*, 1–48. [CrossRef]
24. de Barrios, O.; Meler, A.; Parra, M. MYC's Fine Line Between B Cell Development and Malignancy. *Cells* **2020**, *9*, 523. [CrossRef] [PubMed]
25. Vallespinós, M.; Fernández, D.; Rodríguez, L.; Alvaro-Blanco, J.; Baena, E.; Ortiz, M.; Dukovska, D.; Martínez, D.; Rojas, A.; Campanero, M.R.; et al. B Lymphocyte Commitment Program Is Driven by the Proto-Oncogene *c-Myc*. *J. Immunol.* **2011**, *186*, 6726–6736. [CrossRef] [PubMed]
26. Somasundaram, R.; Jensen, C.T.; Tingvall-Gustafsson, J.; Åhsberg, J.; Okuyama, K.; Prasad, M.; Hagman, J.R.; Wang, X.; Soneji, S.; Strid, T.; et al. EBF1 and PAX5 Control Pro-B Cell Expansion via Opposing Regulation of the Myc Gene. *Blood* **2021**, *137*, 3037–3049. [CrossRef] [PubMed]
27. Habib, T.; Park, H.; Tsang, M.; de Alborán, I.M.; Nicks, A.; Wilson, L.; Knoepfler, P.S.; Andrews, S.; Rawlings, D.J.; Eisenman, R.N.; et al. Myc Stimulates B Lymphocyte Differentiation and Amplifies Calcium Signaling. *J. Cell Biol.* **2007**, *179*, 717–731. [CrossRef]
28. Bisso, A.; Sabò, A.; Amati, B. MYC in Germinal Center-Derived Lymphomas: Mechanisms and Therapeutic Opportunities. *Immunol. Rev.* **2019**, *288*, 178–197. [CrossRef] [PubMed]
29. Luo, W.; Weisel, F.; Shlomchik, M.J. B Cell Receptor and CD40 Signaling Are Rewired for Synergistic Induction of the C-Myc Transcription Factor in Germinal Center B Cells. *Immunity* **2018**, *48*, 313–326.e5. [CrossRef]
30. Finkin, S.; Hartweger, H.; Oliveira, T.Y.; Kara, E.E.; Nussenzweig, M.C. Protein Amounts of the MYC Transcription Factor Determine Germinal Center B Cell Division Capacity. *Immunity* **2019**, *51*, 324–336.e5. [CrossRef]
31. Filip, D.; Mraz, M. The Role of MYC in the Transformation and Aggressiveness of 'Indolent' B-Cell Malignancies. *Leuk. Lymphoma* **2020**, *61*, 510–524. [CrossRef] [PubMed]
32. Dominguez-Sola, D.; Victora, G.D.; Ying, C.Y.; Phan, R.T.; Saito, M.; Nussenzweig, M.C.; Dalla-Favera, R. The Proto-Oncogene MYC Is Required for Selection in the Germinal Center and Cyclic Reentry. *Nat. Immunol.* **2012**, *13*, 1083–1091. [CrossRef]

33. Calado, D.P.; Sasaki, Y.; Godinho, S.A.; Pellerin, A.; Köchert, K.; Sleckman, B.P.; de Alborán, I.M.; Janz, M.; Rodig, S.; Rajewsky, K. The Cell-Cycle Regulator c-Myc Is Essential for the Formation and Maintenance of Germinal Centers. *Nat. Immunol.* **2012**, *13*, 1092–1100. [CrossRef] [PubMed]
34. Sander, S.; Calado, D.P.; Srinivasan, L.; Köchert, K.; Zhang, B.; Rosolowski, M.; Rodig, S.J.; Holzmann, K.; Stilgenbauer, S.; Siebert, R.; et al. Synergy between PI3K Signaling and MYC in Burkitt Lymphomagenesis. *Cancer Cell* **2012**, *22*, 167–179. [CrossRef] [PubMed]
35. Shlomchik, M.J.; Luo, W.; Weisel, F. Linking Signaling and Selection in the Germinal Center. *Immunol. Rev.* **2019**, *288*, 49–63. [CrossRef] [PubMed]
36. Laidlaw, B.J.; Cyster, J.G. Transcriptional Regulation of Memory B Cell Differentiation. *Nat. Rev. Immunol.* **2021**, *21*, 209–220. [CrossRef] [PubMed]
37. Jellusova, J. The Role of Metabolic Checkpoint Regulators in B Cell Survival and Transformation. *Immunol. Rev.* **2020**, *295*, 39–53. [CrossRef]
38. Toboso-Navasa, A.; Gunawan, A.; Morlino, G.; Nakagawa, R.; Taddei, A.; Damry, D.; Patel, Y.; Chakravarty, P.; Janz, M.; Kassiotis, G.; et al. Restriction of Memory B Cell Differentiation at the Germinal Center B Cell Positive Selection Stage. *J. Exp. Med.* **2020**, *217*, e20191933. [CrossRef]
39. Schaub, F.X.; Dhankani, V.; Berger, A.C.; Trivedi, M.; Richardson, A.B.; Shaw, R.; Zhao, W.; Zhang, X.; Ventura, A.; Liu, Y.; et al. Pan-Cancer Alterations of the MYC Oncogene and Its Proximal Network across the Cancer Genome Atlas. *Cell Syst.* **2018**, *6*, 282–300.e2. [CrossRef]
40. Dhanasekaran, R.; Deutzmann, A.; Mahauad-Fernandez, W.D.; Hansen, A.S.; Gouw, A.M.; Felsher, D.W. The MYC Oncogene—The Grand Orchestrator of Cancer Growth and Immune Evasion. *Nat. Rev. Clin. Oncol.* **2022**, *19*, 23–36. [CrossRef]
41. Strasser, A.; Harris, A.W.; Bath, M.L.; Cory, S. Novel Primitive Lymphoid Tumours Induced in Transgenic Mice by Cooperation between Myc and Bcl-2. *Nature* **1990**, *348*, 331–333. [CrossRef]
42. Bissonnette, R.P.; Echeverri, F.; Mahboubi, A.; Green, D.R. Apoptotic Cell Death Induced by C-Myc Is Inhibited by Bcl-2. *Nature* **1992**, *359*, 552–554. [CrossRef]
43. Fanidi, A.; Harrington, E.A.; Evan, G.I. Cooperative Interaction between C-Myc and Bcl-2 Proto-Oncogenes. *Nature* **1992**, *359*, 554–556. [CrossRef] [PubMed]
44. Hoffman, B.; Amanullah, A.; Shafarenko, M.; Liebermann, D.A. The Proto-Oncogene c-Myc in Hematopoietic Development and Leukemogenesis. *Oncogene* **2002**, *21*, 3414–3421. [CrossRef] [PubMed]
45. Robaina, M.C.; Mazzoccoli, L.; Esteves Klumb, C. Germinal Centre B Cell Functions and Lymphomagenesis: Circuits Involving MYC and MicroRNAs. *Cells* **2019**, *8*, 1365. [CrossRef] [PubMed]
46. Adams, J.M.; Harris, A.W.; Pinkert, C.A.; Corcoran, L.M.; Alexander, W.S.; Cory, S.; Palmiter, R.D.; Brinster, R.L. The C-Myc Oncogene Driven by Immunoglobulin Enhancers Induces Lymphoid Malignancy in Transgenic Mice. *Nature* **1985**, *318*, 533–538. [CrossRef] [PubMed]
47. Ferrad, M.; Ghazzaui, N.; Issaoui, H.; Cook-Moreau, J.; Denizot, Y. Mouse Models of C-Myc Deregulation Driven by IgH Locus Enhancers as Models of B-Cell Lymphomagenesis. *Front. Immunol.* **2020**, *11*, 1564. [CrossRef] [PubMed]
48. Benhamou, D.; Labi, V.; Getahun, A.; Benchetrit, E.; Dowery, R.; Rajewsky, K.; Cambier, J.C.; Melamed, D. The C-Myc/MiR17-92/PTEN Axis Tunes PI3K Activity to Control Expression of Recombination Activating Genes in Early B Cell Development. *Front. Immunol.* **2018**, *9*, 2715. [CrossRef] [PubMed]
49. Chung, E.Y.; Psathas, J.N.; Yu, D.; Li, Y.; Weiss, M.J.; Thomas-Tikhonenko, A. CD19 Is a Major B Cell Receptor-Independent Activator of MYC-Driven B-Lymphomagenesis. *J. Clin. Investig.* **2012**, *122*, 2257–2266. [CrossRef] [PubMed]
50. Poe, J.C.; Minard-Colin, V.; Kountikov, E.I.; Haas, K.M.; Tedder, T.F. A C-Myc and Surface CD19 Signaling Amplification Loop Promotes B Cell Lymphoma Development and Progression in Mice. *J. Immunol.* **2012**, *189*, 2318–2325. [CrossRef]
51. Dudley, J.; Mertz, J.; Rajan, L.; Lozano, M.; Broussard, D. What Retroviruses Teach Us about the Involvement of C-Myc in Leukemias and Lymphomas. *Leukemia* **2002**, *16*, 1086–1098. [CrossRef] [PubMed]
52. Meyer, N.; Penn, L.Z. Reflecting on 25 Years with MYC. *Nat. Rev. Cancer* **2008**, *8*, 976–990. [CrossRef] [PubMed]
53. Erikson, J.; Ar-Rushdi, A.; Drwinga, H.L.; Nowell, P.C.; Croce, C.M. Transcriptional Activation of the Translocated C-Myc Oncogene in Burkitt Lymphoma. *Proc. Natl. Acad. Sci. USA* **1983**, *80*, 820–824. [CrossRef] [PubMed]
54. Collins, S.; Groudine, M. Amplification of Endogenous Myc-Related DNA Sequences in a Human Myeloid Leukaemia Cell Line. *Nature* **1982**, *298*, 679–681. [CrossRef] [PubMed]
55. Alitalo, K.; Schwab, M.; Lin, C.C.; Varmus, H.E.; Bishop, J.M. Homogeneously Staining Chromosomal Regions Contain Amplified Copies of an Abundantly Expressed Cellular Oncogene (c-Myc) in Malignant Neuroendocrine Cells from a Human Colon Carcinoma. *Proc. Natl. Acad. Sci. USA* **1983**, *80*, 1707–1711. [CrossRef] [PubMed]
56. Nau, M.M.; Brooks, B.J.; Battey, J.; Sausville, E.; Gazdar, A.F.; Kirsch, I.R.; McBride, O.W.; Bertness, V.; Hollis, G.F.; Minna, J.D. L-Myc, A New Myc-Related Gene Amplified and Expressed in Human Small Cell Lung Cancer. *Nature* **1985**, *318*, 69–73. [CrossRef]
57. Mariani-Costantini, R.; Escot, C.; Theillet, C.; Gentile, A.; Merlo, G.; Lidereau, R.; Callahan, R. In Situ C-Myc Expression and Genomic Status of the c-Myc Locus in Infiltrating Ductal Carcinomas of the Breast. *Cancer Res.* **1988**, *48*, 199–205. [PubMed]

58. Münzel, P.; Marx, D.; Köchel, H.; Schauer, A.; Bock, K.W. Genomic Alterations of the C-Myc Protooncogene in Relation to the Overexpression of c-ErbB2 and Ki-67 in Human Breast and Cervix Carcinomas. *J. Cancer Res. Clin. Oncol.* **1991**, *117*, 603–607. [CrossRef]
59. Augenlicht, H.; Wadler, S.; Pathak, S.; Benson, A.; Hailer, D.; Heerdt, G. Low-Level c-Myc Amplification in Human Colonic Carcinoma Cell Lines and Tumors: A Frequent, P53-Independent Mutation Associated with Improved Outcome in a Randomized Multi-Institutional Trial'. *Cancer Res.* **1997**, *57*, 1769–1775. [PubMed]
60. Kalkat, M.; De Melo, J.; Hickman, K.A.; Lourenco, C.; Redel, C.; Resetca, D.; Tamachi, A.; Tu, W.B.; Penn, L.Z. MYC Deregulation in Primary Human Cancers. *Genes* **2017**, *8*, 151. [CrossRef] [PubMed]
61. He, T.C.; Sparks, A.B.; Rago, C.; Hermeking, H.; Zawel, L.; da Costa, L.T.; Morin, P.J.; Vogelstein, B.; Kinzler, K.W. Identification of C-MYC as a Target of the APC Pathway. *Science* **1998**, *281*, 1509–1512. [CrossRef] [PubMed]
62. Yagi, K.; Furuhashi, M.; Aoki, H.; Goto, D.; Kuwano, H.; Sugamura, K.; Miyazono, K.; Kato, M. C-Myc Is a Downstream Target of the Smad Pathway. *J. Biol. Chem.* **2002**, *277*, 854–861. [CrossRef] [PubMed]
63. Allen-Petersen, B.L.; Sears, R.C. Mission Possible: Advances in MYC Therapeutic Targeting in Cancer. *BioDrugs* **2019**, *33*, 539–553. [CrossRef] [PubMed]
64. Sears, R.; Nuckolls, F.; Haura, E.; Taya, Y.; Tamai, K.; Nevins, J.R. Multiple Ras-Dependent Phosphorylation Pathways Regulate Myc Protein Stability. *Genes Dev.* **2000**, *14*, 2501–2514. [CrossRef] [PubMed]
65. Seitz, V.; Butzhammer, P.; Hirsch, B.; Hecht, J.; Gütgemann, I.; Ehlers, A.; Lenze, D.; Oker, E.; Sommerfeld, A.; von der Wall, E.; et al. Deep Sequencing of MYC DNA-Binding Sites in Burkitt Lymphoma. *PLoS ONE* **2011**, *6*, e26837. [CrossRef]
66. Moyo, T.K.; Wilson, C.S.; Moore, D.J.; Eischen, C.M. Myc Enhances B Cell Receptor Signaling in Precancerous B Cells and Confers Resistance to Btk Inhibition. *Oncogene* **2017**, *36*, 4653–4661. [CrossRef]
67. Psathas, J.N.; Doonan, P.J.; Raman, P.; Freedman, B.D.; Minn, A.J.; Thomas-Tikhonenko, A. The Myc-MiR-17-92 Axis Amplifies B-Cell Receptor Signaling via Inhibition of ITIM Proteins: A Novel Lymphomagenic Feed-Forward Loop. *Blood* **2013**, *122*, 4220–4229. [CrossRef]
68. Musilova, K.; Mraz, M. MicroRNAs in B-Cell Lymphomas: How a Complex Biology Gets More Complex. *Leukemia* **2015**, *29*, 1004–1017. [CrossRef]
69. Quinquenel, A.; Aurran-Schleinitz, T.; Clavert, A.; Cymbalista, F.; Dartigeas, C.; Davi, F.; de Guibert, S.; Delmer, A.; Dilhuydy, M.-S.; Feugier, P.; et al. Diagnosis and Treatment of Chronic Lymphocytic Leukemia: Recommendations of the French CLL Study Group (FILO). *HemaSphere* **2020**, *4*, e473. [CrossRef]
70. Nguyen-Khac, F.; Balogh, Z.; Chauzeix, J.; Veronese, L.; Chapiro, E. Cytogenetics in the Management of Chronic Lymphocytic Leukemia: Guidelines from the Groupe Francophone de Cytogénétique Hématologique (GFCH). *Curr. Res. Transl. Med.* **2023**, *71*, 103410. [CrossRef]
71. Nguyen-Khac, F. "Double-Hit" Chronic Lymphocytic Leukemia, Involving the TP53 and MYC Genes. *Front. Oncol.* **2022**, *11*, 826245. [CrossRef] [PubMed]
72. Largeot, A.; Klapp, V.; Viry, E.; Gonder, S.; Botana, I.F.; Blomme, A.; Benzarti, M.; Pierson, S.; Duculty, C.; Marttila, P.; et al. Inhibition of MYC Translation through Targeting of the Newly Identified PHB-EIF4F Complex as a Therapeutic Strategy in CLL. *Blood* **2023**, *141*, 3166. [CrossRef] [PubMed]
73. Simon-Molas, H.; Montironi, C.; Kabanova, A.; Eldering, E. Metabolic Reprogramming in the CLL TME.; Potential for New Therapeutic Targets. *Semin. Hematol.* **2024**, *61*, 155–162. [CrossRef] [PubMed]
74. Kuzminov, A. Single-Strand Interruptions in Replicating Chromosomes Cause Double-Strand Breaks. *Proc. Natl. Acad. Sci. USA* **2001**, *98*, 8241–8246. [CrossRef] [PubMed]
75. Loeb, L.A.; Preston, B.D. Mutagenesis by Apurinic/Apyrimidinic Sites. *Annu. Rev. Genet.* **1986**, *20*, 201–230. [CrossRef] [PubMed]
76. McNulty, J.M.; Jerkovic, B.; Bolton, P.H.; Basu, A.K. Replication Inhibition and Miscoding Properties of DNA Templates Containing a Site-Specific Cis-Thymine Glycol or Urea Residue. *Chem. Res. Toxicol.* **1998**, *11*, 666–673. [CrossRef] [PubMed]
77. Rouet, P.; Essigmann, J.M. Possible Role for Thymine Glycol in the Selective Inhibition of DNA Synthesis on Oxidized DNA Templates. *Cancer Res.* **1985**, *45*, 6113–6118. [PubMed]
78. Maccabee, M.; Evans, J.S.; Glackin, M.P.; Hatahet, Z.; Wallace, S.S. Pyrimidine Ring Fragmentation Products. Effects of Lesion Structure and Sequence Context on Mutagenesis. *J. Mol. Biol.* **1994**, *236*, 514–530. [CrossRef] [PubMed]
79. Engelward, B.P.; Allan, J.M.; Dreslin, A.J.; Kelly, J.D.; Wu, M.M.; Gold, B.; Samson, L.D. A Chemical and Genetic Approach Together Define the Biological Consequences of 3-Methyladenine Lesions in the Mammalian Genome. *J. Biol. Chem.* **1998**, *273*, 5412–5418. [CrossRef]
80. Merrikh, H.; Zhang, Y.; Grossman, A.D.; Wang, J.D. Replication-Transcription Conflicts in Bacteria. *Nat. Rev. Microbiol.* **2012**, *10*, 449–458. [CrossRef] [PubMed]
81. García-Muse, T.; Aguilera, A. Transcription-Replication Conflicts: How They Occur and How They Are Resolved. *Nat. Rev. Mol. Cell Biol.* **2016**, *17*, 553–563. [CrossRef] [PubMed]
82. Hamperl, S.; Bocek, M.J.; Saldivar, J.C.; Swigut, T.; Cimprich, K.A. Transcription-Replication Conflict Orientation Modulates R-Loop Levels and Activates Distinct DNA Damage Responses. *Cell* **2017**, *170*, 774–786.e19. [CrossRef]
83. St Germain, C.; Zhao, H.; Barlow, J.H. Transcription-Replication Collisions-A Series of Unfortunate Events. *Biomolecules* **2021**, *11*, 1249. [CrossRef] [PubMed]

84. Littler, S.; Sloss, O.; Geary, B.; Pierce, A.; Whetton, A.D.; Taylor, S.S. Oncogenic MYC Amplifies Mitotic Perturbations. *Open Biol.* **2019**, *9*, 190136. [CrossRef] [PubMed]
85. Levine, M.S.; Holland, A.J. The Impact of Mitotic Errors on Cell Proliferation and Tumorigenesis. *Genes Dev.* **2018**, *32*, 620–638. [CrossRef]
86. Vafa, O.; Wade, M.; Kern, S.; Beeche, M.; Pandita, T.K.; Hampton, G.M.; Wahl, G.M. C-Myc Can Induce DNA Damage, Increase Reactive Oxygen Species, and Mitigate P53 Function: A Mechanism for Oncogene-Induced Genetic Instability. *Mol. Cell* **2002**, *9*, 1031–1044. [CrossRef]
87. Ray, S.; Atkuri, K.R.; Deb-Basu, D.; Adler, A.S.; Chang, H.Y.; Herzenberg, L.A.; Felsher, D.W. MYC Can Induce DNA Breaks In Vivo and In Vitro Independent of Reactive Oxygen Species. *Cancer Res.* **2006**, *66*, 6598–6605. [CrossRef]
88. Karlsson, A.; Deb-Basu, D.; Cherry, A.; Turner, S.; Ford, J.; Felsher, D.W. Defective Double-Strand DNA Break Repair and Chromosomal Translocations by MYC Overexpression. *Proc. Natl. Acad. Sci. USA* **2003**, *100*, 9974–9979. [CrossRef]
89. Chiang, Y.-C.; Teng, S.-C.; Su, Y.-N.; Hsieh, F.-J.; Wu, K.-J. C-Myc Directly Regulates the Transcription of the NBS1 Gene Involved in DNA Double-Strand Break Repair. *J. Biol. Chem.* **2003**, *278*, 19286–19291. [CrossRef]
90. Kuppers, R. Molecular Biology of Hodgkin Lymphoma. *Hematology* **2009**, *2009*, 491–496. [CrossRef] [PubMed]
91. Haferlach, C.; Dicker, F.; Schnittger, S.; Kern, W.; Haferlach, T. Comprehensive Genetic Characterization of CLL: A Study on 506 Cases Analysed with Chromosome Banding Analysis, Interphase FISH, IgV(H) Status and Immunophenotyping. *Leukemia* **2007**, *21*, 2442–2451. [CrossRef] [PubMed]
92. Nguyen-Khac, F.; Borie, C.; Callet-Bauchu, E.; Eclache, V.; Struski, S. Cytogenetics in the management of chronic lymphocytic leukemia: An update by the Groupe francophone de cytogénétique hématologique (GFCH). *Ann. Biol. Clin.* **2016**, *74*, 561–567. [CrossRef]
93. Popp, H.D.; Flach, J.; Brendel, S.; Ruppenthal, S.; Kleiner, H.; Seifarth, W.; Schneider, S.; Schulze, T.J.; Weiss, C.; Wenz, F.; et al. Accumulation of DNA Damage and Alteration of the DNA Damage Response in Monoclonal B-Cell Lymphocytosis and Chronic Lymphocytic Leukemia. *Leuk. Lymphoma* **2019**, *60*, 795–804. [CrossRef] [PubMed]
94. Salehi, F.; Behboudi, H.; Kavoosi, G.; Ardestani, S.K. Oxidative DNA Damage Induced by ROS-Modulating Agents with the Ability to Target DNA: A Comparison of the Biological Characteristics of Citrus Pectin and Apple Pectin. *Sci. Rep.* **2018**, *8*, 13902. [CrossRef]
95. Cannan, W.J.; Tsang, B.P.; Wallace, S.S.; Pederson, D.S. Nucleosomes Suppress the Formation of Double-Strand DNA Breaks during Attempted Base Excision Repair of Clustered Oxidative Damages. *J. Biol. Chem.* **2014**, *289*, 19881–19893. [CrossRef] [PubMed]
96. Carew, J.S.; Nawrocki, S.T.; Xu, R.H.; Dunner, K.; McConkey, D.J.; Wierda, W.G.; Keating, M.J.; Huang, P. Increased Mitochondrial Biogenesis in Primary Leukemia Cells: The Role of Endogenous Nitric Oxide and Impact on Sensitivity to Fludarabine. *Leukemia* **2004**, *18*, 1934–1940. [CrossRef]
97. Tomic, J.; Lichty, B.; Spaner, D.E. Aberrant Interferon-Signaling Is Associated with Aggressive Chronic Lymphocytic Leukemia. *Blood* **2011**, *117*, 2668–2680. [CrossRef] [PubMed]
98. Wuerffel, R.; Wang, L.; Grigera, F.; Manis, J.; Selsing, E.; Perlot, T.; Alt, F.W.; Cogne, M.; Pinaud, E.; Kenter, A.L. S-S Synapsis during Class Switch Recombination Is Promoted by Distantly Located Transcriptional Elements and Activation-Induced Deaminase. *Immunity* **2007**, *27*, 711–722. [CrossRef] [PubMed]
99. Péron, S.; Laffleur, B.; Denis-Lagache, N.; Cook-Moreau, J.; Tinguely, A.; Delpy, L.; Denizot, Y.; Pinaud, E.; Cogné, M. AID-Driven Deletion Causes Immunoglobulin Heavy Chain Locus Suicide Recombination in B Cells. *Science* **2012**, *336*, 931–934. [CrossRef]
100. Boutouil, H.; Boyer, F.; Cook-Moreau, J.; Cogné, M.; Péron, S. IgH Locus Suicide Recombination Does Not Depend on NHEJ in Contrast to CSR in B Cells. *Cell. Mol. Immunol.* **2019**, *16*, 201–202. [CrossRef] [PubMed]
101. Dalloul, I.; Boyer, F.; Dalloul, Z.; Pignarre, A.; Caron, G.; Fest, T.; Chatonnet, F.; Delaloy, C.; Durandy, A.; Jeannet, R.; et al. Locus Suicide Recombination Actively Occurs on the Functionally Rearranged IgH Allele in B-Cells from Inflamed Human Lymphoid Tissues. *PLoS Genet.* **2019**, *15*, e1007721. [CrossRef] [PubMed]
102. Du, L.; Oksenych, V.; Wan, H.; Ye, X.; Dong, J.; Ye, A.Y.; Abolhassani, H.; Vlachiotis, S.; Zhang, X.; de la Rosa, K.; et al. Orientation Regulation of Class-Switch Recombination in Human B Cells. *J. Immunol.* **2024**, *213*, 1093–1104. [CrossRef] [PubMed]

cancers

MDPI

Communication

T-Cell Receptor Rearrangements in Early Stages of Mycosis Fungoides May Be Associated with Pronounced Copy Number Variations: A Prognostic Factor?

Carsten Hain [1], Cassandra Cieslak [2], Jörn Kalinowski [1] and Rudolf Stadler [2,*]

[1] Center for Biotechnology (CeBiTec), Bielefeld University, 32429 Bielefeld, Germany; chain@cebitec.uni-bielefeld.de (C.H.); joern@cebitec.uni-bielefeld.de (J.K.)

[2] Department of Dermatology, Johannes Wesling Medical Centre, University Hospital of the Ruhr-University of Bochum (UKRUB), University of Bochum, 32429 Minden, Germany; cassandra.cieslak@ruhr-uni-bochum.de

* Correspondence: rudolf.stadler@ruhr-uni-bochum.de

Simple Summary: This study aims to improve the diagnosis and prognosis of MF by combining two genetic analysis methods: T-cell receptor (TCR) clonality analysis and somatic mutation profiling. This study explores improving MF diagnosis and prognosis by combining T-cell receptor (TCR) clonality analysis and somatic mutation profiling. Key findings include the detection of TCR clonal rearrangements in seven of nine early-stage patients and somatic mutations in two patients. This allowed for the classification of the patients into three molecular profiles. Potential benefits might be enhanced diagnostic accuracy, better prognostic insights, and refined treatment decisions. However, the small sample size limits conclusions. Future research should focus on larger, multi-center studies to validate the results and integrate genetic analyses into MF management.

Abstract: Mycosis fungoides (MF), the most common form of cutaneous T-cell lymphoma, poses significant diagnostic challenges due to its overlap with benign inflammatory skin diseases and the absence of specific symptoms. Accurate early diagnosis and stratification of patients by progression risk are essential for effective treatment. This study proposes a proof-of-concept for integrating T-cell receptor (TCR) clonality analysis with somatic mutation profiling to enhance diagnostic confidence and prognostic accuracy in early-stage MF. This study's methodology comprised the analysis of nine patients with early MF (stages IA/IB) using whole-exome sequencing and TCR repertoire profiling. The analysis revealed the presence of clonal TCR rearrangements in seven patients, while somatic mutations were identified in two. A notable finding was a recurrent chromosome 7 trisomy in these two cases. The patients were stratified into three molecular profiles: (1) somatic mutations with clonal TCR rearrangement (n = 2), (2) clonal TCR rearrangement without somatic mutations (n = 4), and (3) neither somatic mutations nor clonal TCR rearrangement (n = 3). These findings emphasise the heterogeneity of MF and underscore the limitations of relying solely on TCR clonality or mutation burden for diagnosis. This study underscores the potential of somatic mutations as diagnostic markers to distinguish MF from benign conditions and as prognostic indicators for disease progression. A combined genetic approach may refine treatment decisions, particularly for patients with higher tumor cell fractions and pronounced genetic alterations. Despite the limited size of the cohort, the results advocate for larger, multi-center studies to validate these findings and integrate genetic analyses into routine MF management.

Keywords: CTCL; T-cell receptor; diagnostic; genomic aberrations; sequencing

1. Introduction

Mycosis fungoides (MF), the most common form of cutaneous T-cell lymphoma, presents with limited skin lesions and shows highly heterogenous disease progression between patients. Due to often non-specific skin lesions and certain overlap to benign inflammatory skin diseases, MF is a diagnostic challenge, and the current diagnostic procedure combines clinical, histological, immunohistological, and, increasingly, molecular data. Despite this, there is a pronounced diagnostic delay in MF and sometimes the non-specific symptoms are attributed to other diseases such as eczema or parapsoriasis [1,2]. In addition to a confident and timely diagnosis, the identification and differentiation of indolent patients and patients with a high risk of progression are desirable to administer treatment accordingly. The main molecular feature used for diagnosis is the detection of a clonal T-cell receptor (TCR) rearrangement pattern. Exact quantification of the malignant TCR rearrangement is also an accurate prognostic marker. A frequency of the malignant clone in the total T-cell population above 25% indicates an elevated risk of progression into late-stage MF [3]. However, benign inflammatory diseases can also lead to a clonal TCR rearrangement, making a distinction difficult between those and MF purely based on TCR clonality analysis [4]. Therefore, a combination of TCR clonality analysis and the detection of further somatic mutations might be beneficial for a confident diagnosis and an accurate prognosis.

2. Materials and Methods

2.1. Sample Collection

Punch biopsies from 9 patients with MF were obtained with written consent. Patient characteristics are shown in Table 1. Fresh punch biopsies were stored in MACS® Tissue Storage Solution (Miltenyi Biotec, Bergisch Gladbach, Germany). In eight of nine cases, sampling took place shortly after initial diagnosis. In one case (patient 1), the diagnosis of MF had already been established for nine years

Table 1. Clinical and molecular data of 9 MF patients.

ID	Sex	Age	Stage/TNMB	Number SNVs	TCF *TRG*	TCF *TRB*	Therapy	Response
1	M	74	IA/T1bN0M0B0	21	2.5	2.9	UVB 311 nm, Bexaroten, Chlormethin gel 0.1%	CR
2	F	63	IA/T1aN0M0B0	8	14.8	4.5	UVB 311 nm, Mometasonfuroat 0.1%	PR
3	M	74	IB/T2bN0M0B0	248	61.6	19.1	Peg. Interferon 135 µg, PUVA, low-dose radiotherapy	PR
4	M	72	IA/T2bN0M0B1	11	1.0	1.2	UVB 311 nm, Peg. Interferon 135 µg	CR
5	M	61	IIA/T2N1M0B0	9	2.1	1.5	UVB 311 nm, Bexaroten	CR
6	M	64	IA/T1aN0M0B0	15	5.6	11.1	UVB 311 nm	PR
7	M	63	IA/T1aN0M0B0	188	53.7	1.8	Chlormethin gel 0.1%	PR
8	M	60	IA/T1aN0M0B0	16	21.9	41.8	Chlormethin gel 0.1%, PUVA, Clobetasol salve	PR
9	M	68	IA/T1aN0M0B0	11	46.4	10.6	UVB 311 nm	PR

Table legend: Single-nucleotide variants (SNVs); tumor clone frequency (TCF) as determined by T-cell receptor gamma (TRG) or T-cell receptor beta (TRB); patient response is either complete response (CR) or partial response (PR). Additional information is in Table S1.

2.2. Tissue Processing and Whole-Exome Data Generation

The workflow from freshly obtained punch biopsies to somatic variant calls was carried out as described previously [5]. In brief, cells were dissociated and enriched for CD3+ cells and DNA was isolated. A normal control DNA from patient-matched blood was used. From this DNA, whole-exome sequencing libraries were constructed, sequenced, and analyzed from SNVs and CNVs using GATK workflows.

2.3. T-Cell Receptor Sequencing

DNA from CD3+ cells from the punch biopsies was subjected to T-cell receptor profiling using the next-generation sequencing protocol from the Euroclonality consortia with minor modifications [6]. Data analysis was carried out with MixCR [7]. Tumor clone frequency (TCF) was calculated based on the two most frequent alleles for T-cell receptor gamma or the single most frequent allele for T-cell receptor beta.

3. Results

This study aims to provide a proof-of-concept for the combination of different genetic assays in the context of confident diagnosis and prognosis of early-stage MF. To achieve this, we measured somatic mutations and TCR rearrangement frequencies in a skin biopsy using current standard methods like whole-exome sequencing and T-cell receptor sequencing, respectively. The patient cohort consisted of nine patients who were referred to the University Department of Dermatology in Minden after non-response to standard eczematous treatment. The clinical diagnosis of early MF was made based on the clinical picture of each patient with an asymmetric, singular manifestation, the corresponding histological and immunohistological profiles of epidermotropic T cells, and hyperpigmented, mainly CD4-positive T cells, without major spongiosis in the epidermis. Exemplary data for two patients are shown in Figure 1. All cases showed limited MF (stages IA or IB), and while eight cases were early MF, shortly after diagnosis and without prior therapy, one case was an early-stage MF with a seven-year history of the disease. The patients were treated after the diagnosis of early MF according to the EORTC guideline with skin-directed therapies, e.g., chlormethine gel., PUVA, or peg. interferon in combination with Bexarotene, and one case after progression during follow-up to stage IB received low-dose radiotherapy (see Table S1).

For all patients in the cohort, we performed enrichment of CD3+ cells from fresh tumor biopsies. The enriched cells as well as patient-matched normal samples were subjected to whole-exome sequencing as described earlier to detect somatic single-nucleotide variants (SNVs) and copy number variants (CNVs) [5]. Concurrently, the T-cell receptor gamma and beta repertoire and tumor clone frequency (TCF) were assessed in the CD3+ cells using DNA-based amplicon sequencing using a modified Euroclonality workflow [6,8]. The TCF calculation took into account the most frequent (TCR beta chain) or the two most frequent clones (TCR gamma chain). Extensive SNVs and CNVs were detected in two of the nine patients (patients 3 and 7, Figure 2). Of note is the chromosome 7 trisomy present in both samples. The remaining patients showed no confident SNVs or CNVs beyond technical noise. TCR sequencing detected clonal rearrangements for the TCR gamma chain in five patients (patients 2, 3, 7, 8, and 9) and for the TCR beta chain in five patients (patients 2, 3, 6, 8, and 9). With the exception of patients 6 and 7, patients were simultaneously clonal for the gamma and beta TCR chains. Aggregation of somatic variation and TCR rearrangement data separated individual patients into three groups: samples with somatic mutations and a clonal TCR rearrangement (patients 3 and 7) as well as samples without somatic mutations (patients 1, 4, and 5) or with TCR clonality (patients 2, 6, 8, and 9).

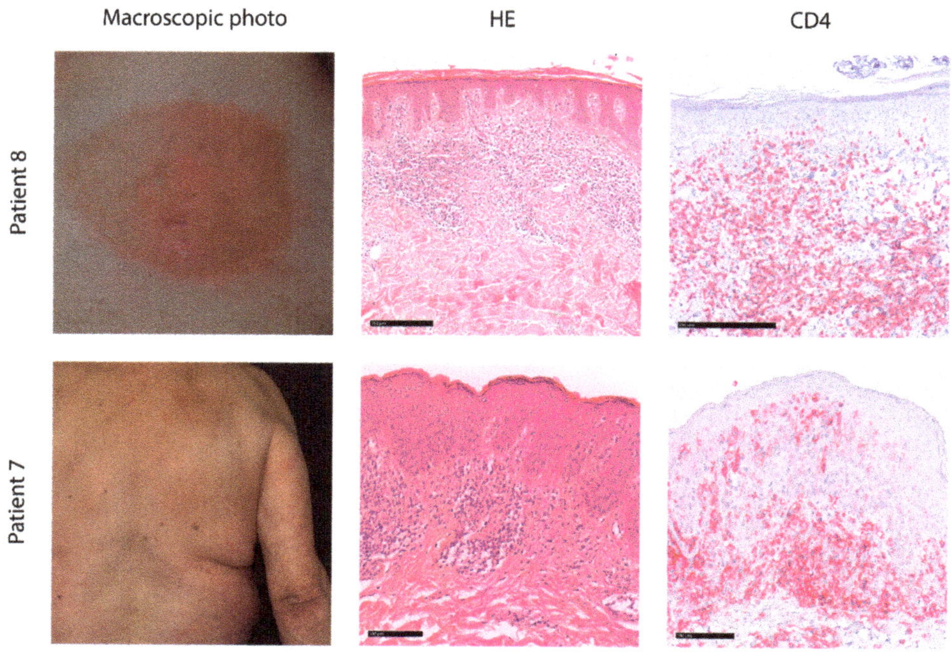

Figure 1. Histological and macroscopic pictures of exemplary patients. Patient 8 presents with a well-circumscribed patch at the lower right back. Histology shows subepidermal bandlike, slightly epidermotropic lymphocytes with a CD4 phenotype. Patient 7 shows disseminated patches on the upper back and arm with a subepidermal dense epidermotropic infiltrate of small lymphocytes with a CD4 phenotype. HE: hematoxylin–eosin staining.

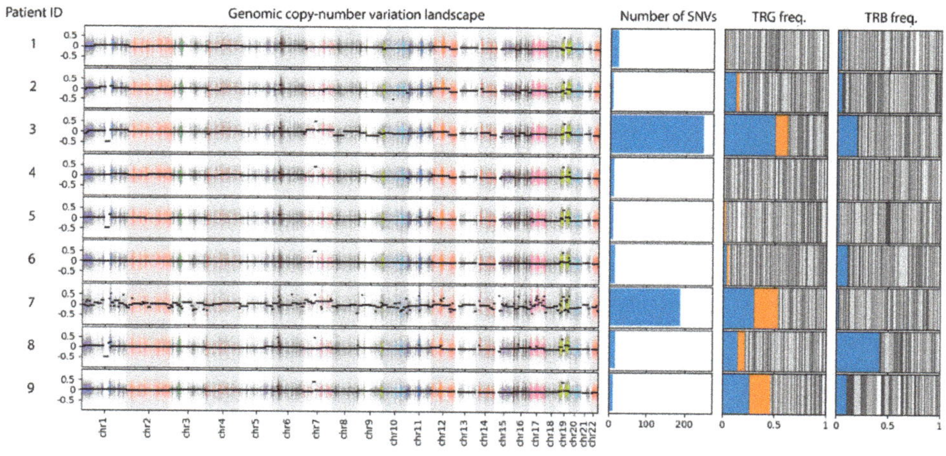

Figure 2. Genetic data of nine early-stage MF patients. For each patient, the copy number landscape, the number of SNVs, and the distribution of T-cell receptor alleles for the gamma (TRG) and beta (TRB) gene is shown. Positive values in the CNV plot indicate amplification, while negative values indicate deletions. Clear CNVs are only visible in samples 3 and 7, best indicated by the chr7 amplification present in both samples. This correlates with the number of SNVs, which is above potential background noise in samples 3 and 7. The T-cell receptor repertoire is depicted as a stacked bar of individual allele frequencies (grey bars). The alleles used for TCF calculation are marked in blue and orange (TRG) and blue (TRB).

4. Discussion

We performed a combined analysis of somatic mutations and TCR rearrangement in a cohort of early-stage MF patients, which stratified the patients into categories based on the presence and absence of somatic mutations and a clonal TCR rearrangement. This dual readout of genetic changes has several advantages. Somatic mutations, especially in the extensiveness observed here in samples 3 and 7, are a clear argument in favor of MF and thus distinguish it from benign inflammatory diseases. In this respect, the somatic mutational burden should be included in the diagnosis of MF. Prospectively, the exact nature of individual mutations might also influence the diagnostic algorithm of MF.

Of special note is the chr7 trisomy found in samples 3 and 7. This alteration is recurrent in various cancers [9–11] and abundant in CTCL, [12–14] has biological consequences like altered expression and nuclear organization [15], and might hint to a general mis-segregation defect in the malignant cells of these cells. With a progressive understanding of the functional consequences of individual mutations, it may be possible to establish a correlation between genetic changes and potential symptoms and the course of the disease.

For the cohort presented here, the performed analysis provides only a snapshot of the current genetic landscape. In this sense, a further examination of the molecular changes, i.e., an analysis of the TCR rearrangements and the somatic mutations during follow-up in the next few years, is particularly interesting. The major question is if the individual groups (somatic mutations yes/no and clonal TCR rearrangement yes/no) develop differently. The cohort used here is too small to answer this question comprehensively and should therefore be understood as a starting point for a future combined analysis of genetic changes in early-stage MF in a multi-center approach. Higher case numbers using such an approach would allow definitive statements to be made which, due to the small cohort size in this study, cannot be made here.

However, it seems that patients with a higher fraction of tumor cells and more pronounced or easy-to-detect somatic mutations require combination therapies and detectable and extensive genetic changes might call for a more active therapy than the current wait-and-see approach for early and limited MF as well as a closer follow-up. In any case, this speculative connection should be observed in further cases. Going forward, a more comprehensive integration of genetic analysis in the diagnosis and prognosis as well as choice of therapy of MF is a necessary aim.

In addition, this study again highlights the technical challenges that accompany the genetic analysis of early-stage MF, namely, the high heterogenity and deceptively low fraction of malignant cells. Striking is that the TCR analysis shows a high tumor clone frequency even in samples that do not show somatic mutations, although the detection limit for SNVs of 5% is well below the measured values for TCR clonality, e.g., in cases 2, 8, and 9 [16]. An incomplete enrichment of the CD3+ cells as well as amplification bias during TCR quantification might explain this discrepancy.

To tackle this issue and reveal the underlying mutational processes in the earliest stages of MF, a more sophisticated preparation of individual cell populations in the tumor biopsy is needed. Additionally the application of single-cell techniques like single-cell RNA [17] or ATAC sequencing [18] to a large cohort of early-stage MF patients might reveal the basic genomic changes in this elusive disease. Generally, progress in the understanding of the fundament mechanisms in MF genetics will help to develop better diagnostic and prognostic avenues.

5. Conclusions

In conclusion, this study provides valuable insights into the genetic landscape of early-stage mycosis fungoides (MF) and highlights the potential benefits of combining

T-cell receptor (TCR) clonality analysis with somatic mutation profiling. The findings underscore the heterogeneity of MF and reveal three distinct molecular profiles among patients:

(1) those with both somatic mutations and clonal TCR rearrangement;
(2) those with TCR clonality but without detectable somatic mutations;
(3) those without either somatic mutations or clonal TCR rearrangement.

The detection of extensive somatic mutations, particularly the recurrent chromosome 7 trisomy, in some patients offers a clear distinction between MF and benign inflammatory conditions [1]. This suggests that incorporating the somatic mutational burden into the diagnostic algorithm for MF could improve diagnostic accuracy.

Furthermore, this study emphasizes the technical challenges in analyzing early-stage MF, such as the low fraction of malignant cells and the discrepancy between TCR clonality and detectable somatic mutations. These findings call for more sophisticated cell preparation techniques and the application of single-cell technologies to better understand the fundamental genomic changes in MF.

While the small cohort size limits definitive conclusions, this research serves as a proof-of-concept and a starting point for future larger-scale, multi-center studies. The integration of a comprehensive genetic analysis in MF diagnosis, prognosis, and treatment selection appears to be a promising avenue for improving patient care.

Ultimately, this study paves the way for a more nuanced approach to MF management, potentially allowing for more personalized treatment strategies based on individual genetic profiles. As our understanding of the molecular mechanisms underlying MF continues to grow, it is likely to lead to better diagnostic tools and more effective therapeutic interventions for patients with this challenging condition.

Supplementary Materials: The following supporting information can be downloaded at https://www.mdpi.com/article/10.3390/cancers17030556/s1, Table S1: Patient Characteristics.

Author Contributions: Conceptualization, R.S. and C.H.; methodology, C.H. and C.C.; investigation, C.H.; resources, J.K. and R.S.; data curation, C.H. and C.C.; writing—original draft preparation, C.H.; writing—review and editing, C.H., J.K. and R.S.; visualization, C.H. All authors have read and agreed to the published version of the manuscript.

Funding: This research received no external funding.

Institutional Review Board Statement: This study was conducted in accordance with the Declaration of Helsinki and approved by the Ethics Committee of the Ruhr University Bochum (date of approval: 23.02.2022).

Informed Consent Statement: Informed consent was obtained from all subjects involved in this study.

Data Availability Statement: Data are available from the authors upon request.

Conflicts of Interest: The authors declare no conflicts of interest.

References

1. Scarisbrick, J.J.; Quaglino, P.; Prince, H.M.; Papadavid, E.; Hodak, E.; Bagot, M.; Servitje, O.; Berti, E.; Ortiz-Romero, P.; Stadler, R.; et al. The PROCLIPI international registry of early-stage mycosis fungoides identifies substantial diagnostic delay in most patients. *Br. J. Dermatol.* **2019**, *181*, 350–357. [CrossRef] [PubMed]
2. Wilcox, R.A. Cutaneous T-cell lymphoma: 2017 update on diagnosis, risk-stratification, and management. *Am. J. Hematol.* **2017**, *92*, 1085–1102. [CrossRef] [PubMed]
3. de Masson, A.; O'Malley, J.T.; Elco, C.P.; Garcia, S.S.; Divito, S.J.; Lowry, E.L.; Tawa, M.; Fisher, D.C.; Devlin, P.M.; Teague, J.E.; et al. High-throughput sequencing of the T cell receptor beta gene identifies aggressive early-stage mycosis fungoides. *Sci. Transl. Med.* **2018**, *10*, eaar5894. [CrossRef] [PubMed]
4. Miyagaki, T. Diagnosis of Early Mycosis Fungoides. *Diagnostics* **2021**, *11*, 1721. [CrossRef] [PubMed]

5. Hain, C.; Stadler, R.; Kalinowski, J. Unraveling the Structural Variations of Early-Stage Mycosis Fungoides-CD3 Based Purification and Third Generation Sequencing as Novel Tools for the Genomic Landscape in CTCL. *Cancers* **2022**, *14*, 4466. [CrossRef] [PubMed]
6. Bruggemann, M.; Kotrova, M.; Knecht, H.; Bartram, J.; Boudjogrha, M.; Bystry, V.; Fazio, G.; Froňková, E.; Giraud, M.; EuroClonality-NGS Working Group; et al. Standardized next-generation sequencing of immunoglobulin and T-cell receptor gene recombinations for MRD marker identification in acute lymphoblastic leukaemia; a EuroClonality-NGS validation study. *Leukemia* **2019**, *33*, 2241–2253. [CrossRef] [PubMed]
7. Bolotin, D.A.; Poslavsky, S.; Mitrophanov, I.; Shugay, M.; Mamedov, I.Z.; Putintseva, E.V.; Chudakov, D.M. MiXCR: Software for comprehensive adaptive immunity profiling. *Nat. Methods* **2015**, *12*, 380–381. [CrossRef] [PubMed]
8. Cieslak, C.; Hain, C.; Ruckert-Reed, C.; Busche, T.; Klages, L.J.; Schaper-Gerhardt, K.; Gutzmer, R.; Kalinowski, J.; Stadler, R. Nanopore Sequencing for T-Cell Receptor Rearrangement Analysis in Cutaneous T-Cell Lymphoma. *Cancers* **2024**, *16*, 3700. [CrossRef] [PubMed]
9. Robb, T.J.; Ward, Z.; Houseman, P.; Woodhouse, B.; Patel, R.; Fitzgerald, S.; Tsai, P.; Lawrence, B.; Parker, K.; Print, C.G.; et al. Chromosomal Aberrations Accumulate during Metastasis of Virus-Negative Merkel Cell Carcinoma. *J. Investig. Dermatol.* **2023**, *143*, 1168–1177.e2. [CrossRef] [PubMed]
10. Mastronikolis, N.S.; Tsiambas, E.; Fotiades, P.P.; Baliou, E.; Karameris, A.; Peschos, D.; Mastronikolis, S.N.; Asimakopoulos, A.D.; Giannakopoulos, X.; Ragos, V. Numerical Imbalances of Chromosome 7 in Oral Squamous Cell Carcinoma. *Anticancer Res.* **2018**, *38*, 2339–2342. [CrossRef] [PubMed]
11. Zhang, F.; Tang, J.M.; Wang, L.; Shen, J.Y.; Zheng, L.; Wu, P.P.; Mastronikolis, S.N.; Asimakopoulos, A.D.; Giannakopoulos, X.; Ragos, V. Phosphorylation of epidermal growth factor receptor and chromosome 7 polysomy in gastric adenocarcinoma. *J. Dig. Dis.* **2012**, *13*, 350–359. [CrossRef] [PubMed]
12. Motamedi, M.; Xiao, M.Z.X.; Iyer, A.; Gniadecki, R. Patterns of Gene Expression in Cutaneous T-Cell Lymphoma: Systematic Review of Transcriptomic Studies in Mycosis Fungoides. *Cells* **2021**, *10*, 1409. [CrossRef] [PubMed]
13. Fanoni, D.; Corti, L.; Alberti-Violetti, S.; Tensen, C.P.; Venegoni, L.; Vermeer, M.; Willemze, R.; Berti, E. Array-based CGH of primary cutaneous CD8+ aggressive EPIDERMO-tropic cytotoxic T-cell lymphoma. *Genes. Chromosomes Cancer* **2018**, *57*, 622–629. [CrossRef] [PubMed]
14. Marcus Muche, J.; Karenko, L.; Gellrich, S.; Karhu, R.; Kytola, S.; Kahkonen, M.; Lukowsky, A.; Sterry, W.; Ranki, A. Cellular coincidence of clonal T cell receptor rearrangements and complex clonal chromosomal aberrations-a hallmark of malignancy in cutaneous T cell lymphoma. *J. Investig. Dermatol.* **2004**, *122*, 574–578. [CrossRef] [PubMed]
15. Braun, R.; Ronquist, S.; Wangsa, D.; Chen, H.; Anthuber, L.; Gemoll, T.; Wangsa, D.; Koparde, V.; Hunn, C.; Habermann, J.K.; et al. Single Chromosome Aneuploidy Induces Genome-Wide Perturbation of Nuclear Organization and Gene Expression. *Neoplasia* **2019**, *21*, 401–412. [CrossRef] [PubMed]
16. Yan, Y.H.; Chen, S.X.; Cheng, L.Y.; Rodriguez, A.Y.; Tang, R.; Cabrera, K.; Zhang, D.Y. Confirming putative variants at </= 5% allele frequency using allele enrichment and Sanger sequencing. *Sci. Rep.* **2021**, *11*, 11640. [CrossRef]
17. Rindler, K.; Jonak, C.; Alkon, N.; Thaler, F.M.; Kurz, H.; Shaw, L.E.; Stingl, G.; Weninger, W.; Halbritter, F.; Bauer, W.M.; et al. Single-cell RNA sequencing reveals markers of disease progression in primary cutaneous T-cell lymphoma. *Mol. Cancer* **2021**, *20*, 124. [CrossRef] [PubMed]
18. Taavitsainen, S.; Engedal, N.; Cao, S.; Handle, F.; Erickson, A.; Prekovic, S.; Wetterskog, D.; Tolonen, T.; Vuorinen, E.M.; Kiviaho, A.; et al. Single-cell ATAC and RNA sequencing reveal pre-existing and persistent cells associated with prostate cancer relapse. *Nat. Commun.* **2021**, *12*, 5307. [CrossRef] [PubMed]

cancers

MDPI

Article

Multi-Omic Data Integration Suggests Putative Microbial Drivers of Aetiopathogenesis in Mycosis Fungoides

Philipp Licht [1] and Volker Mailänder [1,2,*]

[1] Department of Dermatology, University Medical Centre Mainz, 55131 Mainz, Germany; plicht@uni-mainz.de
[2] Max Planck Institute for Polymer Research, 55128 Mainz, Germany
* Correspondence: mailaend@mpip-mainz.mpg.de

Simple Summary: This research aims to better understand the causes of Mycosis fungoides (MF)—a disease where a type of immune cell, the T cell, malignantly transforms into cancer. It is not yet fully understood what triggers MF or how it progresses, partly because it varies so much between patients and because there is debate about whether the disease begins in immature T cells (thymocytes) or in more mature T cells (memory T cells). Recent findings suggest that bacteria living on the skin, particularly a harmful strain of Staphylococcus aureus, may aggravate MF by triggering specific pathways in T cells. To investigate this hypothesis, we explored the gene expression and microbial abundance of MF patients' skin with advanced statistical methods. We found that varied microbial skin colonization between patients may explain why the skin gene expression is so different from patient to patient. We also observed additional evidence that *S. aureus* might indeed trigger pathways in mature T cells that fuel cancer progression. Further, our statistical model suggested that certain viruses, like Epstein–Barr virus, could play a role in starting the disease by disrupting thymocytes (immature T cells). Based on these results, we speculate that both perspectives on the origin of MF could be correct, immature (thymocytes) and mature T cells: Thymocytes undergo malignant transformation, possibly caused by viruses, and bacteria like *S. aureus* fuel the malignant T cells to become prominent cancer cells. Our research could help uncover the complex interplay between bacteria, viruses, and T cells in MF. The findings may pave the way for new treatments targeting the skin microbiome or T cell pathways, offering hope for better management of this challenging disease.

Abstract: Background: Mycosis fungoides (MF) represents the most prevalent entity of cutaneous T cell lymphoma (CTCL). The MF aetiopathogenesis is incompletely understood, due to significant transcriptomic heterogeneity and conflicting views on whether oncologic transformation originates in early thymocytes or mature effector memory T cells. Recently, using clinical specimens, our group showed that the skin microbiome aggravates disease course, mainly driven by an outgrowing, pathogenic *S. aureus* strain carrying the virulence factor spa, which was shown by others to activate the T cell signalling pathway NF-κB. Methods: To explore the role of the skin microbiome in MF aetiopathogenesis, we here performed RNA sequencing, multi-omic data integration of the skin microbiome and skin transcriptome using Multi-Omic Factor Analysis (MOFA), virome profiling, and T cell receptor (TCR) sequencing in 10 MF patients from our previous study group. Results: We observed that inter-patient transcriptional heterogeneity may be largely attributed to differential activation of T cell signalling pathways. Notably, the MOFA model resolved the heterogenous activation pattern of T cell signalling after denoising the transcriptome from microbial influence. The MOFA model suggested that the outgrowing *S. aureus* strain evoked signalling by non-canonical NF-κB and IL-1B, which in turn may have fuelled the aggravated disease course. Further, the MOFA model indicated aberrant pathways of early thymopoiesis alongside enrichment of antiviral innate immunity. In line with this, viral prevalence, particularly of Epstein–Barr virus (EBV), trended higher in both lesional skin and the blood compared to nonlesional skin. Additionally, TCRs in both MF skin lesions and the blood were significantly more likely to recognize EBV peptides involved in latent infection. Conclusions: First, our findings suggest that *S. aureus* with its virulence factor spa fuels MF progression through non-canonical NF-κB and IL-1B signalling. Second, our data provide insights into the potential role of viruses in MF aetiology. Last, we propose a model of microbiome-driven

MF aetiopathogenesis: Thymocytes undergo initial oncologic transformation, potentially caused by viruses. After maturation and skin infiltration, an outgrowing, pathogenic *S. aureus* strain evokes activation and maturation into effector memory T cells, resulting in aggressive disease. Further studies are warranted to verify and extend our data, which are based on computational analyses.

Keywords: cutaneous T cell lymphoma (CTCL); Mycosis fungoides (MF); microbiome; transcriptomics; multi-omics; data integration; bioinformatics; computational biology; disease signalling; NF-κB; IL-1B; *Staphylococcus aureus*; protein A; microbiome-driven pathogenesis

1. Introduction

Cutaneous T cell lymphoma (CTCL) is a heterogeneous group of non-Hodgkin T cell lymphomas with skin homing properties. The most common entity is Mycosis fungoides (MF), with an incidence rate of 4.1 cases per million in the USA [1,2]. MF patients present with several to many cutaneous lesions that are formed by the infiltration of neoplastic T cells and benign reactive lymphocytes. With disease progression, both neoplastic and benign infiltrates accumulate, resulting in inflammatory reddening of skin lesions [1,3,4]. Depending on the degree of lymphocyte infiltration and inflammation, lesions are classified into the stages patch, plaque, and tumour. As MF is a lymphoproliferative disorder which can involve extracutaneous compartments like the blood, a clinical staging system considers these events and classifies patients into stages IA–IVB [1,5]. In early stages (IA–IIA), MF is an indolent disease with a 5-year disease-specific survival of 89%, which, however, dramatically drops to ~20% in the most advanced stages [5]. Because the aetiology and pathogenesis of MF are incompletely understood, treatment options are limited, and a cure is almost not achievable [6,7].

MF is thought to arise from mature, skin-resident CD4+ T cells [8], which resemble the phenotype of tissue-resident effector memory T cells [9]. However, others suggest that the initial oncologic transformation takes place during early thymopoiesis, specifically during the double-negative (DN) stages DN-1 through DN-3. Those "premalignant clones" may then migrate to the skin, where they proliferate [10–15]. The causative agent responsible for the oncogenic transformation of T cells in MF remains uncertain [1]. Viruses are among the potential factors considered due to their involvement in various lymphoma types, including even two other entities of CTCL [16]. Viral involvement in MF is further supported by the elevated risk of MF patients to develop one or more virus-initiated lymphoma types, either simultaneously with MF or later in time [17–23]. However, the role of viruses in the aetiology of MF remains unclear, as findings from different studies have yielded conflicting results [24–26].

Likewise, the molecular drivers of MF pathogenesis remain incompletely understood [27,28]. In benign T cells, three signals orchestrate activity and proliferation: First, an initial T cell response is initiated by antigenic stimulation of the T cell receptor (TCR) together with CD3. Second, co-stimulation is required to augment TCR signalling, which is mediated by various molecules, including CD28 and the tumour necrosis factor receptor superfamily (TNFRSF). Downstream, both TCR signalling and co-stimulation converge on PI3K/AKT, NFAT, and NF-κB [29]. Third, sustained T cell activity is promoted by cytokines, which activate JAK-STAT. While CD8+ T cells require interleukin (IL) 12 and interferon (IFN) α/β to initiate signal three, CD4+ T cells require IL-1 [30,31]. Because these pathways are frequently dysregulated in T cell lymphomas, a "three-signal model" of T cell lymphoma pathogenesis has been proposed [32]. In MF, dysregulation of TCR, TNFRSF/NF-κB, and JAK-STAT pathways is recurrently observed, demonstrating the involvement of all three "T cell lymphoma promoting" signalling pathways [33–38]. However, the MF transcriptome exhibits substantial variability between patients and among lesions within the same patient [27,39], which impedes the identification of a shared pathogenic pathway to date.

It has been proposed that the skin microbiome contributes to or evokes transcriptional heterogeneity [40]. In agreement, we recently identified a subgroup of MF patients with a significantly aggravated disease course and outgrowth of a distinct, pathogenic *S. aureus* strain on plaque lesions. Conversely, another MF patient subgroup presented with a more balanced skin microbiome and a favourable prognosis. Reflecting the differing prevalences of *S. aureus* between the two subgroups, we referred to the subgroup with aggravated disease and *S. aureus* outgrowth as ΔSA-positive, while the other subgroup was termed ΔSA-neutral. Notably, the virulence factor staphylococcal protein A (spa) was highly abundant in the genome of *S. aureus* in the ΔSA-positive subgroup [41]. It has been demonstrated that spa activates the NF-κB pathway [42,43], which is involved in T cell co-stimulation [29], and is a component of the "three-signal model" of T cell lymphoma pathogenesis [32]. Moreover, some studies observed aberrant NF-κB activity in subsets of MF patients with aggressive disease [44–46]. We thus theorized that the skin microbiome shapes MF disease signalling, resulting in exacerbated malignancy.

To investigate our hypothesis, we here performed bulk RNA sequencing (RNAseq), multi-omic data integration of the microbiome and the transcriptome using Multi-Omic Factor Analysis (MOFA) (48), virome profiling, and T cell receptor sequencing (TCRseq). Our analyses yielded three main findings:

First, our data indicated that inter-patient transcriptional heterogeneity was largely driven by differential expression of pathways involved in T cell signalling. Strikingly, denoising the transcriptome from microbial influence using MOFA pronouncedly reduced the heterogeneous activation pattern of T cell signalling pathways. This advocated that the skin microbiome had a substantial impact on MF disease signalling.

Second, the MOFA model suggested that *S. aureus* with its virulence factor spa induced ectopic activity of both non-canonical NF-κB and IL-1B signalling. While non-canonical NF-κB signalling leads to survival and proliferation of naïve T cells and their differentiation into effector memory T cells (49), IL-1B facilitates sustained CD4+ T cell activation [30,31]. Given that CD4+ effector memory T cells are the malignant T cell subset in MF (*9*), spa-bearing *S. aureus* may induce or augment the phenotypic characteristics of MF, resulting in aggressive disease.

Third, the MOFA model implied augmented antiviral immune response along with enriched pathways involved in early thymopoiesis between DN1 and DN3. Viral prevalence, particularly of Epstein–Barr virus (EBV) and human papillomavirus 71 (HPV), trended higher in both lesional skin and the blood. In line with this, TCRs in both MF skin lesions and the blood were significantly more likely to recognize epitopes of EBV compared to TCRs in nonlesional skin. Notably, the most frequently recognized EBV-epitopes were derived from proteins orchestrating latent EBV infection. Collectively, our findings provide evidence to support potential viral involvement in the aetiology of MF, considering that (I) malignant MF T cells infiltrate the skin from the blood (50), (II) the initial oncologic transformation of malignant T cells in MF was mapped between DN1 and DN3 (10–15), and (III) latent EBV infection can induce non-Hodgkin lymphoma, including peripheral T cell lymphoma [47,48].

Taking findings from our preceding and this study together, we present a speculative model delineating microbiome-driven MF aetiopathogenesis: The initial oncologic transformation occurs during early thymopoiesis, possibly induced by viral infection. Following maturation and skin infiltration, outgrowth of a spa-bearing *S. aureus* strain exacerbates disease by activating both non-canonical NF-κB and IL-1B signalling, resulting in the differentiation of naïve T cells into effector memory T cells with sustained activity. Our study sheds light on the critical role of the microbiome in MF aetiopathogenesis.

2. Materials and Methods

2.1. Patients and Clinical Specimens

The patient group comprised a subset of patients included in our previous study [41] that were recruited from the Department of Dermatology, University Medical Centre Mainz,

Germany. Clinical specimens included metagenomic samples from the skin, skin punch biopsies, and peripheral blood (see Table 1). Metagenomic skin samples were obtained using a swab–scrape–swab procedure described earlier [41]. In brief, a pre-moistened swab was brushed over the skin, then the same skin area was gently scraped with a scalpel, and then it was brushed again with the same swab. To address the interindividual profile of skin microbiomes [49], nonlesional skin from the contralateral site of the sampled lesion was used as a control. The skin biopsies were obtained with 4 mm punches from the same MF lesions sampled for the metagenome and were immediately snap frozen and stored at −80 °C until RNA extraction. Peripheral blood was drawn for the isolation of peripheral blood mononuclear cells (PBMCs) and subsequently enriched for the T cell fraction. For full details, please refer to [41].

Table 1. Study group, clinical specimens, and patient metadata. This study group is a subset of the group enrolled in Licht et al. [41], in which we characterized the skin microbiome of MF patients. We demonstrated that *S. aureus* abundance is increased on MF lesions compared to nonlesional skin in a subgroup of patients (ΔSA-positive), while S. aureus abundance does not change in the other subgroup (ΔSA-neutral). ΔSA-positive patients exhibit a poor clinical course compared to ΔSA-neutral patients. Patient IDs in this study match the given patient IDs in Licht et al. [41]. NA = not available; f = female; m = male.

Patient ID	Sex	Age (Years)	Clinical Stage	ΔSA-Subgroup	Stage of Lesion (Body Site)	RNAseq: Tissues Included	TCRseq: Tissues Included
Pat1	f	81	IA	positive	plaque (hip)	lesion, blood	lesion, blood
Pat3	f	47	IB	neutral	patch (thigh)	lesion, nonlesional	NA
Pat4	m	70	IIB	positive	plaque (upper back)	lesion, nonlesional, blood	lesion, nonlesional, blood
Pat5	f	73	IB	neutral	patch (thigh)	lesion, nonlesional	lesion, nonlesional
Pat7	m	54	IB	neutral	plaque (flank)	lesion, nonlesional, blood	lesion, nonlesional
Pat8	m	82	IIB	positive	patch (upper back)	lesion, nonlesional, blood	lesion, blood
Pat9	m	63	IB	neutral	plaque (forearm)	lesion, nonlesional	lesion
Pat10	m	49	IB	neutral	patch (gluteus)	lesion, nonlesional	lesion
Pat13	m	66	IB	neutral	patch (lower leg, back)	lesion, nonlesional	NA
Pat14	m	62	IB	neutral	plaque (lower leg, front)	lesion, nonlesional	lesion, nonlesional, blood

2.2. *RNA Sequencing and Bioinformatic Analysis of the Transcriptome*

Total RNA was extracted from skin biopsies and T cell enriched PBMCs as described previously [41]. Total RNA was sent to Novogene Company Ltd. (Cambridge, UK) for library preparation and sequencing. Briefly, after a quality check with Caliper Life Sciences GX II (Hopkinton, MA, USA), 400 ng RNA was used as input for the NEBNext Ultra RNA Library Prep Kit (New England Biolabs, Ipswich, MA, USA). One sample (nonlesional skin of Pat1) was excluded due to low RNA quality. After the quality check, libraries were sequenced on a NovaSeq 6000 (Illumina, San Diego, CA, USA) at 150 bp paired-end to an average of 78.74 (range 63.82–94.20) million raw reads, with 11.81 (range 9.57–14.13) giga base pairs (Gbp) per sample.

For the transcriptome analysis, raw reads were quality-checked with Fastqc version 0.11.9 [50], and subsequently aligned against the human reference genome Ensembl built GRCh38.p13 using STAR 2.7.9a with the flag –quantMode GeneCounts [51]. The data were loaded into DESeq2 version 1.36.0 [52], filtered for genes with a minimum of ten cumulative counts over all samples, and normalized using the internal DESeq2 method. Subsequently, a differential expression analysis using the Wald test was carried out, adjusted for influences from individual patients which may have been introduced by the paired-sample design. To account for multiple testing, p-values were adjusted using Benjamini–Hochberg and genes with <0.05 adjusted p-value were considered significant. Volcano plots were created using the R package EnhancedVolcano [53]. For principal component analyses, the data were first transformed with the variance-stabilization method (vst) and then analysed using the plotPCA function within DESeq2. The PCA was visualized using ggplot2 [54].

Gene set enrichment analysis was performed using the R package pathfindeR [55]. Log2 fold-changes of significantly deregulated genes determined by DESeq2 were used as input. The Reactome database [56] was used to find enriched pathways. Because we were particularly interested in signalling pathways that might orchestrate MF pathogenesis, we screened the literature for pathways that were described as deregulated in T cell lymphomas or CTCL. The Reactome database was subsequently filtered to retain only pathways that contain at least one of the following terms (case-insensitive): cascade, signal, signalling, signaling, pathway, transition, cycle, regulation, activation, keratinization, cornified, antimicrobial, interferon, IFN, stimulation, stimuli, activate, receptor, TLR. Enriched pathways were clustered using the pathfindeR function cluster_enriched_terms. Plots were created using pathfindeR.

Upregulated pathways in the plaque stage that, according to the literature, may have been elicited by *S. aureus* were further investigated using PCA. Therefore, the influence from the paired-sample design was regressed out from the normalized and rlog-transformed count matrix with the removeBatchEffect function from the limma package [57]. For three differentially regulated pathways of interest in the plaque stage, up- and downregulated genes according to pathfindeR were determined. The corrected count matrix was filtered for the genes of interest and plaque stage samples and parsed into the prcomp R stats function to perform PCA. ggplot2 was used for visualization.

2.3. Multi-Omic Data Integration

We utilized the R implementation of MOFA (Multi-Omic Factor Analysis) version 1.6.0 [58] to integratively analyse the microbiome and transcriptome datasets. MOFA integrates multiple omics datasets by discovering latent factors that capture both shared and omic-specific patterns of variation. These latent factors help to identify key biological signals, denoise the data, and identify relationships between different omics layers.

The microbiome dataset was created in our preceding study where we characterized the MF skin microbiome using shotgun metagenomics and MetaPhlAn 3.0.2 [41,59,60]. To obtain absolute counts of microbial taxa (rather than the relative abundance, which is the default in MetaPhlAn and was used in our previous study), the metagenome was re-profiled including the flag -t rel_ab_w_read_stats using MetaPhlAn 3.0.2 with the intermediate bowtie2 mapping files as input. The resulting profiles were filtered to retain only taxa on the species level and normalized using the wrench method, which accounts for sparse metagenomic count data and is implemented in the R package wrench [61]. The normalized data were log2-transformed with a pseudocount of 1. To select highly variable species for the MOFA model, 50 species with the highest variance were filtered (with the R base function var) and used for all subsequent steps.

The transcriptome dataset was created in this study as described above. After mapping and normalization with STAR 2.7.9a and DESeq2 version 1.36.0 (described above), the data were transformed with the regularized logarithm using the rlog function implemented in DESeq2 [52]. As recommended by the MOFA authors, the large transcriptome dataset was restricted to not overrepresent the smaller microbiome dataset in the MOFA model. Therefore, the top 5000 highly variable genes were selected based on the median absolute deviation calculated with the R base function mad. The MOFA model was trained with a single group, the option scale_views set to TRUE, convergence_mode set to slow, and eight factors.

Gene set enrichment analysis was performed with the MOFA function run_enrichment, which is based on principal component gene set enrichment (PCGSE) [62] using only genes with positive weights as input. The Reactome database, filtered for pathways of interest (see the above section of transcriptome analysis), was used as a background. Pathway enrichment plots were generated with ggplot2.

For the heatmaps in Figure 2A,B, an identical set of genes were plotted using the R package pheatmap (refer to the caption of Figure 2 for more details on gene selection). The heatmap in Figure 2A shows the normalized gene counts, while the heatmap in Figure 2B shows the denoised version of the data, explained by the latent factors of the MOFA model. This means that the MOFA model reconstructs the input data (transcriptome and microbiome) based on the latent factors and shows patterns that are shared between the omics layers and that are omics-specific. The denoised heatmap in Figure 2B was generated with the MOFA function plot_data_heatmap with argument denoise set to TRUE. Internally, plot_data_heatmap uses pheatmap for heatmap plotting, with the MOFA trained data as input. The code used for Figure 2B (and other figures) can be found in the Data Availability Statement.

2.4. Virus Profiling

The virome present on superficial skin layers was profiled via re-analysis of the microbiome dataset generated in our previous investigation [41]. Briefly, human skin was brushed topically with swabs, and whole metagenomic sequencing was performed and subsequently profiled with MetaPhlAn 3.0.2. While viruses were excluded in our previous study, here we activated the flag for viral identification (--add_viruses). The resulting profiles were filtered to retain only viruses.

The virome present in deeper skin layers was profiled using RNAseq reads generated from skin punch biopsies. In contrast to the swabbing procedure, punch biopsies obtain the entire depth of skin layers. Virome profiles from RNAseq reads were generated with the VIRTUS pipeline [63]. Briefly, non-human reads were filtered out through alignment to the human reference genome Ensembl built GRCh38.p13. Next, unmapped (i.e., non-human) reads were aligned against 762 viral genomes. Heatmaps visualizing the virome profile were generated with the R package pheatmap [64].

2.5. Assessment of TCR–Epitope Binding

To investigate whether T cells of MF patients recognize EBV-epitopes, the T cell receptor was sequenced as described previously [41]. Briefly, RNA isolated from skin punch biopsies and T cell enriched PBMCs of MF patients were subjected to library preparation spanning the variable part of the TCR using the NEBNext Immune Sequencing Kit, human (E6320, New England Biolabs, USA), and sequenced on a MiSeq (Illumina, USA) running 300 bp PE. The sequencing data were processed with the pRESTO toolkit [65] (https://usegalaxy.org/u/bradlanghorst/w/presto-nebnext-immune-seq-workflow-v320, accessed on 20 May 2019) and the R package immunearch [66]. EBV-epitopes were obtained from the Immune Epitope Database (IEDB) (https://www.iedb.org/, accessed on 2 April 2019) [67]. The Tool ERGO-II [68] was used to determine binding scores of pairs of TCRs and EBV-epitopes. The closer the binding score is to 1, the higher the probability that a TCR recognizes the epitope [68].

3. Results

3.1. Study Group and Clinical Specimens

The study group consisted of 10 MF patients (mean age at sampling: 64.7 years; range: 47 to 82 years) and was a subset of patients that were enrolled in our preceding investigation where we characterized the MF skin microbiome [41]. To ensure traceability, IDs of included patients in this study were adopted from our previous investigation [41]. In that study, we found that a subgroup of patients presented with an increased abundance of *S. aureus* on MF lesions compared to nonlesional skin of the same patient. We termed this subgroup ΔSA-positive as opposed to the ΔSA-neutral subgroup, where *S. aureus* abundance was consistent between MF lesions and nonlesional skin of the same patient. The ΔSA-positive subgroup had a strong dysbiosis and a significantly aggravated disease course compared to the ΔSA-neutral subgroup [41]. Of the 10 MF patients included in the present study, 3 were in the ΔSA-positive subgroup and 7 were in the ΔSA-neutral subgroup.

Clinical specimens included skin swabs and punch biopsies of lesional skin (5 patch stage and 5 plaque stage) and matched nonlesional skin of the same MF patients as an intra-patient control (10 nonlesional skin samples). For more details about the rationale of using internal controls, please refer to Licht, Dominelli et al. [41]. Additionally, blood was drawn in four patients for the isolation of peripheral blood mononuclear cells (PBMCs). All clinical specimens were taken at the same visit. Skin punch biopsies were used for RNAseq. Where sufficient material was available, skin punch biopsies and PBMCs were used for TCRseq. Shotgun metagenomics from skin swabs were initially carried out in our previous investigation [41] and re-analysed in the present study in a multi-omic data integration approach together with the skin transcriptome. Table 1 summarizes patient metadata, clinical specimens, and data modalities used in the present study.

3.2. Transcriptional Heterogeneity May Be Largely Driven by Differential Activation of Pathways of the "Three-Signal Model" of T Cell Lymphoma Pathogenesis

A number of studies investigated the CTCL transcriptome [27,39] and uncovered various mechanisms of disease progression [33–38]. However, the transcriptome was also found to exhibit significant heterogeneity among patients, complicating the identification of a common pathological mechanism [27,39,46,69–73].

To explore the root of transcriptional heterogeneity, we employed differential gene expression analysis and gene set enrichment analysis (GSEA). Principal component analysis (PCA) of the MF skin transcriptome clearly separated nonlesional skin, patch, and plaque via principal component (PC) 1. However, lesional samples also spread along PC2, demonstrating strong inter-patient transcriptional heterogeneity (Figure 1A). Likewise, GSEA recovered several aberrant pathways known to promote MF exacerbation, such as chemokine signalling [74] (Supplementary Material S2), but also showed substantial activation differences between patients in both patch and plaque lesions (Supplementary Material S3). Notably, pathways of the "three-signal model" of T cell lymphoma pathogenesis exhibited different activation patterns between patch and plaque (Table 2): TCR signalling was enriched in both patch and plaque, whereas interleukin-13 (IL-13) signalling, CD28 co-stimulation, and co-stimulation through death receptors activating NF-κB [44,45] were enriched solely in the plaque stage.

Next, we aimed to assess whether pathways of the "three-signal model" for T cell lymphoma pathogenesis [32] replicate transcriptional heterogeneity. To this end, we compiled a list of genes that are (I) members of the "three-signal model" for T cell lymphomas and (II) were differentially expressed in either or both patch and plaque stages. A clustered heatmap of these genes closely resembled the pattern of the PCA of the entire transcriptome (Figure 2A): Nonlesional skin and plaque samples formed separate clusters, while patch samples exhibited intermediate expression levels. Importantly, plaque samples displayed pronounced heterogeneity in RNA expression levels (Figure 2A). We thus reasoned that inter-patient transcriptional heterogeneity can, at least to some extent, be attributed to differential expression of pathways that orchestrate T cell lymphoma pathogenesis.

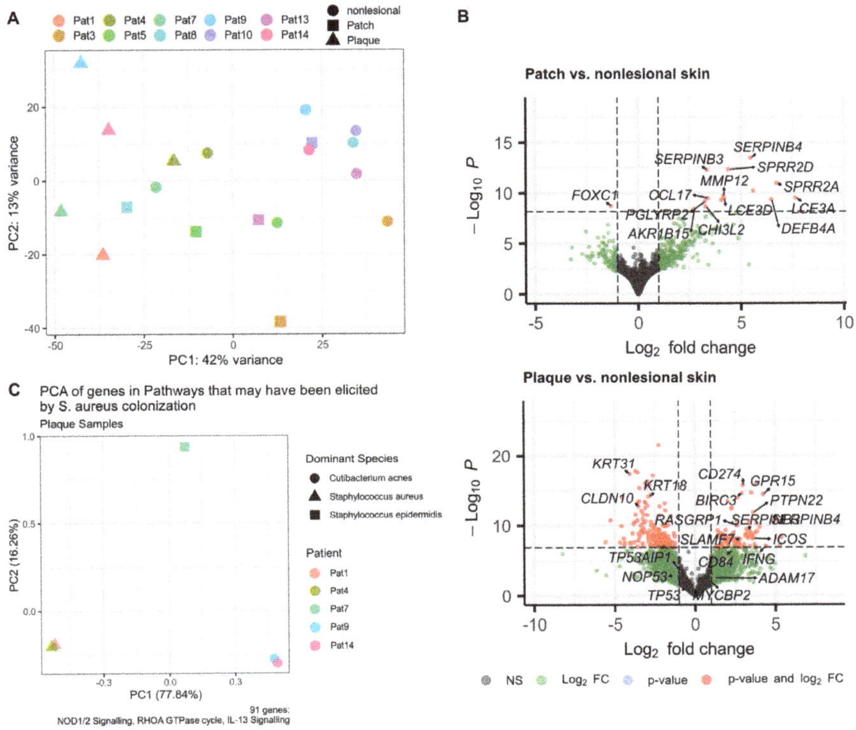

Figure 1. Analysis of the MF skin transcriptome. (**A**) Principal component analysis (PCA) showing inter-heterogeneity of plaque. (**B**) Volcano plots visualizing the distribution of gene fold-changes and statistical significance between patch or plaque and nonlesional skin. For the complete results table of DESeq2 differential gene expression analysis, refer to Supplementary Material S1. (**C**) PCA of genes belonging to upregulated pathways in the plaque stage that may have been elicited by *S. aureus* according to the literature [75–79]. PC = principal component, NS = not significant.

Table 2. Gene set enrichment analysis. Displayed are selected pathways of interest. Please refer to Supplementary Material S2 for the full GSEA list. log2fc = log2 fold-change, NA = not available, padj = *p*-value adjusted to multiple comparisons using Bonferroni.

Pathway	Patch log2fc	padj	Plaque log2fc	padj
Keratinization	6.35	3.6×10^{-5}	1.18	7.1×10^{-8}
Antimicrobial peptides	8.69	1.4×10^{-5}	NA	NA
TCR signalling	4.74	1.3×10^{-7}	1.63	1.4×10^{-14}
CD28 co-stimulation	NA	NA	1.74	1.1×10^{-5}
Non-canonical NF-kB pathway	NA	NA	1.34	4.3×10^{-11}
Interleukin-13 signalling	NA	NA	1.41	6.2×10^{-4}
IFN-γ signalling	8.32	6.7×10^{-5}	1.83	5.8×10^{-3}
Death receptor signalling	NA	NA	1.85	6.4×10^{-21}
Toll-Like Receptor 4 (TLR4) Cascade	2.55	1.5×10^{-2}	1.31	6.4×10^{-19}
NOD1/2 signalling pathway	NA	NA	1.84	4.3×10^{-17}
RHOA GTPase cycle	NA	NA	1.62	4×10^{-29}
Regulation of TP53 activity	NA	NA	1.28	1.8×10^{-10}

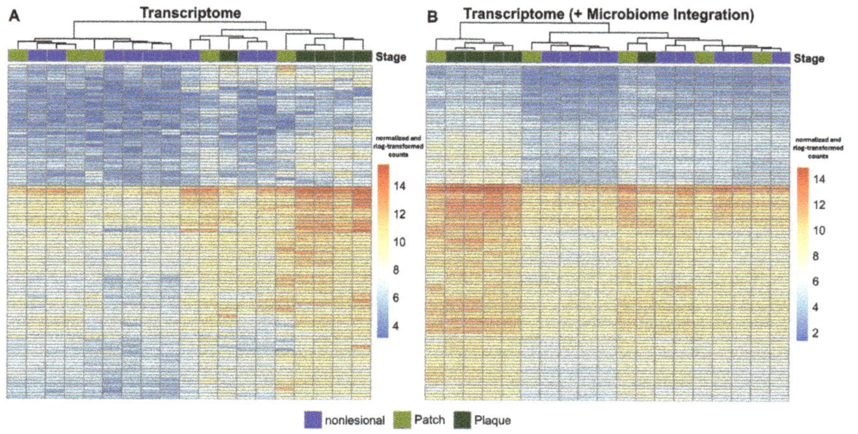

Figure 2. Clustered heatmap of genes involved in the "three-signal model" of T cell lymphoma pathogenesis. The clustered heatmap was created with genes involved in the "three-signal model" of T cell lymphoma pathogenesis that were differentially expressed in the patch and/or plaque stage and were included in the MOFA model (5000 highly variable genes; see Methods and Supplementary Material S6). Raw gene counts were normalized using the median of ratios and rlog-transformed with the DESeq2 package [52]. (**A**) Shown are normalized RNA expression levels before integration of the microbiome. Notably, although only genes of the "three-signal model" of T cell lymphoma pathogenesis were included, the clustering resembled the patterns of the PCA, which was based on the entire transcriptome. Despite clustering together, plaque samples exhibited a high degree of heterogeneity. (**B**) After data integration of the transcriptome with the microbiome, the heterogeneity of plaque samples was largely resolved, suggesting that the microbiome had a strong impact on MF disease signalling. Please refer to the Methods section for information on how the denoised heatmap was generated.

3.3. The MF Skin Transcriptome Shows Responses to Microbial Stimuli

Since differential microbial skin colonization may contribute to transcriptional heterogeneity [40], and we recently identified that the skin microbiome stratifies MF patients and determines the clinical course [41], our next objective was to identify microbial responses in the skin transcriptome.

As expected, we uncovered enriched pathways likely activated by the skin microbiome (Table 2 and Supplementary Material S2). Notably, Toll-Like Receptor 4 (TLR4) was upregulated in patch and showed attenuated activity in plaque. The pathway senses lipopolysaccharides, a cell wall constituent of Gram-negative bacteria, and activates inflammatory responses as well as pyroptosis, which is a form of programmed necrosis. Pyroptosis protects the host from microbial infection but, if overactivated, can also lead to pathological inflammation [80], a typical condition of progressive MF [4]. In accordance, we recently showed that the skin microbiome of patches is strongly dysbiotic, while dysbiosis on plaques was observed only on a subgroup of MF patients [41], providing a potential explanation for the differential TLR4 activation between patch and plaque.

Conversely, several microbe-associated pathways were uniquely upregulated in plaque lesions, potentially driven by the outgrowth of a distinct S. aureus strain harbouring virulence factors such as α-hemolysin and spa [41]. The innate immune sensor NOD2 (nucleotide-binding oligomerization domain 2) recognizes small peptides derived from the peptidoglycan cell wall component of Gram-positive bacteria like *S. aureus*. Importantly, NOD2 mediates protective responses specifically against *S. aureus* through its interaction with the virulence factor α-hemolysin [75,76]. Additionally, we observed enhanced activity of the RHOA GTPase cycle, a pathway that *S. aureus* can exploit via its virulence factor spa to facilitate epithelial invasion [77]. Further, we also noted enrichment of IL-13 signalling, consistent with reports that IL-13 expression is induced in healthy skin upon *S. aureus* exposure [78,79]. Interestingly, malignant T cells in the skin of MF patients express IL-13, whereas malignant T cells found in lymph nodes and blood do not [46]. To further investigate whether the upregulation of NOD1/2 signalling, the RHOA GTPase cycle, and IL-13 signalling might have been elicited by the microbiome, as suggested by the literature, we conducted PCA of genes regulating these pathways (Figure 1C). Plaque samples with *S. aureus* as the dominant species in the microbiome clearly separate from those with other dominant species.

Collectively, we noted several enriched pathways that may be attributed to microbial stimuli, consistent with the microbiome patterns identified on MF lesions in our preceding study [41]. Consequently, our findings indicated that distinct microbial colonization led to differential transcriptomic responses between patients.

3.4. Multi-Omic Data Integration of the Microbiome and the Transcriptome Resolves Transcriptional Heterogeneity

To investigate whether differential skin colonization elicited the heterogeneous expression of genes related to the "three-signal model" for T cell lymphoma pathogenesis, we performed data integration of the microbiome and the transcriptome using Multi-Omic Factor Analysis (MOFA) [58]. Briefly, MOFA can be seen as a multi-omic implementation of PCA and finds latent factors (comparable to principal components in PCA) that capture the main sources of variation across different omic data modalities (which are called views in the MOFA framework). Within the latent factors, weights are allocated to the features of the different views. Consequently, latent factors are characterized by feature weights, which indicate their importance or significance in the variation captured by the latent factor. Downstream, additional analyses such as gene set enrichment analysis (GSEA) can be conducted within each latent factor. The MOFA framework further enables the identification of each view's contribution to the variation captured by a latent factor across the entire multi-omic dataset [58].

Regarding the multi-omic dataset in this study, the weights of the microbes (features of the microbiome) and the weights of the genes (features of the transcriptome) characterize the latent factors. The MOFA model showed a good fit to the multi-omic dataset, as the patterns of the latent factors were congruent to the results observed in both the independent transcriptome analysis (see above) and our previous study on the MF skin microbiome [41] (see Supplementary Material S4). Latent factors 1 to 5 captured shared sources of variation present in both data modalities, indicating a reciprocal influence between the transcriptome and microbiome (Figure 3A). By leveraging the latent factors, MOFA reconstructs the input data to separate shared and specific variations in each data modality, thereby denoising the data and revealing underlying biological signals [58]. Strikingly, the differential expression of genes involved in the "three-signal model" of T cell lymphoma pathogenesis was resolved after integration of the microbiome and transcriptome data modalities (called denoised in MOFA2 terms; Figure 2B). This indicated that the lesion-specific microbiome heavily influences MF disease signalling. Please refer to the Methods section for information how the denoised heatmap was generated.

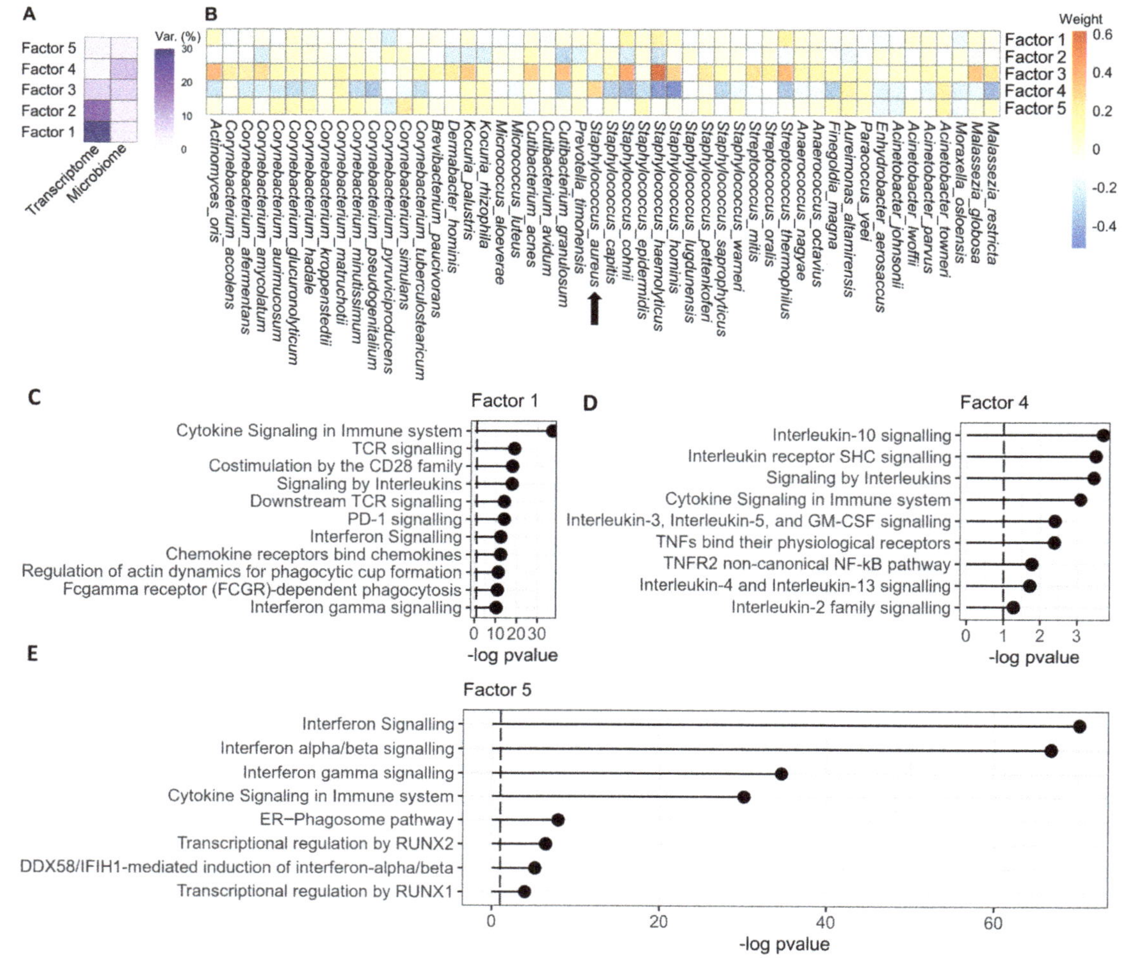

Figure 3. Overview of the latent factors of the MOFA model in the microbiome and transcriptome data modalities. (**A**) Overview of the latent factors capturing variances in the microbiome and transcriptome data modalities. Factor 4 captured the majority of variance in the microbiome, meaning that this factor captured important impacts of the microbiome on the transcriptome. (**B**) Overview of the feature weights of the microbiome per factor. (**C–E**) Gene set enrichment analysis. Factor 1 showed pathomechanisms of high and sustained T cell activity in MF (TCR signalling, CD28 co-stimulation, interleukin signals). Factor 4 showed pathways that were likely evoked by *S. aureus*. Factor 5 showed strong antivirus responses along with enrichment of RUNX1 and RUNX2 signalling, which have pivotal roles in thymopoiesis [81–84].

3.5. Spa-Bearing S. aureus Likely Activates Non-Canonical NF-κB and IL-1B Signalling

We next assessed the impact of the microbiome on MF disease signalling in more detail. **Factor 4** accounted for a substantial proportion of the total variance in the microbiome (Figure 3A), and *S. aureus* was the only microbiome feature with a positive weight. Notably, microbes with anti-*S. aureus* properties, such as *S. hominis* and *S. epidermidis* [85–87], displayed markedly negative weights. Thus, the microbiome feature weights exhibited a pattern similar to plaque lesions of the ΔSA-positive subgroup. These patients presented with *S. aureus* outgrowth and an aggravated clinical course [41].

In the transcriptome, the following pathways controlling T cell activity were enriched: Interleukin signalling, particularly IL-1B, and non-canonical NF-κB signalling (Figure 3D, Supplementary Material S5). While non-canonical NF-κB serves as a co-stimulator and generates and maintains effector memory T cells [88,89], IL-1B signalling facilitates sustained activation of CD4+ T cells [32]. In agreement, the malignant T cell subset in MF is thought to be CD4+ effector memory T cells [9]. Regarding IL-1B signalling, Chng et al. [90] reported that human keratinocytes express IL-1B when challenged with *S. aureus*. This suggests that outgrowth of virulent *S. aureus* may stimulate the tumour microenvironment to activate malignant T cells via IL-1B in a paracrine fashion.

Regarding non-canonical NF-κB signalling, we intriguingly observed that the stimulating ligand, TNF-α [91], displayed an almost neutral weight in factor 4 (Supplementary Material S5), indicating an alternative stimulation route. Instead, we previously identified high abundance of the virulence factor staphylococcal protein A (spa) in the genome of the *S. aureus* strain colonizing ΔSA-positive patients. It was reported that spa can activate the NF-κB pathway [42,43], and some studies associated aggressive CTCL with upregulated NF-κB signalling [44–46]. Notably, several genes emphasized in these studies exhibited positive weights in factor 4 (e.g., *LTA, LTB, BIRC3, TNFSF13, TNFSF14, TNFRSF1B*, and *TNFRSF7*; Supplementary Material S5). Owing to the almost neutral weight of the NF-κB stimulating agent TNF-α and the presence of spa as an alternative stimulator [41], we theorized that outgrowing, spa-bearing *S. aureus* strains upregulated non-canonical NF-κB signalling in MF patients with aggressive disease.

Collectively, the MOFA model indicated that the skin microbiome significantly influences MF disease signalling and is a major source of transcriptional heterogeneity. Further, the MOFA model suggested that *S. aureus* fuelled malignant T cells to promote the aggravated disease course of ΔSA-positive patients via two signalling axes: While spa may activate non-canonical NF-κB signalling, resulting in survival, proliferation, and naïve T cell differentiation into mature, effector memory T cells, the neoplastic T cell subset in MF [9], paracrine IL-1B signalling might facilitate sustained activation.

3.6. Aberrant Signalling of Early Thymopoiesis Alongside Enriched Antiviral Immunity Suggests Viral Involvement in MF Aetiology

Additionally to factor 4, our attention was drawn to the results of **factor 5**. While the microbiome weights showed only minimal signals (Figure 3B), GSEA identified significant upregulation of host defence signals against viruses (Figure 3E), indicating that factor 5 represented a source of variation independent of bacteria. In particular, GSEA showed upregulation of interferon alpha and beta (IFN-α/β) signalling, which is the first line of innate immune defence upon viral infection. Downstream, IFN-α/β signalling stimulates genes that inhibit the replication machinery of viruses at various mechanistic levels [92]. Several of these interferon-stimulated genes exhibited high weights in factor 5, for example *MX1, IFIT1, IFIT3, OAS1, OAS2, IFI27*, and *OASL* (Supplementary Material S5). *IFIT3*, which displayed the highest weight in factor 5, was shown to specifically boost antiviral signalling by IFN-α/β [93]. Further, GSEA found an enrichment of the ER–phagosome pathway (Figure 3E), which can be hijacked by viruses for their own translation, replication, and particle budding in order to spread into other host cells [94]. Notably, the phagosome-pathway genes *TAP1* and *TAP2*, which displayed high weights in factor 5 (Supplementary Material S5), are highly expressed by Epstein–Barr virus (EBV)-infected lymphocytes [95], and interact with Epstein–Barr-nuclear antigen 1 (EBNA1) [96].

Remarkably, EBV infection can lead to *RUNX1* expression [97,98], and signalling by *RUNX1* and *RUNX2* was significantly enriched in factor 5 (Figure 3E). The RUNX family is a frequent target of retroviral insertion [99–101], resulting in the development of several T cell lymphoma entities [102,103]. *RUNX1* regulates the expansion of mature CD4+ T cells [82], which constitutes the neoplastic T cell subset in MF [9]. Interestingly, both *RUNX1* and *RUNX2* are important regulators of early thymopoiesis during the double-negative stages of T lymphocytes [81–84], a developmental step that has been mapped to the initial oncologic transformation of T cells in MF [10–15].

Since the transcriptome dataset included in the MOFA model was restricted to 5000 genes (see Methods section), we screened the entire transcriptomic dataset to investigate signals of aberrant thymopoiesis in more detail. It has been reported that ectopic expression of *RUNX2* strongly expands immature thymocytes during the double-negative stages, resulting in a preneoplastic state of thymocytes characterized by low proliferation rates [81]. However, concomitant overexpression of *MYC* rescues proliferation and facilitates differentiation. Additionally, *MYC* and *RUNX* collaboratively inhibit the tumour suppressor p53, resulting in decreased apoptosis of malignant T cells. Together, this ultimately leads to the accumulation of mature, neoplastic T cells [81,104]. In agreement with these reports, several genes were aberrantly expressed in the MF transcriptome (Supplementary Material S1): Besides *RUNX1* and *RUNX2*, we identified enriched *MYCBP2* (Figure 1B), which is a member of the c-myc family with the function to increase c-myc activity [105]. Further, although the pathway to regulate p53 activity was upregulated (Table 2), important components of the p53 machinery were downregulated (Figure 1B): These included the p53 gene *TP53* itself, the p53 stabilizing protein *NOP53* [106], and TP53AIP1, which is regulated by p53 and induces apoptosis [107]. A complete results table of DESeq2 differential gene expression analysis is available in Supplementary Material S1.

In summary, our data show dysregulation of pathways involved in early thymopoiesis, probably representing the initial oncologic transformation of T cells in MF. Moreover, factor 5 suggested a connection between viruses and MF aetiology given the concomitant upregulation of host responses to viruses and *RUNX1/2* signalling.

3.7. Increased Viral Prevalence and EBV-Epitope Recognition in MF Skin Lesions

In MF, malignant T cells circulate in the blood and infiltrate the upper dermis but have not been described to be present in the most superficial layers of the skin [108]. To determine whether viruses may be involved in in MF aetiology, we investigated viral prevalence in different layers of the skin and the blood of MF patients. To this end, we examined whole metagenomic sequencing (WMS) and RNAseq data. While the WMS data were generated from skin swabs and thus only contain material from superficial skin layers, the RNAseq data were generated from both the blood and skin punch biopsies representing the whole skin, including deeper skin layers like the epidermis and dermis.

In whole skin samples, total viral load trended higher in MF lesions compared to nonlesional skin (Figure 4B). In contrast, total viral load did not differ in superficial skin layers (Figure 4A). Among all viruses detected in MF lesions of whole skin samples, EBV and human papillomavirus 71 (HPV71) displayed the highest prevalence, albeit HPV71 was also frequently present in nonlesional skin. Remarkably, EBV and HPV71 were also present in the blood of some patients (Figure 4B). Both EBV and various human papillomaviruses are implicated in the development of cancer, including T cell lymphomas [109–111]. Since MF is characterized by the infiltration of circulating T cells from the blood into the skin [108], the concomitant presence of viruses in both compartments indicates a potential viral involvement in MF.

Next, we evaluated whether T cells of whole skin samples or T cells in the blood were directed against known EBV-epitopes obtained from the Immune Epitope Database (IEDB) [67]. We sequenced the TCRs of MF lesions, nonlesional skin, and blood and utilized ERGO-II [68] to calculate binding scores for the most abundant TCRs of a given sample with each of the EBV-epitopes. TCRs in MF lesions and the blood showed a significantly higher probability to recognize EBV-epitopes than T cells in nonlesional skin (Figure 4C), which agreed with the increased viral prevalence of MF lesions in whole skin samples. Interestingly, latent membrane protein (LMP) 1, LMP2, and Epstein–Barr-nuclear antigen (EBNA) 1 were the most frequently recognized EBV-epitopes by TCRs in the skin and blood (Figure 4D). An expression pattern of LMP1, LMP2, and EBNA1 represents a characteristic EBV gene profile of latency type I/II, commonly observed in EBV-induced

T cell lymphomas [112,113]. In NK/T cell lymphoma, LMP1 upregulates *CD274* (PD-L1) via the NF-κB axis [114], which agreed with our results of enriched *CD274* and NF-κB signalling in the plaque stage (Table 2, Figure 1B).

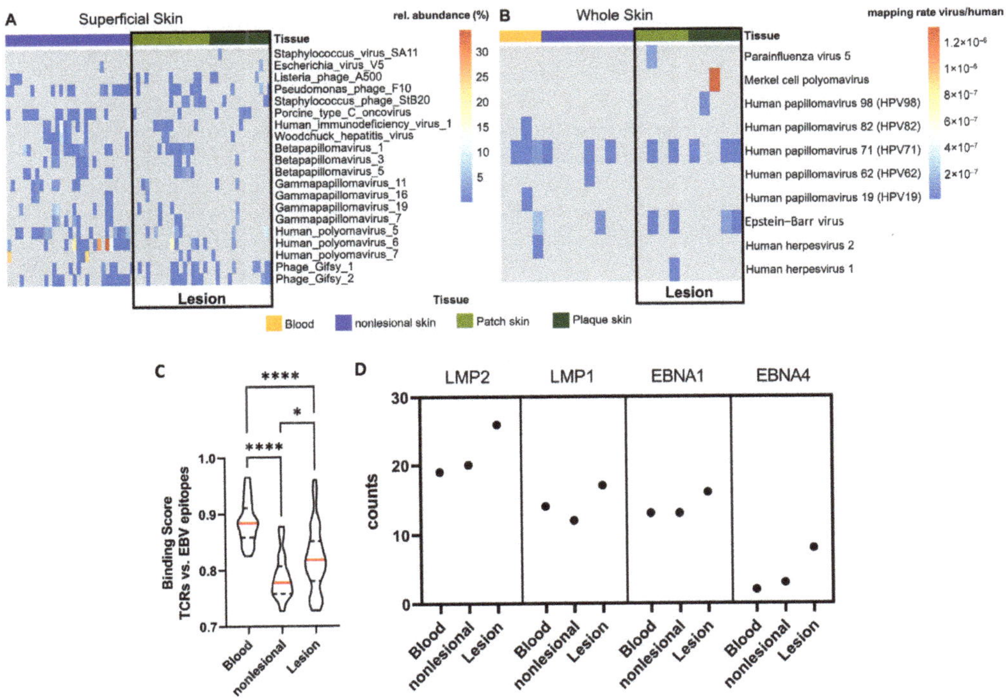

Figure 4. Viral association with MF. (**A**,**B**), Viral prevalence identified in WMS reads of superficial skin (**A**) and RNAseq reads of whole skin (**B**). Viral prevalence in MF lesions of whole skin trended higher compared to nonlesional skin, specifically EBV and HPV71 (**B**), whereas no difference was present on superficial skin (**A**). In (**A**), the top 20 viruses with the highest variation between MF lesions and nonlesional skin are shown. In (**B**), all viruses detected are shown. In (**C**,**D**), T cells in MF lesions and the blood specifically recognize EBV-epitopes. (**C**) The probability of TCRs in blood, nonlesional skin, and MF lesions to recognize EBV-epitopes was assessed using ERGO-II [68]. Binding scores (1 = perfect binding, 0 = no binding) were calculated for each TCR with each EBV-epitope. n = 175; displayed are violin plots with the median (red), 1st quartile, and 3rd quartile. Kruskal–Wallis test: * for $p \leq 0.05$, **** for $p \leq 0.0001$. (**D**) Displayed are the EBV-epitopes that were recognized most often by TCRs of MF patients.

Together, the trend of increased EBV and HPV71 prevalence in whole skin samples along with a significantly higher probability of EBV-epitope recognition by TCRs in MF lesions and the blood compared to nonlesional skin suggest viral involvement in MF aetiology.

4. Discussion

The clinical course of MF varies greatly, with some patients having only minor progress, whereas others suffer from fast progression and high disease burden [5,115]. Likewise, the transcriptome is tremendously heterogeneous, exhibiting patient and even lesion specificity [27,39,73]. Moreover, ambiguity exists about the initial oncologic transformation of malignant T cells, as studies from different groups mapped the event to either early thymocytes [10–15] or mature, effector memory T cells [9]. Consequently, a common pathogenic mechanism of all patients or patient subgroups is still in doubt, resulting in a non-optimal treatment regimen [27]. We recently demonstrated that the skin microbiome of MF patients is altered, and identified a subgroup of patients that was overgrown with a distinct, pathogenic *S. aureus* strain. This subgroup exhibited a significantly aggravated disease course, possibly owing to the virulence factor spa, which was present in the *S. aureus* genome [41]. In line with this, others showed that spa can activate the NF-κB axis [42], which is recurrently deregulated in MF patients with aggressive disease [34,44–46,116]. We thus theorized that (I) the lesion-specific microbiome determines transcriptomic response, thereby contributing to or causing heterogeneity [40], and that (II) spa-bearing *S. aureus* elicits NF-κB signalling to fuel MF. Therefore, we investigated how the skin microbiome affects the skin transcriptome in a subset of 10 MF patients that were enrolled in our preceding study [41]. Using RNAseq, multi-omic data integration, virome profiling, and TCRseq, we obtained novel insights into the role of the microbiome in both the aetiology and pathogenesis of MF.

First, we recovered substantial transcriptomic heterogeneity on both gene and pathway levels and observed that this heterogeneity may be largely driven by differential expression of T cell signalling pathways. Strikingly, our results suggested that the differential T cell signalling pattern may have been caused by the skin microbiome, since denoising the transcriptome with the microbiome using MOFA considerably reduced heterogeneity (Figure 2A,B).

Second, latent factor 4 implied that *S. aureus* elicited upregulation of both non-canonical NF-κB and IL-1B signalling, which could explain the aggravated disease course of MF patients overgrown with a spa-bearing *S. aureus* strain. Non-canonical NF-κB signalling is known to promote survival and proliferation of thymocytes and mature T cells, induce differentiation of naïve T cells into effector memory T cells, and support clonal expansion [88,89], which are all common characteristics of malignant T cells in MF [1,8,9,117]. NF-κB signalling is typically initiated by the interaction of TNFα with either TNFRSF1A (also known as TNFR1) or TNFRSF1B (also known as TNFR2) [88,89]. However, the enrichment of non-canonical NF-κB signalling in factor 4 appeared to be independent of TNFα, since it displayed a very low weight. Instead, we propose that the *S. aureus* virulence factor spa induced non-canonical NF-κB signalling since Gómez et al. showed that spa activates TNFRSF1A [42,43]. In general, TNFRSF1A induces the canonical form of NF-κB signalling leading to apoptosis, while TNFRSF1B induces non-canonical NF-κB signalling resulting in survival and proliferation. However, there is some level of crosstalk between the two forms of NF-κB pathways. In highly activated T cells, such as the malignant T cells in MF, activation of TNFRSF1A results in survival rather than apoptosis [88,89,118]. Further, the extracellular domains of TNFRSF1A and TNFRSF1B exhibit a high degree of structural similarity [119], possibly allowing spa to activate both receptors. Additionally, while TNFRSF1A is expressed nearly ubiquitously across various cell types throughout the body, TNFRSF1B expression is considerably more restricted, including thymocytes and T cells [89]. Thus, spa may activate either TNFRSF1A, TNFRSF1B, or both to initiate non-canonical NF-κB signalling to fuel MF progression.

In agreement with our findings, a clinical study reported that systemic inhibition of NF-κB in CTCL patients induced a skin response in 30.4% of patients, whereas the blood response was mixed [120–122] Further, Shin et al. identified ectopic NF-κB signalling in 30.6% of MF patients and that these patients had an aggravated disease course compared to patients without ectopic NF-κB signalling [44]. We previously identified that 31.3% of MF patients were overgrown with the virulent, spa-bearing *S. aureus* strain and had an aggravated disease course [41]. Considering the comparable prevalences of the specified subgroups and the presumed NF-κB-activating effect of spa, this may suggest a potential association between skin colonization by a spa-carrying *S. aureus* strain and exacerbation of the disease in the aforementioned studies. Additional research on the spa-NF-κB interaction including mechanistic assays with bigger patient groups is warranted.

Regarding our finding of upregulated IL-1B signalling in factor 4, it was shown that *S. aureus* induces secretion of IL-1B by eosinophils [123], which infiltrate MF lesions [124]. Furthermore, human keratinocytes, skin-derived dendritic cells, and lymphocytes produce several cytokines, including IL-1A, IL-1B, and IL-4, when challenged with *S. aureus* or spa [90,125,126]. This suggests a paracrine stimulation of malignant T cells by the tumour microenvironment via cytokine signalling in response to *S. aureus*.

Last, we discovered evidence indicating viral involvement in the aetiology of MF. Latent factor 5 captured a concomitant upregulation of innate antivirus defence mechanisms and aberrant *RUNX1/2* signalling. The latter is known to coordinate thymopoiesis during double-negative (DN) stages DN1 through DN3 [81,82,127], which have been mapped to the initial oncologic transformation in MF [10–15]. We further observed that viral load, particularly of HPV71 and EBV, trended higher in both deeper skin layers of MF lesions and the blood compared to nonlesional skin, whereas superficial skin layers of MF lesions showed no difference. HPV71 was shown to degrade p53 [128], which can result in neoplasia [129]. Nevertheless, while certain papillomaviruses are categorized as high-risk factors for the onset of solid cancers [130] and have also been loosely linked to an elevated risk of lymphomas [131], HPV71 is generally regarded as having low oncogenic potential [132]. In contrast, EBV is a well-known oncovirus [133–135]. Remarkably, we found that T cells residing in MF lesions or in the blood were significantly more affinitive to EBV-epitopes than T cells of nonlesional skin (Figure 4C). Moreover, the most frequently recognized EBV-epitopes were epitopes of LMP1, LMP2, and EBNA1, which are genes commonly expressed by EBV in latent infection [113]. Although EBV has a strong B cell tropism, leading to B cell Hodgkin's lymphoma (HL) and non-Hodgkin's lymphoma (nHL) [133–135], the virus also infects T cells, thereby causing some entities of T cell nHL [47,48]. Notably, CTCL patients have an increased risk of developing secondary HL and nHL [17–20]. It has been proposed that a single precursor malignant T cell can trigger HL, CTCL, and lymphomatoid papulosis, a benign T cell neoplasm, in the same patient [21,22]. Others reported the simultaneous presence of HL and CTCL within the same lymph node and theorized that the two malignancies arise from distinct B and T cells [23], indicating that a common trigger induced oncogenesis.

To the best of our knowledge, we were the first to investigate viral prevalence in different skin compartments and found that EBV and HPV71 trended higher solely in deeper skin layers of MF lesions, where malignant T cells in MF reside [108]. Collectively, our data indicate that viruses, probably EBV and/or HPV71, might play a role in MF aetiology. However, the sample size was rather small and not all lesional and blood samples were positive for EBV and/or HPV71. Moreover, despite a significantly higher EBV-epitope recognition of these tissues defined by computational estimation, mechanistic assays are warranted for a definitive conclusion. Thus, hypotheses on viral involvement in MF need to be considered speculative at this point. Additional investigations with more participants and longitudinal viral monitoring across tissues and different skin compartments are necessary to determine whether viruses are the aetiologic agent in MF.

An open question remains whether the age distribution of our patient group may have influenced the skin condition/clinical appearance, since 7 out of 10 patients were aged 60 or older (4 ΔSA-neutral, 3 ΔSA-neutral). In skin-disease-free subjects, several skin integrity parameters such as hydration, sebum production, or cytokine production decrease with age [136,137]. Studies in MF have indicated that skin integrity and moisture levels are lower in comparison to age-matched healthy subjects, which could lead to an increased susceptibility to bacterial infection [138–140]. This implies that older MF patients could experience a more severe skin barrier defect than younger MF patients and hence could experience infections more frequently. Notably, a cut-point of 60 years of age has been identified as a clinically prognostic outcome factor. Given that 70% of the patients in our study group were aged 60 and older and were evenly distributed between the two ΔSA groups, this indicates appropriate age matching within our study group. However, bigger patient groups with a more diverse population are warranted.

Further limitations of our study are the small patient group size, the mono-centric nature of our study, and the lack of mechanistic assays to confirm our findings, which are based on computational analyses from primary clinical specimens.

Additionally, the skewed distribution of patients between the two ΔSA groups could have introduced bias. To mitigate this concern, we specifically opted to utilize MOFA2, which is an unsupervised method agnostic of grouping variables. Analogous to PCA, MOFA2 generates clusters based on integrated multi-omic profiles [58]. Therefore, the input to MOFA2 entails the entire group rather than a comparison of two (potentially imbalanced) groups. Importantly, despite the smaller size of the ΔSA-positive group compared to the ΔSA-neutral group, the MOFA2 model identified substantial upregulation of both non-canonical NF-κB and IL-1B signalling, which was largely disregarded by conventional GSEA (Supplementary Material S2, plaque vs. nonlesional skin). Nonetheless, our findings are based on a small patient group derived from our preceding study [41]. To assess the robustness of the MOFA2 model, we performed a thorough evaluation of (a.) the stability of our model throughout iterations and (b.) the robustness of our model against technical noise. Our assessment suggests that MOFA robustly detects the same latent factors and patterns with and without noise. Please refer to Supplementary Material S7 for the full stability assessment. The current study represents an in-depth investigation of a rather limited patient group, and further research is required for validation and generalization.

5. Conclusions

In this study, we provided novel insights into the role of the microbiome in MF aetiopathogenesis. First, our results indicated that the skin microbiome largely contributes to transcriptional heterogeneity. Second, our data suggest that a spa-bearing *S. aureus* strain, which overgrows a subgroup of MF patients with aggravated disease course, might evoke non-canonical NF-κB and IL-1B signalling in the skin. Third, our data collectively suggested that viruses, particularly EBV and HPV71, may be the aetiologic agent in MF.

Together with the results from our preceding study, these findings led us to propose a model of microbiome-driven MF aetiopathogenesis (Figure 5): The initial oncologic transformation could emerge during early thymopoiesis triggered by aberrant *RUNX* expression, potentially caused by viral infection such as with EBV and/or HPV71. Malignant T cells infiltrate the skin, leading the microenvironment to release AMPs. Eventually, this process eliminates microbes, resulting in skin dysbiosis and a diminished epithelial barrier. Additionally, certain AMPs attract CD4+ T cells, which could include malignant T cells, thereby augmenting the infiltration of (malignant) T cells into MF lesions. Over time, some microbes acquire resistance to AMPs and recolonize the lesions. In the ΔSA-neutral subgroup, microbes with anti-*S. aureus* properties accumulate, resulting in a balanced microbiome and favourable clinical course. However, in the ΔSA-positive subgroup, a virulent *S. aureus* strain carrying the virulence factor spa outgrows and activates non-canonical NF-κB and IL-1B signalling, resulting in the generation of mature effector memory T cells and poor outcome. Hence, both perspectives on the cells of origin in MF may apply: early T cell progenitors [10–14] and mature, effector memory T cells [8,9].

Additional research is warranted to validate and generalize our findings, which are based on a rather small patient group. Furthermore, research allowing mechanistic insights is needed to understand how the microbiome affects malignant T cells in MF.

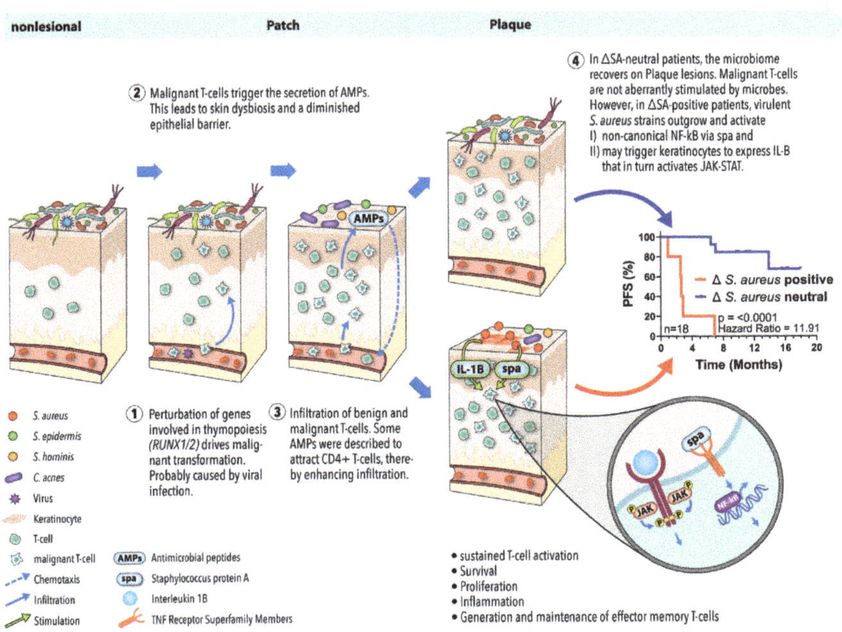

Figure 5. Proposed model of microbiome-driven MF aetiopathogenesis. T cell precursors undergo initial oncologic transformation due to aberrant expression of RUNX1 and RUNX2, probably caused by viruses like EBV and/or HPV71. After maturation and skin infiltration, the malignant T cells trigger the microenvironment to secrete AMPs. AMPs kill most of the physiological skin microbiota, resulting in dysbiosis and a diminished epithelial barrier. In addition, AMPs may recruit benign and malignant CD4+ T cells, establishing a loop of sustained dysbiosis and T cell infiltration. Since AMP levels remain constant over the disease course, some microbes eventually acquire resistance and recolonize the lesions. In the ΔSA-neutral subgroup, microbes with anti-S. aureus activity accumulate, resulting in a balanced microbiome that does not fuel the disease. However, in the ΔSA-positive subgroup, virulent S. aureus strains bearing the virulence factor spa overgrow and activate non-canonical NF-κB as well as IL-1B signalling. The two pathways cause inflammation, sustained activity, survival, and proliferation of T cells, as well as the generation and maintenance of effector memory T cells. These characteristics are hallmarks of malignant T cells in MF and may explain the significantly aggravated clinical course of the ΔSA-positive subgroup.

Supplementary Materials: The following supporting information can be downloaded at https://www.mdpi.com/article/10.3390/cancers16233947/s1: Supplementary Material S1: Differentially expressed genes in patch and plaque vs. nonlesional skin; Supplementary Material S2: Gene set enrichment analysis of patch and plaque vs. nonlesional skin; Supplementary Material S3: Heterogeneous activation pattern of pathways between nonlesional skin and patch or plaque, respectively; Supplementary Material S4: Characterization of the fitted MOFA2 model; Supplementary Figure S1: Total variance explained in the transcriptome and the microbiome data modalities by the MOFA model. Supplementary Material S5: Gene set enrichment analysis of each MOFA2 factor; Supplementary Material S6: Transcriptome input for the MOFA2 model. Supplementary Material S7: Stability Assessment of the MOFA model.

Author Contributions: Conceptualization, P.L. and V.M.; methodology, P.L.; validation, P.L. and V.M.; formal analysis, P.L.; investigation, P.L.; resources, V.M.; data curation, P.L.; writing—original draft preparation, P.L.; writing—review and editing, P.L. and V.M.; visualization, P.L.; supervision, V.M.; project administration, P.L. and V.M.; funding acquisition, P.L. and V.M. All authors have read and agreed to the published version of the manuscript.

Funding: This research received no external funding.

Institutional Review Board Statement: All procedures in this study were conducted in accordance with the Declaration of Helsinki from 1964 and its later amendments. The local ethics committee, the Medical Association of Rhineland-Palatine (Ethik-Kommisssion der Landesaerztekammer Rheinland-Pfalz), Germany, assigned ethical approval under number 2020-14813.

Informed Consent Statement: Informed consent was obtained from all subjects involved in this study.

Data Availability Statement: The code used to generate the plots in Figures 1, 2, 3 and 4A,B, Supplementary Figure S1, and Supplementary Material S5 was deposited on GitHub (https://github.com/phlicht/Multi-Omic_Data_Integration_in_MF). All datasets used in this study (i.e., RNAseq, WGS, and TCRseq) were deposited on the Gene Expression Omnibus (GEO) under the SuperSeries GSE221150 (https://www.ncbi.nlm.nih.gov/geo/query/acc.cgi?acc=GSE221150): RNA sequencing data and associated analysis files can be accessed under GSE221148 (https://www.ncbi.nlm.nih.gov/geo/query/acc.cgi?acc=GSE221148). This includes viral profiles from both whole skin biopsies and the blood. Microbiome sequencing data and associated analysis files can be accessed under GSE221149 (https://www.ncbi.nlm.nih.gov/geo/query/acc.cgi?acc=GSE221149). This includes re-analysed microbiome profiles with absolute counts and viral profiles from superficial skin layers. TCR Sequencing data and associated analysis files can be accessed under GSE218874 (https://www.ncbi.nlm.nih.gov/geo/query/acc.cgi?acc=GSE218874).

Acknowledgments: We thank Stefan Kindel, University Medical Centre Mainz, for the creation of the figure summarizing our proposed model of microbial-driven MF aetiopathogenesis.

Conflicts of Interest: The authors declare no conflicts of interest.

References

1. Dummer, R.; Vermeer, M.H.; Scarisbrick, J.J.; Kim, Y.H.; Stonesifer, C.; Tensen, C.P.; Geskin, L.J.; Quaglino, P.; Ramelyte, E. Cutaneous T Cell Lymphoma. *Nat. Rev. Dis. Prim.* **2021**, *7*, 61. [CrossRef]
2. Bradford, P.T.; Devesa, S.S.; Anderson, W.F.; Toro, J.R. Cutaneous Lymphoma Incidence Patterns in the United States: A Population-Based Study of 3884 Cases. *Blood* **2009**, *113*, 5064–5073. [CrossRef] [PubMed]
3. Nielsen, P.R.; Eriksen, J.O.; Sørensen, M.D.; Wehkamp, U.; Lindahl, L.M.; Bzorek, M.; Iversen, L.; Woetman, A.; Ødum, N.; Litman, T.; et al. Role of B-Cells in Mycosis Fungoides. *Acta Derm. Venereol.* **2021**, *101*, adv00413. [CrossRef] [PubMed]
4. Girardi, M.; Heald, P.W.; Wilson, L.D. The Pathogenesis of Mycosis Fungoides. *N. Engl. J. Med.* **2004**, *350*, 1978–1988. [CrossRef]
5. Olsen, E.; Vonderheid, E.; Pimpinelli, N.; Willemze, R.; Kim, Y.; Knobler, R.; Zackheim, H.; Duvic, M.; Estrach, T.; Lamberg, S.; et al. Revisions to the Staging and Classification of Mycosis Fungoides and Sezary Syndrome: A Proposal of the International Society for Cutaneous Lymphomas (ISCL) and the Cutaneous Lymphoma Task Force of the European Organization of Research and Treatment of Ca. *Blood* **2007**, *110*, 1713–1722. [CrossRef]
6. Whittaker, S.; Hoppe, R.; Prince, H.M. How I Treat Mycosis Fungoides and Sezary Syndrome. *Blood* **2016**, *127*, 3142–3153. [CrossRef] [PubMed]
7. Ansell, S.M. Non-Hodgkin Lymphoma: Diagnosis and Treatment. *Mayo Clin. Proc.* **2015**, *90*, 1152–1163. [CrossRef] [PubMed]
8. Kirsch, I.R.; Watanabe, R.; O'Malley, J.T.; Williamson, D.W.; Scott, L.-L.; Elco, C.P.; Teague, J.E.; Gehad, A.; Lowry, E.L.; LeBoeuf, N.R.; et al. TCR Sequencing Facilitates Diagnosis and Identifies Mature T Cells as the Cell of Origin in CTCL. *Sci. Transl. Med.* **2015**, *7*, 308ra158. [CrossRef]
9. Campbell, J.J.; Clark, R.A.; Watanabe, R.; Kupper, T.S. Sézary Syndrome and Mycosis Fungoides Arise from Distinct T-Cell Subsets: A Biologic Rationale for Their Distinct Clinical Behaviors. *Blood* **2010**, *116*, 767–771. [CrossRef]
10. Iyer, A.; Hennessey, D.; O'Keefe, S.; Patterson, J.; Wang, W.; Salopek, T.; Wong, G.K.-S.; Gniadecki, R. Clonotypic Heterogeneity in Cutaneous T-Cell Lymphoma (Mycosis Fungoides) Revealed by Comprehensive Whole-Exome Sequencing. *Blood Adv.* **2019**, *3*, 1175–1184. [CrossRef]
11. Hamrouni, A.; Fogh, H.; Zak, Z.; Ødum, N.; Gniadecki, R. Clonotypic Diversity of the T-Cell Receptor Corroborates the Immature Precursor Origin of Cutaneous T-Cell Lymphoma. *Clin. Cancer Res.* **2019**, *25*, 3104–3114. [CrossRef] [PubMed]
12. Iyer, A.; Hennessey, D.; O'Keefe, S.; Patterson, J.; Wang, W.; Wong, G.K.-S.; Gniadecki, R. Skin Colonization by Circulating Neoplastic Clones in Cutaneous T-Cell Lymphoma. *Blood* **2019**, *134*, 1517–1527. [CrossRef] [PubMed]
13. Iyer, A.; Hennessey, D.; O'Keefe, S.; Patterson, J.; Wang, W.; Wong, G.K.-S.; Gniadecki, R. Branched Evolution and Genomic Intratumor Heterogeneity in the Pathogenesis of Cutaneous T-Cell Lymphoma. *Blood Adv.* **2020**, *4*, 2489–2500. [CrossRef] [PubMed]
14. Iyer, A.; Hennessey, D.; Gniadecki, R. Clonotype Pattern in T-Cell Lymphomas Map the Cell of Origin to Immature Lymphoid Precursors. *Blood Adv.* **2022**, *6*, 2334–2345. [CrossRef]

15. Ren, J.; Qu, R.; Rahman, N.-T.; Lewis, J.M.; King, A.L.O.; Liao, X.; Mirza, F.N.; Carlson, K.R.; Huang, Y.; Gigante, S.; et al. Integrated Transcriptome and Trajectory Analysis of Cutaneous T-Cell Lymphoma Identifies Putative Precancer Populations. *Blood Adv.* **2022**, *7*, 445–457. [CrossRef]
16. Mesri, E.A.; Feitelson, M.A.; Munger, K. Human Viral Oncogenesis: A Cancer Hallmarks Analysis. *Cell Host Microbe* **2014**, *15*, 266–282. [CrossRef]
17. Väkevä, L.; Ranki, A.; Pukkala, E. Increased Risk of Secondary Cancers in Patients with Primary Cutaneous T Cell Lymphoma. *J. Investig. Dermatol.* **2000**, *115*, 62–65. [CrossRef]
18. Kantor, A.F.; Curtis, R.E.; Vonderheid, E.C.; van Scott, E.J.; Fraumeni, J.F. Risk of Second Malignancy after Cutaneous T-Cell Lymphoma. *Cancer* **1989**, *63*, 1612–1615. [CrossRef]
19. Amber, K.T.; Bloom, R.; Nouri, K. Second Primary Malignancies in CTCL Patients from 1992 to 2011: A SEER-Based, Population-Based Study Evaluating Time from CTCL Diagnosis, Age, Sex, Stage, and CD30+ Subtype. *Am. J. Clin. Dermatol.* **2016**, *17*, 71–77. [CrossRef]
20. Brownell, I.; Etzel, C.J.; Yang, D.J.; Taylor, S.H.; Duvic, M. Increased Malignancy Risk in the Cutaneous T-Cell Lymphoma Patient Population. *Clin. Lymphoma Myeloma* **2008**, *8*, 100–105. [CrossRef]
21. Davis, T.H.; Morton, C.C.; Miller-Cassman, R.; Balk, S.P.; Kadin, M.E. Hodgkin's Disease, Lymphomatoid Papulosis, and Cutaneous T-Cell Lymphoma Derived from a Common T-Cell Clone. *N. Engl. J. Med.* **1992**, *326*, 1115–1122. [CrossRef] [PubMed]
22. Wood, G.S.; Crooks, C.F.; Uluer, A.Z. Lymphomatoid Papulosis and Associated Cutaneous Lymphoproliferative Disorders Exhibit a Common Clonal Origin. *J. Investig. Dermatol.* **1995**, *105*, 51–55. [CrossRef]
23. Steinhoff, M. Cutaneous T Cell Lymphoma and Classic Hodgkin Lymphoma of the B Cell Type within a Single Lymph Node: Composite Lymphoma. *J. Clin. Pathol.* **2004**, *57*, 329–331. [CrossRef]
24. Mirvish, E.D.; Pomerantz, R.G.; Geskin, L.J. Infectious Agents in Cutaneous T-Cell Lymphoma. *J. Am. Acad. Dermatol.* **2011**, *64*, 423–431. [CrossRef]
25. Mirvish, J.J.; Pomerantz, R.G.; Falo, L.D.; Geskin, L.J. Role of Infectious Agents in Cutaneous T-Cell Lymphoma: Facts and Controversies. *Clin. Dermatol.* **2013**, *31*, 423–431. [CrossRef]
26. Dulmage, B.O.; Feng, H.; Mirvish, E.; Geskin, L. Black Cat in a Dark Room: The Absence of a Directly Oncogenic Virus Does Not Eliminate the Role of an Infectious Agent in Cutaneous T-Cell Lymphoma Pathogenesis. *Br. J. Dermatol.* **2015**, *172*, 1449–1451. [CrossRef] [PubMed]
27. García-Díaz, N.; Piris, M.Á.; Ortiz-Romero, P.L.; Vaqué, J.P. Mycosis Fungoides and Sézary Syndrome: An Integrative Review of the Pathophysiology, Molecular Drivers, and Targeted Therapy. *Cancers* **2021**, *13*, 1931. [CrossRef] [PubMed]
28. Yumeen, S.; Girardi, M. Insights into the Molecular and Cellular Underpinnings of Cutaneous T Cell Lymphoma. *Yale J. Biol. Med.* **2020**, *93*, 111–121.
29. Smith-Garvin, J.E.; Koretzky, G.A.; Jordan, M.S. T Cell Activation. *Annu. Rev. Immunol.* **2009**, *27*, 591–619. [CrossRef]
30. Curtsinger, J.M.; Mescher, M.F. Inflammatory Cytokines as a Third Signal for T Cell Activation. *Curr. Opin. Immunol.* **2010**, *22*, 333–340. [CrossRef]
31. Santarlasci, V.; Cosmi, L.; Maggi, L.; Liotta, F.; Annunziato, F. IL-1 and T Helper Immune Responses. *Front. Immunol.* **2013**, *4*, 182. [CrossRef] [PubMed]
32. Wilcox, R.A. A Three-Signal Model of T-Cell Lymphoma Pathogenesis. *Am. J. Hematol.* **2016**, *91*, 113–122. [CrossRef] [PubMed]
33. McGirt, L.Y.; Jia, P.; Baerenwald, D.A.; Duszynski, R.J.; Dahlman, K.B.; Zic, J.A.; Zwerner, J.P.; Hucks, D.; Dave, U.; Zhao, Z.; et al. Whole-Genome Sequencing Reveals Oncogenic Mutations in Mycosis Fungoides. *Blood* **2015**, *126*, 508–519. [CrossRef] [PubMed]
34. Ungewickell, A.; Bhaduri, A.; Rios, E.; Reuter, J.; Lee, C.S.; Mah, A.; Zehnder, A.; Ohgami, R.; Kulkarni, S.; Armstrong, R.; et al. Genomic Analysis of Mycosis Fungoides and Sézary Syndrome Identifies Recurrent Alterations in TNFR2. *Nat. Genet.* **2015**, *47*, 1056–1060. [CrossRef]
35. Da Silva Almeida, A.C.; Abate, F.; Khiabanian, H.; Martinez-Escala, E.; Guitart, J.; Tensen, C.P.; Vermeer, M.H.; Rabadan, R.; Ferrando, A.; Palomero, T. The Mutational Landscape of Cutaneous T Cell Lymphoma and Sézary Syndrome. *Nat. Genet.* **2015**, *47*, 1465–1470. [CrossRef]
36. Choi, J.; Goh, G.; Walradt, T.; Hong, B.S.; Bunick, C.G.; Chen, K.; Bjornson, R.D.; Maman, Y.; Wang, T.; Tordoff, J.; et al. Genomic Landscape of Cutaneous T Cell Lymphoma. *Nat. Genet.* **2015**, *47*, 1011–1019. [CrossRef]
37. Damsky, W.E.; Choi, J. Genetics of Cutaneous T Cell Lymphoma: From Bench to Bedside. *Curr. Treat. Options Oncol.* **2016**, *17*, 33. [CrossRef]
38. Pérez, C.; Mondéjar, R.; García-Díaz, N.; Cereceda, L.; León, A.; Montes, S.; Durán Vian, C.; Pérez Paredes, M.G.; González-Morán, A.; Miguel, V.; et al. Advanced-stage Mycosis Fungoides: Role of the Signal Transducer and Activator of Transcription 3, Nuclear Factor-κB and Nuclear Factor of Activated T Cells Pathways. *Br. J. Dermatol.* **2020**, *182*, bjd.18098. [CrossRef]
39. Dulmage, B.O.; Geskin, L.J. Lessons Learned from Gene Expression Profiling of Cutaneous T-Cell Lymphoma. *Br. J. Dermatol.* **2013**, *169*, 1188–1197. [CrossRef]
40. Licht, P.; Mailänder, V. Transcriptional Heterogeneity and the Microbiome of Cutaneous T-Cell Lymphoma. *Cells* **2022**, *11*, 328. [CrossRef]
41. Licht, P.; Dominelli, N.; Kleemann, J.; Pastore, S.; Müller, E.-S.; Haist, M.; Hartmann, K.S.; Stege, H.; Bros, M.; Meissner, M.; et al. The Skin Microbiome Stratifies Patients with Cutaneous T Cell Lymphoma and Determines Event-Free Survival. *NPJ Biofilms Microbiomes* **2024**, *10*, 74. [CrossRef] [PubMed]

42. Gómez, M.I.; Lee, A.; Reddy, B.; Muir, A.; Soong, G.; Pitt, A.; Cheung, A.; Prince, A. Staphylococcus Aureus Protein A Induces Airway Epithelial Inflammatory Responses by Activating TNFR1. *Nat. Med.* **2004**, *10*, 842–848. [CrossRef] [PubMed]
43. Gómez, M.I.; O'Seaghdha, M.; Magargee, M.; Foster, T.J.; Prince, A.S. Staphylococcus Aureus Protein A Activates TNFR1 Signaling through Conserved IgG Binding Domains. *J. Biol. Chem.* **2006**, *281*, 20190–20196. [CrossRef]
44. Shin, J.; Monti, S.; Aires, D.J.; Duvic, M.; Golub, T.; Jones, D.A.; Kupper, T.S. Lesional Gene Expression Profiling in Cutaneous T-Cell Lymphoma Reveals Natural Clusters Associated with Disease Outcome. *Blood* **2007**, *110*, 3015–3027. [CrossRef]
45. Tracey, L.; Villuendas, R.; Dotor, A.M.; Spiteri, I.; Ortiz, P.; García, J.F.; Peralto, J.L.R.; Lawler, M.; Piris, M.A. Mycosis Fungoides Shows Concurrent Deregulation of Multiple Genes Involved in the TNF Signaling Pathway: An Expression Profile Study. *Blood* **2003**, *102*, 1042–1050. [CrossRef]
46. Rindler, K.; Bauer, W.M.; Jonak, C.; Wielscher, M.; Shaw, L.E.; Rojahn, T.B.; Thaler, F.M.; Porkert, S.; Simonitsch-Klupp, I.; Weninger, W.; et al. Single-Cell RNA Sequencing Reveals Tissue Compartment-Specific Plasticity of Mycosis Fungoides Tumor Cells. *Front. Immunol.* **2021**, *12*, 666935. [CrossRef] [PubMed]
47. Gru, A.A.; Haverkos, B.H.; Freud, A.G.; Hastings, J.; Nowacki, N.B.; Barrionuevo, C.; Vigil, C.E.; Rochford, R.; Natkunam, Y.; Baiocchi, R.A.; et al. The Epstein-Barr Virus (EBV) in T Cell and NK Cell Lymphomas: Time for a Reassessment. *Curr. Hematol. Malig. Rep.* **2015**, *10*, 456–467. [CrossRef]
48. Heslop, H.E. Biology and Treatment of Epstein-Barr Virus–Associated Non-Hodgkin Lymphomas. *Hematology* **2005**, *2005*, 260–266. [CrossRef] [PubMed]
49. Byrd, A.L.; Belkaid, Y.; Segre, J.A. The Human Skin Microbiome. *Nat. Rev. Microbiol.* **2018**, *16*, 143–155. [CrossRef]
50. Andrews, S.; Babraham, I. FastQC: A Quality Control Tool for High Throughput Sequence Data. Available online: https://www.bioinformatics.babraham.ac.uk/projects/fastqc/ (accessed on 8 October 2019).
51. Dobin, A.; Davis, C.A.; Schlesinger, F.; Drenkow, J.; Zaleski, C.; Jha, S.; Batut, P.; Chaisson, M.; Gingeras, T.R. STAR: Ultrafast Universal RNA-Seq Aligner. *Bioinformatics* **2013**, *29*, 15–21. [CrossRef]
52. Love, M.I.; Huber, W.; Anders, S. Moderated Estimation of Fold Change and Dispersion for RNA-Seq Data with DESeq2. *Genome Biol.* **2014**, *15*, 550. [CrossRef] [PubMed]
53. Blighe, K.; Rana, S.; Lewis, M. EnhancedVolcano: Publication-Ready Volcano Plots with Enhanced Colouring and Labeling. Available online: https://github.com/kevinblighe/EnhancedVolcano (accessed on 8 October 2019).
54. Wickham, H. *Ggplot2: Elegant Graphics for Data Analysis*; Springer: New York, NY, USA, 2016; ISBN 978-3-319-24277-4.
55. Ulgen, E.; Ozisik, O.; Sezerman, O.U. PathfindR: An R Package for Comprehensive Identification of Enriched Pathways in Omics Data Through Active Subnetworks. *Front. Genet.* **2019**, *10*, 858. [CrossRef] [PubMed]
56. Jassal, B.; Matthews, L.; Viteri, G.; Gong, C.; Lorente, P.; Fabregat, A.; Sidiropoulos, K.; Cook, J.; Gillespie, M.; Haw, R.; et al. The Reactome Pathway Knowledgebase. *Nucleic Acids Res.* **2019**, *48*, D498–D503. [CrossRef] [PubMed]
57. Smyth, G.K.; Ritchie, M.; Thorne, N.; Wettenhall, J.; Shi, W.; Hu, Y. Limma: Linear Models for Microarray and RNA-Seq Data User's Guide. Available online: https://www.bioconductor.org/packages/release/bioc/vignettes/limma/inst/doc/usersguide.pdf (accessed on 12 November 2024).
58. Argelaguet, R.; Velten, B.; Arnol, D.; Dietrich, S.; Zenz, T.; Marioni, J.C.; Buettner, F.; Huber, W.; Stegle, O. Multi-Omics Factor Analysis—A Framework for Unsupervised Integration of Multi-omics Data Sets. *Mol. Syst. Biol.* **2018**, *14*, e8124. [CrossRef] [PubMed]
59. Truong, D.T.; Franzosa, E.A.; Tickle, T.L.; Scholz, M.; Weingart, G.; Pasolli, E.; Tett, A.; Huttenhower, C.; Segata, N. MetaPhlAn2 for Enhanced Metagenomic Taxonomic Profiling. *Nat. Methods* **2015**, *12*, 902–903. [CrossRef]
60. Beghini, F.; McIver, L.J.; Blanco-Míguez, A.; Dubois, L.; Asnicar, F.; Maharjan, S.; Mailyan, A.; Manghi, P.; Scholz, M.; Thomas, A.M.; et al. Integrating Taxonomic, Functional, and Strain-Level Profiling of Diverse Microbial Communities with BioBakery 3. *Elife* **2021**, *10*, e65088. [CrossRef]
61. Kumar, M.S.; Slud, E.V.; Okrah, K.; Hicks, S.C.; Hannenhalli, S.; Corrada Bravo, H. Analysis and Correction of Compositional Bias in Sparse Sequencing Count Data. *BMC Genom.* **2018**, *19*, 799. [CrossRef]
62. Frost, H.R.; Li, Z.; Moore, J.H. Principal Component Gene Set Enrichment (PCGSE). *BioData Min.* **2015**, *8*, 25. [CrossRef]
63. Yasumizu, Y.; Hara, A.; Sakaguchi, S.; Ohkura, N. VIRTUS: A Pipeline for Comprehensive Virus Analysis from Conventional RNA-Seq Data. *Bioinformatics* **2021**, *37*, 1465–1467. [CrossRef]
64. Kolde, R. Pheatmap: Pretty Heatmaps. Available online: https://cran.r-project.org/package=pheatmap (accessed on 5 October 2018).
65. Vander Heiden, J.A.; Yaari, G.; UdumanSegre, M.; Stern, J.N.H.; O'Connor, K.C.; Hafler, D.A.; Vigneault, F.; Kleinstein, S.H. PRESTO: A Toolkit for Processing High-Throughput Sequencing Raw Reads of Lymphocyte Receptor Repertoires. *Bioinformatics* **2014**, *30*, 1930–1932. [CrossRef]
66. ImmunoMind Team Immunarch: An R Package for Painless Bioinformatics Analysis of T-Cell and B-Cell Immune Repertoires 2019. Available online: https://zenodo.org/records/10840553 (accessed on 20 November 2024).
67. Vita, R.; Mahajan, S.; Overton, J.A.; Dhanda, S.K.; Martini, S.; Cantrell, J.R.; Wheeler, D.K.; Sette, A.; Peters, B. The Immune Epitope Database (IEDB): 2018 Update. *Nucleic Acids Res.* **2019**, *47*, D339–D343. [CrossRef] [PubMed]
68. Springer, I.; Tickotsky, N.; Louzoun, Y. Contribution of T Cell Receptor Alpha and Beta CDR3, MHC Typing, V and J Genes to Peptide Binding Prediction. *Front. Immunol.* **2021**, *12*, 664514. [CrossRef]

69. Gaydosik, A.M.; Tabib, T.; Geskin, L.J.; Bayan, C.-A.; Conway, J.F.; Lafyatis, R.; Fuschiotti, P. Single-Cell Lymphocyte Heterogeneity in Advanced Cutaneous T-Cell Lymphoma Skin Tumors. *Clin. Cancer Res.* **2019**, *25*, 4443–4454. [CrossRef]
70. Buus, T.B.; Willerslev-Olsen, A.; Fredholm, S.; Blümel, E.; Nastasi, C.; Gluud, M.; Hu, T.; Lindahl, L.M.; Iversen, L.; Fogh, H.; et al. Single-Cell Heterogeneity in Sézary Syndrome. *Blood Adv.* **2018**, *2*, 2115–2126. [CrossRef]
71. Borcherding, N.; Voigt, A.P.; Liu, V.; Link, B.K.; Zhang, W.; Jabbari, A. Single-Cell Profiling of Cutaneous T-Cell Lymphoma Reveals Underlying Heterogeneity Associated with Disease Progression. *Clin. Cancer Res.* **2019**, *25*, 2996–3005. [CrossRef]
72. Herrera, A.; Cheng, A.; Mimitou, E.P.; Seffens, A.; George, D.D.; Bar-Natan, M.; Heguy, A.; Ruggles, K.V.; Scher, J.U.; Hymes, K.; et al. Multimodal Single-Cell Analysis of Cutaneous T Cell Lymphoma Reveals Distinct Sub-Clonal Tissue-Dependent Signatures. *Blood* **2021**, *138*, 1456–1464. [CrossRef] [PubMed]
73. Litvinov, I.V.; Tetzlaff, M.T.; Thibault, P.; Gangar, P.; Moreau, L.; Watters, A.K.; Netchiporouk, E.; Pehr, K.; Prieto, V.G.; Rahme, E.; et al. Gene Expression Analysis in Cutaneous T-Cell Lymphomas (CTCL) Highlights Disease Heterogeneity and Potential Diagnostic and Prognostic Indicators. *Oncoimmunology* **2017**, *6*, e1306618. [CrossRef] [PubMed]
74. Stolearenco, V.; Namini, M.R.J.; Hasselager, S.S.; Gluud, M.; Buus, T.B.; Willerslev-Olsen, A.; Ødum, N.; Krejsgaard, T. Cellular Interactions and Inflammation in the Pathogenesis of Cutaneous T-Cell Lymphoma. *Front. Cell Dev. Biol.* **2020**, *8*, 851. [CrossRef]
75. Hruz, P.; Zinkernagel, A.S.; Jenikova, G.; Botwin, G.J.; Hugot, J.-P.; Karin, M.; Nizet, V.; Eckmann, L. NOD2 Contributes to Cutaneous Defense against Staphylococcus Aureus through α-Toxin-Dependent Innate Immune Activation. *Proc. Natl. Acad. Sci. USA* **2009**, *106*, 12873–12878. [CrossRef]
76. Deshmukh, H.S.; Hamburger, J.B.; Ahn, S.H.; McCafferty, D.G.; Yang, S.R.; Fowler, V.G. Critical Role of NOD2 in Regulating the Immune Response to Staphylococcus Aureus. *Infect. Immun.* **2009**, *77*, 1376–1382. [CrossRef]
77. Soong, G.; Martin, F.J.; Chun, J.; Cohen, T.S.; Ahn, D.S.; Prince, A. Staphylococcus Aureus Protein A Mediates Invasion across Airway Epithelial Cells through Activation of RhoA GTPase Signaling and Proteolytic Activity. *J. Biol. Chem.* **2011**, *286*, 35891–35898. [CrossRef] [PubMed]
78. Nakatsuji, T.; Chen, T.H.; Two, A.M.; Chun, K.A.; Narala, S.; Geha, R.S.; Hata, T.R.; Gallo, R.L. Staphylococcus Aureus Exploits Epidermal Barrier Defects in Atopic Dermatitis to Trigger Cytokine Expression. *J. Investig. Dermatol.* **2016**, *136*, 2192–2200. [CrossRef] [PubMed]
79. Byrd, A.L.; Deming, C.; Cassidy, S.K.B.; Harrison, O.J.; Ng, W.-I.; Conlan, S.; Belkaid, Y.; Segre, J.A.; Kong, H.H. Staphylococcus Aureus and Staphylococcus Epidermidis Strain Diversity Underlying Pediatric Atopic Dermatitis. *Sci. Transl. Med.* **2017**, *9*, eaal4651. [CrossRef] [PubMed]
80. Bergsbaken, T.; Fink, S.L.; Cookson, B.T. Pyroptosis: Host Cell Death and Inflammation. *Nat. Rev. Microbiol.* **2009**, *7*, 99–109. [CrossRef] [PubMed]
81. Vaillant, F.; Blyth, K.; Andrew, L.; Neil, J.C.; Cameron, E.R. Enforced Expression of Runx2 Perturbs T Cell Development at a Stage Coincident with β-Selection. *J. Immunol.* **2002**, *169*, 2866–2874. [CrossRef]
82. Egawa, T.; Tillman, R.E.; Naoe, Y.; Taniuchi, I.; Littman, D.R. The Role of the Runx Transcription Factors in Thymocyte Differentiation and in Homeostasis of Naive T Cells. *J. Exp. Med.* **2007**, *204*, 1945–1957. [CrossRef]
83. Ichikawa, M.; Asai, T.; Saito, T.; Yamamoto, G.; Seo, S.; Yamazaki, I.; Yamagata, T.; Mitani, K.; Chiba, S.; Hirai, H.; et al. AML-1 Is Required for Megakaryocytic Maturation and Lymphocytic Differentiation, but Not for Maintenance of Hematopoietic Stem Cells in Adult Hematopoiesis. *Nat. Med.* **2004**, *10*, 299–304. [CrossRef]
84. Growney, J.D.; Shigematsu, H.; Li, Z.; Lee, B.H.; Adelsperger, J.; Rowan, R.; Curley, D.P.; Kutok, J.L.; Akashi, K.; Williams, I.R.; et al. Loss of Runx1 Perturbs Adult Hematopoiesis and Is Associated with a Myeloproliferative Phenotype. *Blood* **2005**, *106*, 494–504. [CrossRef]
85. Cogen, A.L.; Yamasaki, K.; Sanchez, K.M.; Dorschner, R.A.; Lai, Y.; MacLeod, D.T.; Torpey, J.W.; Otto, M.; Nizet, V.; Kim, J.E.; et al. Selective Antimicrobial Action Is Provided by Phenol-Soluble Modulins Derived from Staphylococcus Epidermidis, a Normal Resident of the Skin. *J. Investig. Dermatol.* **2010**, *130*, 192–200. [CrossRef]
86. Nakatsuji, T.; Chen, T.H.; Narala, S.; Chun, K.A.; Two, A.M.; Yun, T.; Shafiq, F.; Kotol, P.F.; Bouslimani, A.; Melnik, A.V.; et al. Antimicrobials from Human Skin Commensal Bacteria Protect against Staphylococcus Aureus and Are Deficient in Atopic Dermatitis. *Sci. Transl. Med.* **2017**, *9*, eaah4680. [CrossRef]
87. Williams, M.R.; Costa, S.K.; Zaramela, L.S.; Khalil, S.; Todd, D.A.; Winter, H.L.; Sanford, J.A.; O'Neill, A.M.; Liggins, M.C.; Nakatsuji, T.; et al. Quorum Sensing between Bacterial Species on the Skin Protects against Epidermal Injury in Atopic Dermatitis. *Sci. Transl. Med.* **2019**, *11*, eaat8329. [CrossRef] [PubMed]
88. Sun, S.-C. The Non-Canonical NF-KB Pathway in Immunity and Inflammation. *Nat. Rev. Immunol.* **2017**, *17*, 545–558. [CrossRef] [PubMed]
89. Faustman, D.; Davis, M. TNF Receptor 2 Pathway: Drug Target for Autoimmune Diseases. *Nat. Rev. Drug Discov.* **2010**, *9*, 482–493. [CrossRef]
90. Chng, K.R.; Tay, A.S.L.; Li, C.; Ng, A.H.Q.; Wang, J.; Suri, B.K.; Matta, S.A.; McGovern, N.; Janela, B.; Wong, X.F.C.C.; et al. Whole Metagenome Profiling Reveals Skin Microbiome-Dependent Susceptibility to Atopic Dermatitis Flare. *Nat. Microbiol.* **2016**, *1*, 16106. [CrossRef]
91. Wang, X.; Lin, Y. Tumor Necrosis Factor and Cancer, Buddies or Foes? *Acta Pharmacol. Sin.* **2008**, *29*, 1275–1288. [CrossRef]
92. Katze, M.G.; He, Y.; Gale, M. Viruses and Interferon: A Fight for Supremacy. *Nat. Rev. Immunol.* **2002**, *2*, 675–687. [CrossRef]

93. Liu, X.-Y.; Chen, W.; Wei, B.; Shan, Y.-F.; Wang, C. IFN-Induced TPR Protein IFIT3 Potentiates Antiviral Signaling by Bridging MAVS and TBK1. *J. Immunol.* **2011**, *187*, 2559–2568. [CrossRef] [PubMed]
94. Roy, C.R.; Salcedo, S.P.; Gorvel, J.-P.E. Pathogen–Endoplasmic-Reticulum Interactions: In through the out Door. *Nat. Rev. Immunol.* **2006**, *6*, 136–147. [CrossRef]
95. Murray, P.G.; Constandinou, C.M.; Crocker, J.; Young, L.S.; Ambinder, R.F. Analysis of Major Histocompatibility Complex Class I, TAP Expression, and LMP2 Epitope Sequence in Epstein-Barr Virus–Positive Hodgkin's Disease. *Blood* **1998**, *92*, 2477–2483. [CrossRef]
96. Wang, Y.; Finan, J.E.; Middeldorp, J.M.; Hayward, S.D. P32/TAP, a Cellular Protein That Interacts with EBNA-1 of Epstein–Barr Virus. *Virology* **1997**, *236*, 18–29. [CrossRef]
97. Zhao, B.; Zou, J.; Wang, H.; Johannsen, E.; Peng, C.; Quackenbush, J.; Mar, J.C.; Morton, C.C.; Freedman, M.L.; Blacklow, S.C.; et al. Epstein-Barr Virus Exploits Intrinsic B-Lymphocyte Transcription Programs to Achieve Immortal Cell Growth. *Proc. Natl. Acad. Sci. USA* **2011**, *108*, 14902–14907. [CrossRef] [PubMed]
98. Gunnell, A.; Webb, H.M.; Wood, C.D.; McClellan, M.J.; Wichaidit, B.; Kempkes, B.; Jenner, R.G.; Osborne, C.; Farrell, P.J.; West, M.J. RUNX Super-Enhancer Control through the Notch Pathway by Epstein-Barr Virus Transcription Factors Regulates B Cell Growth. *Nucleic Acids Res.* **2016**, *44*, 4636–4650. [CrossRef] [PubMed]
99. Stewart, M.; Terry, A.; Hu, M.; O'Hara, M.; Blyth, K.; Baxter, E.; Cameron, E.; Onions, D.E.; Neil, J.C. Proviral Insertions Induce the Expression of Bone-Specific Isoforms of PEBP2αA (CBFA1): Evidence for a New Myc Collaborating Oncogene. *Proc. Natl. Acad. Sci. USA* **1997**, *94*, 8646–8651. [CrossRef]
100. Suzuki, T.; Shen, H.; Akagi, K.; Morse, H.C.; Malley, J.D.; Naiman, D.Q.; Jenkins, N.A.; Copeland, N.G. New Genes Involved in Cancer Identified by Retroviral Tagging. *Nat. Genet.* **2002**, *32*, 166–174. [CrossRef] [PubMed]
101. Wotton, S.; Stewart, M.; Blyth, K.; Vaillant, F.; Kilbey, A.; Neil, J.C.; Cameron, E.R. Proviral Insertion Indicates a Dominant Oncogenic Role for Runx1/AML-1 in T-Cell Lymphoma. *Cancer Res.* **2002**, *62*, 7181–7185.
102. Blyth, K.; Cameron, E.R.; Neil, J.C. The Runx Genes: Gain or Loss of Function in Cancer. *Nat. Rev. Cancer* **2005**, *5*, 376–387. [CrossRef]
103. Dutta, A.; Zhao, B.; Love, P.E. New Insights into TCR β-Selection. *Trends Immunol.* **2021**, *42*, 735–750. [CrossRef]
104. Blyth, K.; Vaillant, F.; Hanlon, L.; Mackay, N.; Bell, M.; Jenkins, A.; Neil, J.C.; Cameron, E.R. Runx2 and MYC Collaborate in Lymphoma Development by Suppressing Apoptotic and Growth Arrest Pathways in Vivo. *Cancer Res.* **2006**, *66*, 2195–2201. [CrossRef]
105. Liang, J.; Zhou, W.; Sakre, N.; DeVecchio, J.; Ferrandon, S.; Ting, A.H.; Bao, S.; Bissett, I.; Church, J.; Kalady, M.F. Epigenetically Regulated MiR-1247 Functions as a Novel Tumour Suppressor via MYCBP2 in Methylator Colon Cancers. *Br. J. Cancer* **2018**, *119*, 1267–1277. [CrossRef]
106. Lee, S.; Kim, J.-Y.; Kim, Y.-J.; Seok, K.-O.; Kim, J.-H.; Chang, Y.-J.; Kang, H.-Y.; Park, J.-H. Nucleolar Protein GLTSCR2 Stabilizes P53 in Response to Ribosomal Stresses. *Cell Death Differ.* **2012**, *19*, 1613–1622. [CrossRef]
107. Oda, K.; Arakawa, H.; Tanaka, T.; Matsuda, K.; Tanikawa, C.; Mori, T.; Nishimori, H.; Tamai, K.; Tokino, T.; Nakamura, Y.; et al. P53AIP1, a Potential Mediator of P53-Dependent Apoptosis, and Its Regulation by Ser-46-Phosphorylated P53. *Cell* **2000**, *102*, 849–862. [CrossRef] [PubMed]
108. Hwang, S.T.; Janik, J.E.; Jaffe, E.S.; Wilson, W.H. Mycosis Fungoides and Sézary Syndrome. *Lancet* **2008**, *371*, 945–957. [CrossRef] [PubMed]
109. Shannon-Lowe, C.; Rickinson, A.B.; Bell, A.I. Epstein-Barr Virus-Associated Lymphomas. *Philos. Trans. R. Soc. B Biol. Sci.* **2017**, *372*, 20160271. [CrossRef]
110. Rollison, D.E.; Engels, E.A.; Halsey, N.A.; Shah, K.V.; Viscidi, R.P.; Helzlsouer, K.J. Prediagnostic Circulating Antibodies to JC and BK Human Polyomaviruses and Risk of Non-Hodgkin Lymphoma. *Cancer Epidemiol. Biomark. Prev.* **2006**, *15*, 543–550. [CrossRef] [PubMed]
111. Du-Thanh, A.; Foulongne, V.; Guillot, B.; Dereure, O. Recently Discovered Human Polyomaviruses in Lesional and Non-Lesional Skin of Patients with Primary Cutaneous T-Cell Lymphomas. *J. Dermatol. Sci.* **2013**, *71*, 140–142. [CrossRef]
112. Ryan, J.L.; Morgan, D.R.; Dominguez, R.L.; Thorne, L.B.; Elmore, S.H.; Mino-Kenudson, M.; Lauwers, G.Y.; Booker, J.K.; Gulley, M.L. High Levels of Epstein–Barr Virus DNA in Latently Infected Gastric Adenocarcinoma. *Lab. Investig.* **2009**, *89*, 80–90. [CrossRef]
113. Kutok, J.L.; Wang, F. Spectrum of Epstein-Barr Virus-Associated Diseases. *Annu. Rev. Pathol.* **2006**, *1*, 375–404. [CrossRef]
114. Bi, X.; Wang, H.; Zhang, W.; Wang, J.; Liu, W.; Xia, Z.; Huang, H.; Jiang, W.; Zhang, Y.; Wang, L. PD-L1 Is Upregulated by EBV-Driven LMP1 through NF-KB Pathway and Correlates with Poor Prognosis in Natural Killer/T-Cell Lymphoma. *J. Hematol. Oncol.* **2016**, *9*, 109. [CrossRef]
115. Kim, Y.H.; Liu, H.L.; Mraz-Gernhard, S.; Varghese, A.; Hoppe, R.T. Long-Term Outcome of 525 Patients with Mycosis Fungoides and Sézary Syndrome. *Arch. Dermatol.* **2003**, *139*, 857–866. [CrossRef]
116. Wang, L.; Ni, X.; Covington, K.R.; Yang, B.Y.; Shiu, J.; Zhang, X.; Xi, L.; Meng, Q.; Langridge, T.; Drummond, J.; et al. Genomic Profiling of Sézary Syndrome Identifies Alterations of Key T Cell Signaling and Differentiation Genes. *Nat. Genet.* **2015**, *47*, 1426–1434. [CrossRef]
117. Horna, P.; Moscinski, L.C.; Sokol, L.; Shao, H. Naïve/Memory T-cell Phenotypes in Leukemic Cutaneous T-cell Lymphoma: Putative Cell of Origin Overlaps Disease Classification. *Cytom. Part B Clin. Cytom.* **2019**, *96*, 234–241. [CrossRef] [PubMed]

118. Kim, E.Y.; Priatel, J.J.; Teh, S.-J.; Teh, H. TNF Receptor Type 2 (P75) Functions as a Costimulator for Antigen-Driven T Cell Responsesa In Vivo. *J. Immunol.* **2006**, *176*, 1026–1035. [CrossRef] [PubMed]
119. Mukai, Y.; Nakamura, T.; Yoshikawa, M.; Yoshioka, Y.; Tsunoda, S.; Nakagawa, S.; Yamagata, Y.; Tsutsumi, Y. Solution of the Structure of the TNF-TNFR2 Complex. *Sci. Signal.* **2010**, *3*, ra83. [CrossRef] [PubMed]
120. Nicolay, J.P.; Müller-Decker, K.; Schroeder, A.; Brechmann, M.; Möbs, M.; Géraud, C.; Assaf, C.; Goerdt, S.; Krammer, P.H.; Gülow, K. Dimethyl Fumarate Restores Apoptosis Sensitivity and Inhibits Tumor Growth and Metastasis in CTCL by Targeting NF-KB. *Blood* **2016**, *128*, 805–815. [CrossRef]
121. Nicolay, J.P.; Albrecht, J.D.; Assaf, C.; Dippel, E.; Stadler, R.; Wehkamp, U.; Wobser, M.; Guelow, K.; Goerdt, S.; Krammer, P.H. Dimethyl Fumarate (DMF) Therapy in CTCL: Results from a Clinical Phase II Study. *Eur. J. Cancer* **2021**, *156*, S21–S22. [CrossRef]
122. Nicolay, J.P.; Melchers, S.; Albrecht, J.D.; Assaf, C.; Dippel, E.; Stadler, R.; Wehkamp, U.; Wobser, M.; Zhao, J.; Burghaus, I.; et al. Dimethyl Fumarate Treatment in Relapsed and Refractory Cutaneous T Cell Lymphoma—A Multicenter Phase II Study. *Blood J.* **2023**, *142*, 794–805. [CrossRef]
123. Hosoki, K.; Nakamura, A.; Nagao, M.; Hiraguchi, Y.; Tanida, H.; Tokuda, R.; Wada, H.; Nobori, T.; Suga, S.; Fujisawa, T. Staphylococcus Aureus Directly Activates Eosinophils via Platelet-Activating Factor Receptor. *J. Leukoc. Biol.* **2012**, *92*, 333–341. [CrossRef]
124. Fredholm, S.; Gjerdrum, L.M.R.; Willerslev-Olsen, A.; Petersen, D.L.; Nielsen, I.Ø.; Kauczok, C.-S.; Wobser, M.; Ralfkiaer, U.; Bonefeld, C.M.; Wasik, M.A.; et al. STAT3 Activation and Infiltration of Eosinophil Granulocytes in Mycosis Fungoides. *Anticancer Res.* **2014**, *34*, 5277–5286.
125. Aufiero, B.; Guo, M.; Young, C.; Duanmu, Z.; Talwar, H.; Lee, H.K.; Murakawa, G.J. Staphylococcus Aureus Induces the Expression of Tumor Necrosis factor-alpha in Primary Human Keratinocytes. *Int. J. Dermatol.* **2007**, *46*, 687–694. [CrossRef]
126. Tufano, M.A.; Cipollaro de l'Ero, G.; Ianniello, R.; Galdiero, M.; Galdiero, F. Protein A and Other Surface Components of Staphylococcus Aureus Stimulate Production of IL-1 Alpha, IL-4, IL-6, TNF and IFN-Gamma. *Eur. Cytokine Netw.* **1991**, *2*, 361–366.
127. Matthijssens, F.; Sharma, N.D.; Nysus, M.; Nickl, C.K.; Kang, H.; Perez, D.R.; Lintermans, B.; Van Loocke, W.; Roels, J.; Peirs, S.; et al. RUNX2 Regulates Leukemic Cell Metabolism and Chemotaxis in High-Risk T Cell Acute Lymphoblastic Leukemia. *J. Clin. Investig.* **2021**, *131*, e141566. [CrossRef] [PubMed]
128. Fu, L.; Van Doorslaer, K.; Chen, Z.; Ristriani, T.; Masson, M.; Travé, G.; Burk, R.D. Degradation of P53 by Human Alphapapillomavirus E6 Proteins Shows a Stronger Correlation with Phylogeny than Oncogenicity. *PLoS ONE* **2010**, *5*, e12816. [CrossRef]
129. Martinez-Zapien, D.; Ruiz, F.X.; Poirson, J.; Mitschler, A.; Ramirez, J.; Forster, A.; Cousido-Siah, A.; Masson, M.; Pol, S.V.; Podjarny, A.; et al. Structure of the E6/E6AP/P53 Complex Required for HPV-Mediated Degradation of P53. *Nature* **2016**, *529*, 541–545. [CrossRef] [PubMed]
130. Egawa, N.; Egawa, K.; Griffin, H.; Doorbar, J. Human Papillomaviruses; Epithelial Tropisms, and the Development of Neoplasia. *Viruses* **2015**, *7*, 3863–3890. [CrossRef] [PubMed]
131. Intaraphet, S.; Farkas, D.K.; Johannesdottir Schmidt, S.A.; Cronin-Fenton, D.; Søgaard, M. Human Papillomavirus Infection and Lymphoma Incidence Using Cervical Conization as a Surrogate Marker: A Danish Nationwide Cohort Study. *Hematol. Oncol.* **2017**, *35*, 172–176. [CrossRef]
132. Van Doorslaer, K.; Burk, R.D. Evolution of Human Papillomavirus Carcinogenicity. *Adv. Virus Res.* **2010**, *77*, 41–62.
133. Farrell, P.J. Epstein–Barr Virus and Cancer. *Annu. Rev. Pathol. Mech. Dis.* **2019**, *14*, 29–53. [CrossRef]
134. Faulkner, G.C.; Krajewski, A.S.; Crawford, D.H. The Ins and Outs of EBV Infection. *Trends Microbiol.* **2000**, *8*, 185–189. [CrossRef]
135. Gandhi, M.K.; Tellam, J.T.; Khanna, R. Epstein-Barr Virus-Associated Hodgkin's Lymphoma. *Br. J. Haematol.* **2004**, *125*, 267–281. [CrossRef]
136. Luebberding, S.; Krueger, N.; Kerscher, M. Age-related Changes in Skin Barrier Function—Quantitative Evaluation of 150 Female Subjects. *Int. J. Cosmet. Sci.* **2013**, *35*, 183–190. [CrossRef]
137. Choi, E.H. Aging of the Skin Barrier. *Clin. Dermatol.* **2019**, *37*, 336–345. [CrossRef] [PubMed]
138. Suga, H.; Sugaya, M.; Miyagaki, T.; Ohmatsu, H.; Kawaguchi, M.; Takahashi, N.; Fujita, H.; Asano, Y.; Tada, Y.; Kadono, T.; et al. Skin Barrier Dysfunction and Low Antimicrobial Peptide Expression in Cutaneous T-Cell Lymphoma. *Clin. Cancer Res.* **2014**, *20*, 4339–4348. [CrossRef] [PubMed]
139. Nakajima, R.; Miyagaki, T.; Kamijo, H.; Oka, T.; Shishido-Takahashi, N.; Suga, H.; Sugaya, M.; Sato, S. Decreased Progranulin Expression in Mycosis Fungoides: A Possible Association with the High Frequency of Skin Infections. *Eur. J. Dermatol.* **2018**, *28*, 790–794. [CrossRef]
140. Wolk, K.; Mitsui, H.; Witte, K.; Gellrich, S.; Gulati, N.; Humme, D.; Witte, E.; Gonsior, M.; Beyer, M.; Kadin, M.E.; et al. Deficient Cutaneous Antibacterial Competence in Cutaneous T-Cell Lymphomas: Role of Th2-Mediated Biased Th17 Function. *Clin. Cancer Res.* **2014**, *20*, 5507–5516. [CrossRef] [PubMed]

cancers

MDPI

Article

High-Throughput Molecular Characterization of the Microbiome in Breast Implant-Associated Anaplastic Large Cell Lymphoma and Peri-Implant Benign Seromas

Evelina Rogges [1,2], Giorgio Bertolazzi [3], Davide Vacca [4], Marina Borro [5,6,7], Gianluca Lopez [8], Maurizio Simmaco [5,6,7], Anna Scattone [9], Guido Firmani [10], Michail Sorotos [10], Fabio Santanelli di Pompeo [10], Niccolò Noccioli [8], Emanuele Savino [11], Andrea Vecchione [2] and Arianna Di Napoli [2,*]

1 Department of Medical and Surgical Sciences and Translational Medicine, PhD School in Translational Medicine and Oncology, Sapienza University of Rome, Faculty of Medicine and Psychology, 00189 Rome, Italy; evelina.rogges@uniroma1.it
2 Pathology Unit, Department of Clinical and Molecular Medicine, Sant'Andrea University Hospital, Sapienza University of Rome, 00189 Rome, Italy; andrea.vecchione@uniroma1.it
3 Department of Medicine and Surgery, Kore University of Enna, 94100 Enna, Italy; giorgio.bertolazzi@unikore.it
4 Tumor Immunology Unit, Department of Health Sciences, University of Palermo, 90127 Palermo, Italy; davide.vacca@outlook.com
5 Center for Precision Medicine (CPM), Sant'Andrea University Hospital, 00189 Rome, Italy; marina.borro@uniroma1.it (M.B.); maurizio.simmaco@uniroma1.it (M.S.)
6 Department of Neuroscience, Mental Health and Sensory Organs (NESMOS), Faculty of Medicine and Psychology, Sapienza University of Rome, 00189 Rome, Italy
7 Unit of Laboratory Analysis and Advanced Molecular Diagnostics, Department of Diagnostic Sciences, Sant'Andrea University Hospital, 00189 Rome, Italy
8 Morphologic and Molecular Pathology Unit, St. Andrea University Hospital, Sapienza University of Rome, 00189 Rome, Italy; gianluca.lopez10@gmail.com (G.L.); nocciolin@gmail.com (N.N.)
9 Department of Pathology, IRCCS Istituto Tumori 'Giovanni Paolo II', 70124 Bari, Italy; a.scattone@oncologico.bari.it
10 Plastic Surgery Unit, Department of Neuroscience, Mental Health, and Sense Organs (NESMOS), Faculty of Medicine and Psychology, Sapienza University of Rome, 00189 Rome, Italy; guido.firmani@uniroma1.it (G.F.); michail.sorotos@uniroma1.it (M.S.); fabio.santanelli@uniroma1.it (F.S.d.P.)
11 Advanced Pathology Laboratory, IFOM ETS-The AIRC Institute of Molecular Oncology, 20139 Milan, Italy; emanuele.savino@ifom.eu
* Correspondence: arianna.dinapoli@uniroma1.it

Simple Summary

Breast implant-associated anaplastic large cell lymphoma (BIA-ALCL) is a rare type of T-cell lymphoma linked to textured breast implants. A leading hypothesis suggests that chronic inflammation, possibly triggered by bacteria forming biofilms on the implant surface, may play a role in its development. However, previous studies based on breast implant capsules have yielded inconsistent results. In our study, we investigated the microbiota in late-onset seromas from 10 BIA-ALCL patients and 12 with a non-neoplastic effusion. Using two different next-generation sequencing methods, we found no single bacterial species or family specifically associated with BIA-ALCL. Instead, BIA-ALCL samples showed a higher abundance of non-aerobic bacteria, which aligns with prior evidence of a hypoxic (low-oxygen) tumor environment. Our findings suggest that while bacteria may foster chronic inflammation, BIA-ALCL likely stems from a combination of factors, rather than a specific infectious cause.

Abstract

Background: Breast implant-associated anaplastic large cell lymphoma (BIA-ALCL) is a mature T-cell lymphoma linked to textured breast implants. A leading hypothesis suggests that chronic inflammation, combined with immunological and genetic factors, drives its pathogenesis. Two previous studies investigating bacterial biofilms on breast implant capsules have produced conflicting results, particularly regarding the enrichment of *Ralstonia* spp. **Methods**: We analyzed the microbiota profiles in seroma samples from 10 BIA-ALCL patients and 12 patients with non-neoplastic effusion, subclassified into acute-, mixed-, and chronic-type based on cellular composition. We used two metagenomic approaches: 16S rRNA gene sequencing and Nanopore sequencing with the "What's in My Pot?" (WIMP) taxonomic classifier. Our analyses included alpha and beta diversity metrics, as well as comparisons of Gram status and oxygen requirements. **Results**: Both sequencing methods identified *Staphylococcaceae*, *Propionibacteriaceae*, and *Bradyrhizobiaceae* as the most prevalent bacterial families in both BIA-ALCL and benign seroma samples. Notably, the *Burkholderiaceae* family was more abundant in some of the benign seromas according to the 16S rRNA sequencing, but *Ralstonia* spp. were not detected. BIA-ALCL showed higher richness (based on Nanopore data) and higher evenness (based on 16S rRNA data) compared to acute-type seromas, indicating a more homogenous representation of the different taxa identified. BIA-ALCL seromas did not cluster together based on Nanopore data, but they did form a distinct cluster with 16S rRNA data. This cluster was differentiated from the other two clusters by a relatively balanced presence of multiple families without overt dominance. We observed no significant differences in Gram staining between BIA-ALCL and benign samples using either method. However, non-aerobic bacterial families were enriched in BIA-ALCL cases only when analyzed with the Nanopore pipeline. **Conclusions**: Overall, our findings did not identify a distinctive microbial signature specifically associated with BIA-ALCL.

Keywords: BIA-ALCL; microbiome; NGS

1. Introduction

Breast implant-associated anaplastic large cell lymphoma (BIA-ALCL) is an uncommon mature T-cell lymphoma that arises in relation to a textured breast implant [1]. This CD30-positive, anaplastic lymphoma kinase (ALK)-negative malignancy has been increasingly recognized, with the Food and Drug Administration (FDA) reporting a total of 1380 cases of BIA-ALCL with 64 known deaths as of 8 July 2025 [2]. The initial sign of disease is typically a late-onset peri-implant seroma, whereas tumor mass or regional lymphadenopathy are less frequent presentations. Diagnostic protocols for aspirated fluid involve cultural analysis, cytomorphological examination, immunocytochemistry for CD30 expression, and eventual T-cell receptor gene rearrangement analysis [3,4]. However, most late seromas are benign, often linked to infection, trauma, or implant rupture. Based on the cellular composition, we previously sub-classified these late-onset non-neoplastic seromas as acute-type (>50% neutrophils), mixed-type (5 to 50% neutrophils admixed with monocytes and lymphocytes), and chronic-type (when monocytes or lymphocytes predominate) [3].

To date, the etiopathogenesis of BIA-ALCL remains unclear but is considered multifactorial. A prominent hypothesis suggests that in genetically predisposed individuals, implant-associated compounds together with bacterial biofilm on the implant surface may chronically stimulate immune cells in hypoxic conditions [5,6]. This persistent stimulation can lead to proliferating T-cells accumulating genomic alterations, eventually transforming into lymphoma cells. This hypothesis is further sustained by genomic studies identifying somatic genetic alterations of genes involved in the JAK/STAT3 inflammatory signaling,

TP53, or in epigenetic modifiers [7–13]. The constitutive activation of the JAK/STAT3 inflammatory pathway is also consistently evidenced by the expression of phosphorylated STAT3 and transcriptional activation of its target genes [9,12,14,15].

According to the most recent edition of the International Organization for Standardization (ISO) 14607:2024 [16], mammary implant surfaces are classified as either smooth or intentionally textured. Smooth surfaces have a surface roughness (Ra) of 0 μm, while intentionally textured surfaces are further subdivided into microtextured (0 μm < Ra < 50 μm) and macrotextured (Ra ≥ 50 μm) [16]. Recently, a moderate level of evidence linked the development of BIA-ALCL to textured implants, particularly in relation to implants with an intermediate to high surface roughness [17] (Scientific Committee on Health, Environmental and Emerging Risks, SCHEER 2021), primarily due to the higher global incidence of BIA-ALCL cases reported in patients with such devices [1,2]. Highly textured implants have also been associated with increased macrophage activity, scar tissue formation, higher levels of inflammatory T cells [18–20], and greater bacterial colonization [21–24], all contributing to chronic inflammation.

Regarding the bacterial biofilm, Gram-positive bacteria such as *Staphylococcus aureus* and *Staphylococcus epidermidis* are established pathogens in breast implant infections [25]. Conversely, Gram-negative bacterial lipopolysaccharide (LPS) has shown greater efficacy in triggering the activation and proliferation of BIA-ALCL tumor cells in vitro compared to Staphylococcal superantigen enterotoxin A (SEA) [26]. Previous studies on the microbiome of implant capsules have yielded conflicting results. Hu et al., by pyrosequencing the V1–V3 region of the 16S rRNA gene, reported a significantly greater proportion of the Gram-negative *Ralstonia* spp. in 19 BIA-ALCL capsular specimens compared with 12 nontumoral capsules, where a higher number of *Staphylococcus* spp. were found [27]. In contrast, Walker et al. found no significant differences in the relative abundance of Gram-negative bacteria between seven BIA-ALCL capsules and control specimens (i.e., implants, contralateral breast, skin, and one seroma). They also detected low abundance of *Ralstonia* spp. in their samples, attributing this discrepancy to different sequencing methods (next generation sequencing (NGS) vs. pyrosequencing) of the V1–V3 region of 16S rRNA gene [28]. This controversy leaves open the discussion on the precise role of bacterial biofilm in BIA-ALCL pathogenesis. In the current study, we profile for the first time the microbiota of the effusions in a cohort of patients with both neoplastic and benign seromas by employing two distinct metagenomics methods: the 16S rRNA gene NGS and the Nanopore sequencing, a third-generation NGS, coupled with the "What's in my Pot?" (WIMP) data analysis pipeline. We focused on fresh-frozen seroma fluid, as the seroma represents the primary biological microenvironment in which BIA-ALCL develops. Seroma fluid captures the active peri-implant niche more directly than capsular tissue and provides a highly biologically relevant substrate for microbiome characterization. Indeed, fresh-frozen seroma samples not only preserve microbial DNA with minimal degradation but also avoid the artifacts introduced by the processing and storage of formalin-fixed and paraffin-embedded (FFPE) capsule tissue.

2. Materials and Methods

2.1. Sample Collection

Twenty-two late-onset fresh seroma samples from 22 patients with a diagnosis of BIA-ALCL (10 samples) or benign effusion (12 samples) were collected at Sant'Andrea University Hospital of Rome, Italy, from 2014 to 2020. Patients with non-neoplastic effusions were further subclassified into three types according to their cellular composition as previously described: acute-type (4 samples), mixed-type (3 samples), and chronic-type (5 samples) [3]. All patients were women and had textured implants. Patients with non-neoplastic effusions had a mean age of 53 years (range 33–69), and 7 out of 12 (58.3%) had a history of breast cancer. For patients diagnosed as BIA-ALCL, the mean age was 55 years (range 38 to 76), and their implants were placed for either reconstructive (6 out of 10, 60%) or cosmetic reasons (4 out of 10, 40%). The BIA-ALCL cases were staged according to the MD Anderson TNM staging system [29] as follows: stage IA (6 patients), stage IB (1 patient), stage IIA (2 patients), and stage IIB (1 patient). The clinical characteristics of both the benign and BIA-ALCL samples, along with the pathological stages for BIA-ALCL patients, are detailed in Tables 1 and 2. All fresh seroma samples were immediately frozen at −80 °C until DNA extraction. DNA was extracted using the QIAamp ® DNA Mini Kit (Qiagen, Hilden, Germany) and quantification was assessed using Qubit™ dsDNA Quantification High Sensitivity Assay Kits (ThermoFisher Scientific, Waltham, MA, USA). Two distinct DNA aliquots were taken for Nanopore and 16S rRNA gene sequencing.

Table 1. Clinical characteristics of the patients with benign effusions.

CASE ID	Type of Seroma	Age	Reason of the Implant
Acute-01	Acute	42	Aesthetic
Acute-02	Acute	43	Reconstruction after breast cancer
Acute-03	Acute	59	Reconstruction after breast cancer
Acute-04	Acute	33	Aesthetic
Mixed-01	Mixed	56	Aesthetic
Mixed-02	Mixed	50	Reconstruction after breast cancer
Mixed-03	Mixed	63	Aesthetic
Chronic-01	Chronic	60	Reconstruction after breast cancer
Chronic-02	Chronic	38	Aesthetic
Chronic-03	Chronic	69	Reconstruction after breast cancer
Chronic-04	Chronic	69	Reconstruction after breast cancer
Chronic-05	Chronic	52	Reconstruction after breast cancer

Table 2. Clinical characteristic and pathological stage of the BIA-ALCL patients.

CASE ID	Age (Years)	Reason of the Implant	Time from the Implant (Years)	MD Anderson Stage	Therapy	Outcome
BIA-01	53	Aesthetic	3	IA	Radical en-bloc capsulectomy	CR
BIA-02	67	Reconstruction after breast cancer	13	IIA	Radical en-bloc Capsulectomy+RT	CR
BIA-03	66	Reconstruction after breast cancer	11	IIB	Radical en-bloc capsulec-tomy+CHOEP	CR
BIA-04	55	Aesthetic	10	IA	Radical en-bloc capsulectomy	CR
BIA-05	49	Reconstruction after breast cancer	8	IA	Radical en-bloc capsulectomy	CR
BIA-06	76	Reconstruction after breast cancer	15	IIA	Radical en-bloc capsulectomy+RT	CR
BIA-07	54	Reconstruction after breast cancer	13	IA	Radical en-bloc capsulectomy	CR

Table 2. *Cont.*

CASE ID	Age (Years)	Reason of the Implant	Time from the Implant (Years)	MD Anderson Stage	Therapy	Outcome
BIA-08	51	Reconstruction after breast cancer	NA	IB	Radical en-bloc capsulectomy	CR
BIA-09	40	Aesthetic	12	IA	Radical en-bloc capsulectomy	CR
BIA-10	38	Aesthetic	6	IA	Radical en-bloc capsulectomy	CR

2.2. Nanopore Sequencing and Data Analysis

The libraries were prepared on 100 ng input DNA extracted from each seroma by using the polymerase chain reaction (PCR) barcoding protocols (EXP-SQK-PBK004, Version: PBK_9073_v1_revM) (Oxford Nanopore Technologies (ONT), Oxford, UK). Sequencing was performed on an ONT MinION flow cell (FLO-MIN106 R9 Version) connected to a Mk1B device (ONT Ltd.; MIN-101B, Oxford Nanopore Technologies, Oxford, UK) following the manufacturer's instructions [30]. The sequencing was run up to 48 h each time, using Nanopore software MinKNOW v3.1.19 (for the first and second runs) and MinKNOW v4.0.4 for the last run (Oxford Nanopore Technologies Ltd., UK), which collects the sequencing data and converts it into base-called reads. For the post-base-calling analysis, the cloud-based EPI2ME software v.3.4.2.1259 [31] (https://epi2me.nanoporetech.com/about/ (accessed on 15 March 2019, 20 March 2019, and 26 August 2020) with "What's in my pot?" (WIMP) (quantitative analysis tool for real-time species identification—a part of EPI2ME) was used for the evaluation of species diversity. WIMP filters fastq files with a mean q-score below a minimal threshold and the centrifuge classification engine assigns each read to a taxon in the National Center for Biotechnology Information (NCBI) taxonomy and reference database [32–34].

This method covers bacterial, viral, and fungal genomes [34,35]. We used the taxname function from the *taxize* R package (R version 4.5.0) [36] to retrieve family, genus, and species names from the NCBI database and assign them to the organisms identified by WIMP; patients were divided into three runs composed of 4, 8, and 10 DNA samples each (Supplementary Figure S1). In each run, a control sample extracted from a Human Herpesvirus 8 (HHV8)-positive primary effusion lymphoma (PEL) was added, and the nuclease-free water was used as a blank test. All fastq files are available on request to the corresponding author. We did not observe any batch effect among the samples from different runs (Supplementary Figure S1). Indeed, the clusters derived from the beta diversity analysis using the Bray–Curtis matrix were formed independently of the sequencing run.

2.3. Microbiome Sequencing by 16S rRNA Gene Sequencing and Data Analysis

The Ion16S™ Metagenomics Kit (ThermoFisher Scientific, Waltham, MA, USA) was used to obtain the library from the DNA extracted from 21 samples. The DNA of patient Chronic-05 was excluded as it was not sufficient for the analysis. The kit includes two sets of primers to amplify the V2, 4, 8 and the V3, 5, 6, 7, 9 hypervariable regions of the bacterial 16S rRNA gene. Template preparation and chip loading were carried out using an IonS5 Ion Chef™ System (ThermoFisher Scientific, Waltham, MA, USA), and sequencing was performed using an IonS5 System (ThermoFisher Scientific, Waltham, MA, USA). After sequencing, the fastq files were processed using a custom script based on the Quantitative Insights Into Microbial Ecology 2 (QIIME2) software (version QIIME2 2023.5).

Quality control retained sequences with a length between 140 and 400 base pairs (bps) and a mean sequence quality score > 20, while sequences with homopolymers > 7 bps and mismatched primers were omitted. To calculate downstream diversity measures, 16SrRNA Operational Taxonomic Units (OTUs) were defined at 100% sequence homology; OTUs not encompassing at least 2 sequences of the same sample were removed. All reads were classified to the lowest possible taxonomic rank using QIIME2 software and a reference dataset from the Curated MicroSEQ(R) 16S Reference Library v2013.1 (ThermoFisher Scientific) and Curated Greengenes v13.5 [37]. Fastq files are available on request to the corresponding author.

2.4. Statistical Analysis

For hierarchical clustering analysis of the samples, the Bray–Curtis distance metric was considered, and the Ward-D2 aggregation method was used for building the dendrogram within the R package hclust. The relative counts were considered for the bar plot representation. The microbiome differential abundance analysis between BIA-ALCL and non-BIA-ALCL groups was performed by applying the ancombc2 function of the Analysis of Compositions of Microbiomes with Bias Correction (ANCOMBC) R package [38] to the original count matrix. The diversity indices were calculated in R based on their formula [39,40]. Fractions and indices were compared between the BIA-ALCL and non-BIA-ALCL groups through the non-parametric bootstrap t-test. All statistical analyses have been performed using R statistical software v4.5.0 [41].

3. Results

3.1. Abundance Analysis of BIA-ALCL and Benign Samples

Using Nanopore's WIMP workflow, we identified a total of 85 unique bacterial, 4 viral, and 6 fungal families across all samples. Specifically, 55 bacterial, 4 viral, and 4 fungal families were detected in BIA-ALCL samples, while benign seromas contained 56 bacterial, 6 fungal, and 3 viral families (Table 3 and Supplementary Table S1). In contrast, 16S rRNA sequencing, which targets only bacterial genomes, identified a total of 222 bacterial families. Of these, 130 were found in BIA-ALCL samples and 212 in benign seromas (Table 4 and Supplementary Table S1).

Table 3. Number of microorganism families identified in each seroma group by Nanopore sequencing.

Sample Type	Microorganisms (Total *n*)	Bacterial Families (*n*)	Fungal Families (*n*)	Viral Families (*n*)
BIA-ALCL	63	55	4	4
Benign seroma	65	56	6	3
Acute-type	15	12	2	1
Mixed-type	29	26	2	1
Chronic-type	43	37	4	2

Table 4. Number of bacterial families identified in each seroma group by 16S rRNA sequencing.

Sample Type	Bacterial Families (*n*)
BIA-ALCL	130
Benign seroma	212
Acute-type	175
Mixed-type	180
Chronic-type	119

Based on Nanopore sequencing, the most represented families, in both BIA-ALCL and non-neoplastic seromas, were *Propionibacteriaceae and Staphylococcaceae* (Figure 1, upper panel and Supplementary Table S2). Conversely, 16S rRNA gene sequencing revealed that *Bradyrhizobiaceae* and *Staphylococcaceae* predominated (Figure 1, lower panel and Supplementary Table S2).

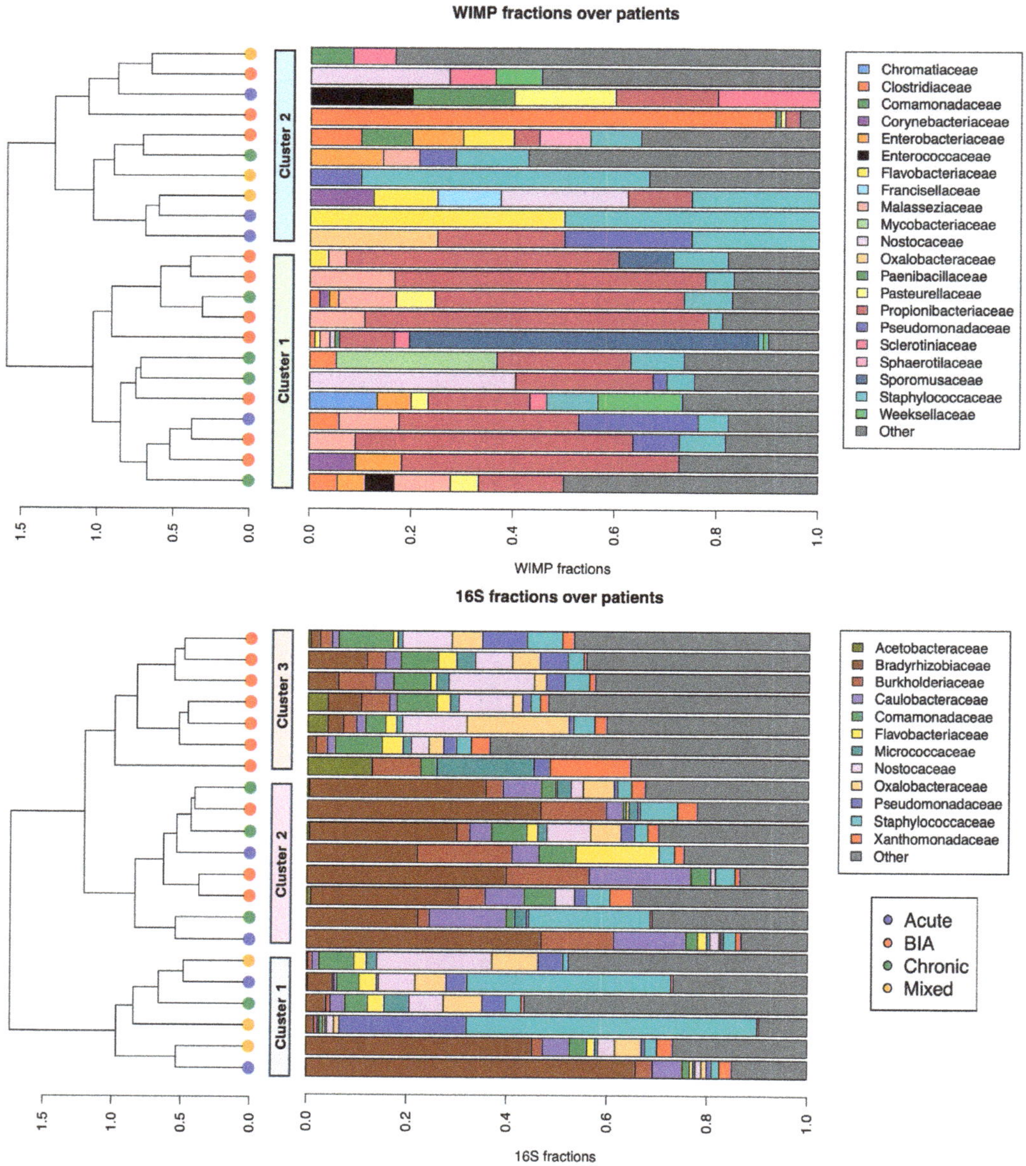

Figure 1. Relative frequencies of organism families across patients, ordered by an unsupervised cluster analysis based on the Bray–Curtis dissimilarity matrix. Using Nanopore sequencing data, samples were segregated into two clusters, independent of the type of effusion (**upper** panel). In contrast, using the 16S rRNA sequencing data, most BIA-ALCL samples clustered together in cluster 3 (**lower** panel).

We then calculated the beta diversity to examine differences in the types and proportions of microbes present across different samples, using the unsupervised Bray–Curtis dissimilarity matrix (Figure 1). Nanopore sequencing showed no significant difference between benign and BIA-ALCL seromas. However, patients, regardless of the type of seromas, clustered in two distinct groups (Figure 1, upper panel). The major families driving the diversity between these two groups were *Propionibacteriaceae* ($p < 0.001$), *Malasseziaceae* ($p = 0.006$), *Pasteurellaceae* ($p = 0.009$), and *Sclerotiniaceae* ($p = 0.02$), all of which were more represented in cluster 1.

Using 16S rRNA sequencing data, 7 out of 10 BIA-ALCL seromas clustered together (cluster 3) (Figure 1, lower panel). Unlike clusters 1 and 2, where single taxa often dominated the community, samples in cluster 3 shared a relatively balanced presence of multiple families without overt dominance. Notably, numerous families were significantly underrepresented in cluster 3, including *Burkholderiaceae* (log fold change (logFC) = −1.235, $p = 0.005$), which encompasses *Ralstonia* spp. that were not detected.

Due to the overall low abundance and impact of viruses and fungi on the clustering of samples, our further analyses on Nanopore data focused solely on bacterial families. We then investigated whether any microbial families were recurrently associated with BIA-ALCL or benign seroma patients by performing a Fisher's exact test on each family, based on its presence or absence across samples. Using both sequencing approaches, we found no microbial family significantly enriched in the BIA-ALCL group (adjusted p-values > 0.05, Supplementary Table S3).

To gain a deeper understanding of the microbiome characteristics within the BIA-ALCL environment, we performed a differential abundance analysis comparing the microbiomes of BIA-ALCL and benign groups (see Methods Section). Only a limited number of bacterial families were found to be differentially abundant between the two conditions (Supplementary Table S4). Specifically, just one family (*Clostridiaceae*) showed higher abundance in the BIA-ALCL group based on Nanopore sequencing. However, this trend was not consistent across sequencing methods. Indeed, in the 16S rRNA dataset, *Clostridiaceae* exhibited greater abundance in benign samples. The 16S analysis identified a greater number of differentially abundant families (log2FC > 0.58, adjusted p-value < 0.05; see Supplementary Table S4). Nonetheless, all these families displayed a low average relative abundance (<10%) in the BIA-ALCL group and were also present in the benign seroma group (Supplementary Table S2). These findings suggest that no specific bacterial families are uniquely or strongly enriched in the BIA-ALCL microbiome.

To assess whether the abundance differences showed a similar pattern between the outputs of WIMP and 16S, we selected a broader group of families that show a differential abundance behavior (adjusted p-value < 0.1, no threshold on $\log_2$FC). Then we evaluated the concordance between the two sets of identified families. As shown in the Venn diagram (Figure 2), no families are displaying a consistent trend across the two sequencing methods. This result is consistent with the lack of a predominant family, as indicated by the results of Fisher's exact test (Supplementary Table S3). In our view, what we have observed is likely due to the absence of families with a marked difference in abundance that could truly characterize the BIA-ALCL patient group.

3.2. Bacterial Family Differences Within Samples by Using Alpha Diversity Assays

To uncover eventual differences within each benign and BIA-ALCL group, we calculated alpha diversity, which provides information on the diversity of the microbial community within individual samples by quantifying the richness (number of different microbial species) and the evenness (distribution of those species). Alpha diversity was assessed using richness alone (absolute number of distinct families) or both richness and evenness (Shannon index and inverse Simpson index).

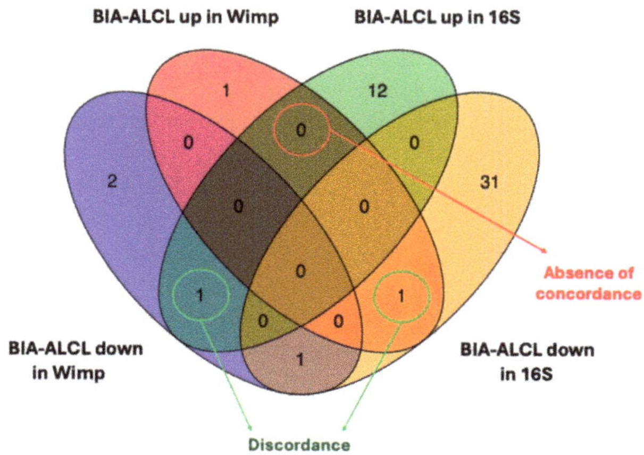

Figure 2. Venn diagram showing the overlap of microbial families identified as differentially abundant in BIA-ALCL patients across two sequencing methods (WIMP and 16S rRNA, adjusted *p*-values < 0.1). Each section represents families that were either increased ("BIA-up") or decreased ("BIA-down") in the BIA-ALCL group, as determined by ANCOM-BC. The absence of overlap among methods indicates a lack of concordance in the identified differentially abundant families. The central area with no shared families highlights the non-concordant results (red circle).

Based on Nanopore sequencing, BIA-ALCL samples did not show any significant differences when compared to all benign seromas. However, when specifically compared to benign acute-type seromas, BIA-ALCL had a significantly higher median value of richness ($p = 0.014$) (Figure 3), suggesting BIA-ALCL had a higher number of different bacterial families.

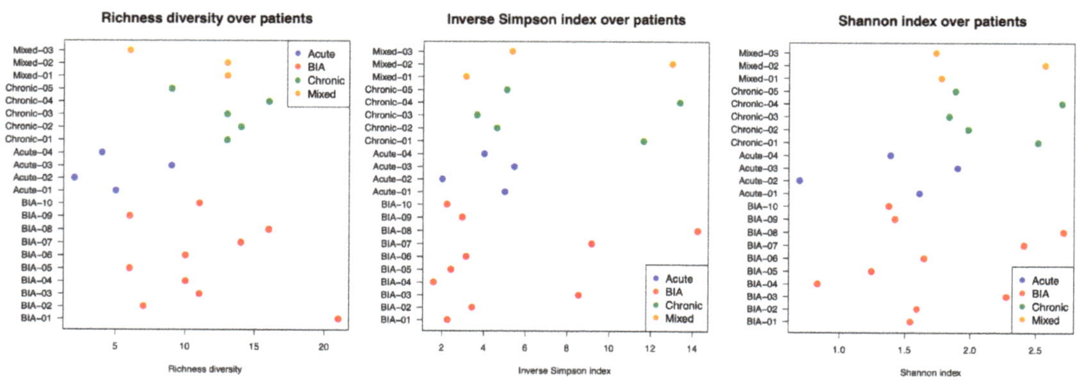

mean.BIA	other.group	mean	pval	sign
	Acute	5.0	0.014	*
11.2	Chronic	13.0	0.35	
	Mixed	10.7	0.83	

mean.BIA	other.group	mean	pval	sign
	Acute	4.1	0.57	
4.9	Chronic	7.6	0.28	
	Mixed	7.1	0.53	

mean.BIA	other.group	mean	pval	sign
	Acute	1.3	0.36	
1.7	Chronic	2.2	0.092	
	Mixed	2.0	0.38	

Figure 3. Alpha diversity, measured at the family level using Nanopore sequencing data, was compared across patient groups. BIA-ALCL had a significantly (<0.05 *) higher median value of richness when compared to acute-type seromas ($p = 0.014$).

Moreover, the inverse Simpson index evaluated on 16S rRNA gene sequencing data (Figure 4) revealed that the BIA-ALCL group had significantly higher microbial diversity than the benign acute-type group ($p = 0.0072$). This difference suggests a loss of evenness, with the dominance of a few taxa, during acute infection. Conversely, BIA-ALCL seromas host more compositionally diverse communities, which is more consistent with a chronic inflammatory process.

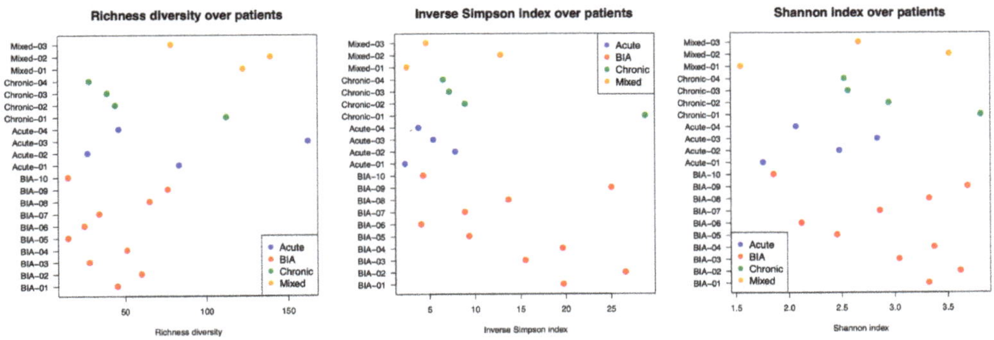

mean.BIA	other.group	mean	pval	sign
41.4	Acute	79.5	0.32	
	Chronic	55.7	0.52	
	Mixed	113.0	0.067	

mean.BIA	other.group	mean	pval	sign
14.65	Acute	4.819	0.0072	**
	Chronic	12.837	0.754	
	Mixed	6.642	0.121	

mean.BIA	other.group	mean	pval	sign
2.96	Acute	2.3	0.067	
	Chronic	3.0	0.99	
	Mixed	2.6	0.53	

Figure 4. Alpha diversity, measured at the family level using 16S rRNA gene sequencing data, was compared across patient groups. The inverse Simpson index revealed significantly (<0.01 **) higher microbial diversity in BIA-ALCL samples compared to the benign acute-type group ($p = 0.0072$).

3.3. Impact of Bacteria GRAM Stain in Benign and BIA-ALCL Seromas

We successfully classified most bacteria in our cohort (99%) as either Gram-positive (GP) or Gram-negative (GN). This classification relied on three different databases: BacDive [42], UK Culture Collections [43], and BactoTraits [44]. Upon assessing their distribution across samples and using both sequencing datasets, we found no significant differences in the proportion of GP and GN between BIA-ALCL and benign seromas (Figure 5).

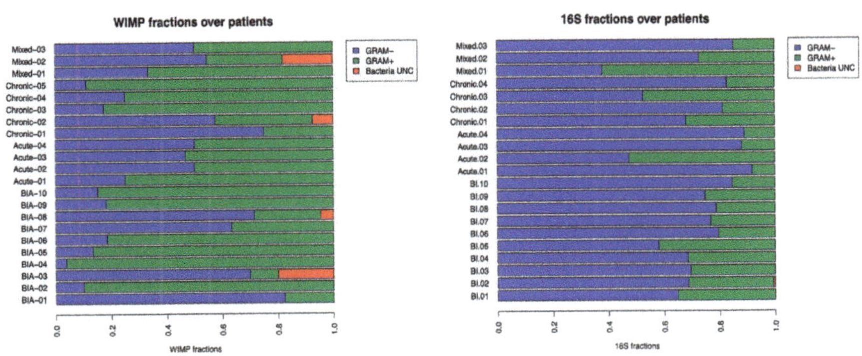

Class	mean.BIA	mean.benign	pval
GRAM-	0.366	0.413	0.68
GRAM+	0.609	0.566	0.72
Bacteria UNC	0.025	0.021	0.93

Class	mean.BIA	mean.benign	pval
GRAM-	0.724	0.724	1.0
GRAM+	0.276	0.276	0.99
Bacteria UNC	0.001	0.000	0.47

Figure 5. Relative frequencies of Gram-positive (GP, green bars) and Gram-negative (GN, blue bars) bacteria across patients with BIA-ALCL and benign seromas. No statistically significant differences were found in the proportion of GP and GN bacteria between BIA-ALCL and benign samples using either sequencing method. Red bars indicate the non-classified bacteria.

3.4. Oxygen Tolerance of the Bacteria in Benign and BIA-ALCL Seromas

Considering the known hypoxic microenvironment of peri-implant seromas [6], we classified the bacteria according to their oxygen requirements using the same three databases: BacDive [42], UK Culture Collections [43], and BactoTraits [44]. We were able to classify most of the bacteria of our cohort (95.4%) based on their oxygen tolerance. Nanopore data revealed a significant enrichment of non-aerobic (anaerobic and microaerophilic) bacteria in BIA-ALCL samples compared to benign ones (p = 0.025). Although 16S rRNA sequencing data showed a trend towards more non-aerobic bacteria in BIA-ALCL compared to the benign seromas, this difference did not reach statistical significance (p = 0.087) (Figure 6).

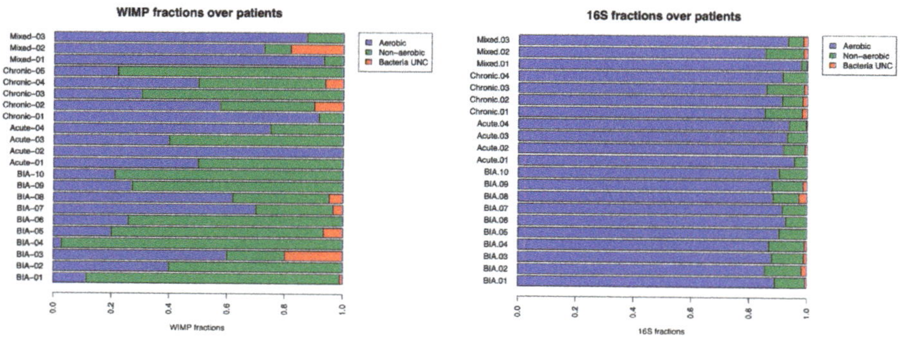

Class	mean.BIA	mean.no-BIA	pval	sign
Aerobic	0.341	0.642	0.016	*
Non-aerobic	0.624	0.329	0.025	*
Bacteria UNC	0.036	0.029	0.789	

Class	mean.BIA	mean.no-BIA	pval	sign
Aerobic	0.89	0.913	0.123	
Non-aerobic	0.101	0.078	0.087	
Bacteria UNC	0.009	0.009	0.947	

Figure 6. Relative frequencies of aerobic and anaerobic bacteria compared across patients with BIA-ALCL and benign seromas. BIA-ALCL samples were significantly (<0.05 *) enriched in non-aerobic bacterial families, but this was statistically significant only with WIMP data.

4. Discussion

This study provides the first comprehensive characterization of the microbiota in late-onset textured breast implant seromas, encompassing both BIA-ALCL and non-neoplastic effusions, by utilizing seroma fluid samples and a dual sequencing approach: Nanopore sequencing with the WIMP workflow and 16S rRNA gene sequencing. Both methods failed in identifying a microbiome profile specifically associated with BIA-ALCL.

Previous studies on the BIA-ALCL bacterial community have focused more on breast implant capsules than seroma samples; however, they present conflicting perspectives. Both Walker et al. and Hu et al. employed 16S rRNA gene sequencing (V1–V3 regions) with the same taxonomic classifier but different sequencing methods (Illumina NGS vs. FLX pyrosequencing) [27,28]. Hu et al. identified *Ralstonia* spp. as a major driver of diversity in BIA-ALCL [27], while Walker et al. found no consistent differences and low abundance of *Ralstonia* spp., primarily in non-ALCL controls [28]. Walker et al. also identified *Staphylococcaceae* and *Propinobacteriaceae* as common colonizers of the breast environment rather than disease-specific pathogens. Our findings corroborate those reported by Walker et al. Regardless of the sequencing method applied, *Staphylococcaceae*, *Propionibacteriaceae*, and *Bradyrhizobiaceae* emerged as among the most prevalent bacterial families in both BIA-ALCL and benign seroma samples. In addition, *Burkholderiaceae*—the family encompassing *Ralstonia* spp.—was detected predominantly in non-BIA-ALCL samples, with no evidence of *Ralstonia* spp. themselves.

We also showed the presence of higher richness and evenness in BIA-ALCL compared with acute seromas. This observation supports a biological model in which BIA-ALCL arises within a chronic inflammatory environment rather than an acute infectious setting. Specifically, acute seromas typically show low evenness due to the overgrowth of a single pathogen, whereas BIA-ALCL displays a more ecologically balanced microbial structure, consistent with chronicity.

In general, Nanopore sequencing detected a lower number of bacteria families when compared to 16S rRNA sequencing, potentially due to the method's lack of an amplification step. Nevertheless, our positive control (PEL) showed a high number of HHV8 reads, confirming the detection of a known disease-driving pathogen. Furthermore, *Pseudomonas aeruginosa* was successfully cultured from one benign mixed-type seroma in a previous study [3] and consistently confirmed by both WIMP and 16S rRNA sequencing in the present study. These observations suggest that disease-defining organisms tend to be more abundant and readily detected, implying that a single microorganism is unlikely to be solely responsible for BIA-ALCL pathogenesis.

While beta diversity analysis with the WIMP workflow failed to distinguish BIA-ALCL from benign effusions, the 16S rRNA sequencing data revealed a tendency for BIA-ALCL samples to form a more distinct cluster (Figure 1 cluster 3). This cluster, notably, exhibited a highly even community composition, differentiating it from the other two clusters, where individual taxa dominated. Indeed, cluster 3 does not reflect the presence of specific pathogens or a unique taxonomic signature. Instead, this clustering pattern arises from a compositional profile that is distinct (more even) but not dominated by any particular taxon. As shown by the enrichment analysis (Supplementary Table S3), while non-neoplastic seromas were significantly enriched in several taxa, no taxa showed significant enrichment in the BIA-ALCL group. In keeping with this result, BIA-ALCL samples showed higher richness (based on Nanopore data) and a higher evenness (based on 16S rRNA data) compared to acute-type seromas, indicative of a more homogenous representation of the different taxa identified. This pattern of community structure further supports the hypothesis that the singular overgrowth of a specific pathogen does not characterize BIA-ALCL.

Mempin et al. highlighted a potential role for Gram-negative bacteria in BIA-ALCL pathobiology, observing that BIA-ALCL tumor cells responded more to LPS stimulation in vitro compared to other mitogenic agents [26]. In contrast, our study found no statistically significant differences in the abundance of Gram-negative and Gram-positive bacteria in BIA-ALCL when compared to benign seromas, a result also corroborated by Walker et al. [28].

Intriguingly, Oishi et al.'s gene expression profiling of BIA-ALCL identified an up-regulation of hypoxia signaling genes, including carbonic anhydrase-9 (CA9), which can stimulate BIA-ALCL cell lines in vitro [6]. Consistent with this, our Nanopore data showed anaerobic and microaerophilic bacteria were significantly more abundant in BIA-ALCL, with 16S rRNA sequencing showing a similar trend. The enrichment of non-aerobic bacteria observed in BIA-ALCL seromas—significant in Nanopore data and trending in 16S—should, however, be interpreted cautiously. This association is correlative and does not establish a causal link between the hypoxic tumor microenvironment and microbial composition. Rather, it may reflect ecological compatibility with the biochemical milieu of peri-implant seromas, without implying selective microbial participation in lymphomagenesis. Establishing a true causal role for anaerobic bacteria in BIA-ALCL would require functional assays and in-depth biological investigations that could also provide additional insights into microbial metabolism and potential virulence. As our primary goal was to clarify whether the disease is associated with a dominant pathogen or rather with a diverse microbial milieu, functional predictions were beyond the scope of this study and should be pursued in future research with larger and functionally oriented datasets.

A limitation of our study was the lack of direct concordance between Nanopore-based metagenomics (using the WIMP workflow) and 16S rRNA sequencing at the individual sample level. This discrepancy is attributable mainly to inherent technical differences in sample processing, sequencing targets, and bioinformatics. Shotgun sequencing offers broad detection of DNA but has lower sensitivity for low-abundance taxa due to a lack of amplification [45]. Conversely, while 16S rRNA sequencing is more focused and sensitive for bacterial detection, it can introduce amplification biases. Consequently, taxonomic profiles generated by each method can differ significantly per sample. Additionally, the different classification databases employed (NCBI RefSeq for WIMP vs. Greengenes/MicroSEQ for 16S rRNA) may further contribute to these discrepancies. The WIMP workflow performs taxonomic assignment by aligning raw reads against the comprehensive NCBI RefSeq database. While this broad genomic reference enables species-level resolution when complete genomes are available, it may underperform when target organisms are underrepresented or poorly annotated in the RefSeq database. In contrast, 16S rRNA gene sequencing relies on amplification of specific hypervariable regions of the 16S rRNA gene, with taxonomic classification based on curated reference libraries such as Greengenes or MicroSEQ [45]. Classification rate and accuracy and taxonomic read classification have been reported to be heavily impacted by the reference database choice, with NCBI RefSeq showing a lower performance when compared to other databases, particularly at low taxonomic ranks [46,47]. These methodological and database differences likely explain why in our work certain families were differentially represented across the two methods and why some were exclusively detected by one method. Despite these individual sample variations, both platforms successfully captured overlapping microbial trends at the group level. This demonstrates the value of multi-modal sequencing for investigating complex microbiota in breast implant-associated effusions and confirms that BIA-ALCL is associated with complex ecological patterns rather than a single pathogen.

The lack of concordance between Nanopore-based metagenomics and 16S rRNA amplicon sequencing precludes any direct, sample-level taxonomic comparison between the two platforms. This discrepancy is expected, as shotgun sequencing and amplicon-based approaches differ substantially in sensitivity, amplification biases, and reference database structure. An example is *Clostridiaceae*, a family found more abundant in the BIA-ALCL group based on Nanopore sequencing and in benign samples by 16S rRNA sequencing. Despite this methodological discordance, both techniques converged on the same overarching conclusion: BIA-ALCL seromas do not harbor a reproducible or disease-specific microbial signature. Therefore, the robustness of our findings lies not in taxonomic agreement between methods, but in the consistent group-level pattern observed across two independent sequencing strategies. Of note, the objective of this study was not to characterize the microbiome of each individual patient, but rather to determine whether BIA-ALCL seromas shared a dominant bacterial family at the group level. If such a predominant family were truly present, the two sequencing platforms might classify it differently at the taxonomic level, but it would still emerge as a reproducible signal across samples. Instead, no such dominant family was detected by either method. This indicates that the discordance observed between techniques concerns taxonomic assignment, not the presence or absence of a shared microbial signature. The fact that two technologically distinct approaches both failed to identify a predominant family strengthens the validity of this negative result and is fully consistent with the absence of a disease-specific pathogen previously reported by Walker et al.

Another limitation of our study is the relatively small sample size, which reflects the rarity of BIA-ALCL and the inherent difficulties in assembling larger cohorts even at referral centers. However, this limit is shared with the other two available microbiome studies in BIA-ALCL, which included 26 [27] and 7 [28] BIA-ALCL cases, respectively. Despite this limitation, the consistent negative findings across two independent sequencing platforms we employed provide a solid foundation for our conclusion that no dominant microbial signature characterizes BIA-ALCL seromas.

5. Conclusions

In conclusion, our data did not reveal a distinctive microbial signature associated with BIA-ALCL when compared to non-neoplastic seromas, suggesting a lack of direct causality between specific microbial taxa and BIA-ALCL pathogenesis. The development of BIA-ALCL is more plausibly a complex process involving multiple genetic and non-genetic factors, warranting future studies to investigate their interplay.

Supplementary Materials: The following supporting information can be downloaded at https://www.mdpi.com/article/10.3390/cancers17233839/s1, Figure S1: Evaluation of batch effect among samples using the Bray–Curtis matrix; Table S1: List of all the organisms identified by both Nanopore and 16S rRNA sequencing; Table S2: Family ranking based on the average fractions over samples; Table S3: Fisher's exact test for comparing microbial presence/absence between BIA-ALCL and benign groups; Table S4: Differential abundance analysis between BIA-ALCL and benign samples.

Author Contributions: Conceptualization, A.D.N.; methodology, A.D.N. and E.R.; software, G.B. and D.V.; validation, A.D.N., E.R. and G.B.; formal analysis, G.B., D.V., M.S. (Maurizio Simmaco), M.B., G.L. and E.S.; investigation, G.B., D.V., G.L., M.S. (Maurizio Simmaco) and M.B.; resources, A.D.N., A.V., A.S., N.N., F.S.d.P., M.S. (Michai Sorotos) and G.F.; data curation, A.D.N. and E.R.; writing—original draft preparation, A.D.N. and E.R.; writing—review and editing, A.D.N. and E.R.; visualization, A.D.N., E.R. and G.B.; supervision, A.D.N.; project administration, A.D.N.; funding acquisition, A.D.N. All authors have read and agreed to the published version of the manuscript.

Funding: This work has been supported by Sapienza University project Ateneo 2020 to Arianna Di Napoli (RP120172AB1F2C84).

Institutional Review Board Statement: The study was conducted in accordance with the Declaration of Helsinki and approved by the Ethics Committee of Sant'Andrea Hospital/University "Sapienza" of Rome (EC n. 82 SA_2017, approved on 30/03/2017).

Informed Consent Statement: Informed consent was obtained from all subjects involved in the study.

Data Availability Statement: All data is available upon request to the corresponding author.

Acknowledgments: E.S. is a student within the European School of Molecular Medicine (SEMM).

Conflicts of Interest: Fabio Santanelli di Pompeo (FSDP) and Michail Sorotos (MS) would like to disclose that the NESMOS Department with which they are affiliated received research funds from Motiva Establishment Labs in 2017 and from GC Aesthetics in 2018 and 2020. The NESMOS Department has also received mini-implants from Establishment Labs, GC Aesthetics, and Sebbin for research purposes. FSDP was a paid consultant for BellaSeno GmbH from 2021 until 2023; he received reimbursements for travel/lodging expenses from ICEAG in 2015 and SCHEER-WG in 2019, 2020, and 2021; and is a member of Notified Body 0373, part of the Superior Institute of Health, which carried out CE Mark certification activities for the Italian Ministry of Health for the year 2021. FSDP received a speaker fee from Establishment Labs in 2023 and 2024. MS received a speaker fee from Establishment Labs in 2024. They have no ownerships or investments to disclose. All other authors hereby certify that to the best of their knowledge, no financial support or benefits have been received, neither by themselves directly, nor by any member of their immediate family or any individual or entity with whom or with which they may have a significant relationship, from any commercial source which is related directly or indirectly to the scientific work which is reported on in the article.

None of the authors has a financial interest in any of the products, devices, or drugs mentioned in this manuscript.

Abbreviations

The following abbreviations are used in this manuscript:

BIA-ALCL	Breast Implant-Associated Anaplastic Large Cell Lymphoma
FDA	Food and Drug Administration
ISO	International Organization for Standardization
SCHEER	Scientific Committee on Health, Emerging and Environmental Risks
NGS	Next Generation Sequencing
WIMP	What's in my Pot?
ALK	Anaplastic Lymphoma Kinase
HHV8	Herpes Virus 8
QIIME	Quantitative Insights Into Microbial Ecology 2
ANCOMBC	Analysis of Compositions of Microbiomes with Bias Correction
ONT	Oxford Nanopore Technologies
NCBI	National Center for Biotechnology Information

References

1. Alaggio, R.; Amador, C.; Anagnostopoulos, I.; Attygalle, A.D.; Araujo, I.B.d.O.; Berti, E.; Bhagat, G.; Borges, A.M.; Boyer, D.; Calaminici, M.; et al. The 5th Edition of the World Health Organization Classification of Haematolymphoid Tumours: Lymphoid Neoplasms. *Leukemia* **2022**, *36*, 1720–1748. [CrossRef]
2. U.S. Food and Drug Administration. Medical Device Reports of Breast Implant-Associated Anaplastic Large Cell Lymphoma. Available online: https://www.fda.gov/medical-devices/breast-implants/medical-device-reports-breast-implant-associated-anaplastic-large-cell-lymphoma (accessed on 2 August 2025).
3. Di Napoli, A.; Pepe, G.; Giarnieri, E.; Cippitelli, C.; Bonifacino, A.; Mattei, M.; Martelli, M.; Falasca, C.; Cox, M.C.; Santino, I.; et al. Cytological Diagnostic Features of Late Breast Implant Seromas: From Reactive to Anaplastic Large Cell Lymphoma. *PLoS ONE* **2017**, *12*, e0181097. [CrossRef]
4. Jaffe, E.S.; Ashar, B.S.; Clemens, M.W.; Feldman, A.L.; Gaulard, P.; Miranda, R.N.; Sohani, A.R.; Stenzel, T.; Yoon, S.W. Best Practices Guideline for the Pathologic Diagnosis of Breast Implant-Associated Anaplastic Large-Cell Lymphoma. *J. Clin. Oncol.* **2020**, *38*, 1102–1111. [CrossRef]
5. DeCoster, R.C.; Clemens, M.W.; Di Napoli, A.; Lynch, E.B.; Bonaroti, A.R.; Rinker, B.D.; Butterfield, T.A.; Vasconez, H.C. Cellular and Molecular Mechanisms of Breast Implant-Associated Anaplastic Large Cell Lymphoma. *Plast. Reconstr. Surg.* **2021**, *147*, 30e–41e. [CrossRef]
6. Oishi, N.; Hundal, T.; Phillips, J.L.; Dasari, S.; Hu, G.; Viswanatha, D.S.; He, R.; Mai, M.; Jacobs, H.K.; Ahmed, N.H.; et al. Molecular Profiling Reveals a Hypoxia Signature in Breast Implant-Associated Anaplastic Large Cell Lymphoma. *Haematologica* **2021**, *106*, 1714–1724. [CrossRef]
7. Di Napoli, A.; Jain, P.; Duranti, E.; Margolskee, E.; Arancio, W.; Facchetti, F.; Alobeid, B.; Santanelli di Pompeo, F.; Mansukhani, M.; Bhagat, G. Targeted next Generation Sequencing of Breast Implant-Associated Anaplastic Large Cell Lymphoma Reveals Mutations in JAK/STAT Signalling Pathway Genes, TP53 and DNMT3A. *Br. J. Haematol.* **2018**, *180*, 741–744. [CrossRef] [PubMed]
8. Blombery, P.; Thompson, E.R.; Jones, K.; Arnau, G.M.; Lade, S.; Markham, J.F.; Li, J.; Deva, A.; Johnstone, R.W.; Khot, A.; et al. Whole Exome Sequencing Reveals Activating JAK1 and STAT3 Mutations in Breast Implant-Associated Anaplastic Large Cell Lymphoma Anaplastic Large Cell Lymphoma. *Haematologica* **2016**, *101*, e387–e390. [CrossRef] [PubMed]
9. Oishi, N.; Brody, G.S.; Ketterling, R.P.; Viswanatha, D.S.; He, R.; Dasari, S.; Mai, M.; Benson, H.K.; Sattler, C.A.; Boddicker, R.L.; et al. Genetic Subtyping of Breast Implant-Associated Anaplastic Large Cell Lymphoma. *Blood* **2018**, *132*, 544–547. [CrossRef]
10. Letourneau, A.; Maerevoet, M.; Milowich, D.; Dewind, R.; Bisig, B.; Missiaglia, E.; de Leval, L. Dual JAK1 and STAT3 Mutations in a Breast Implant-Associated Anaplastic Large Cell Lymphoma. *Virchows Arch.* **2018**, *473*, 505–511. [CrossRef]
11. Laurent, C.; Nicolae, A.; Laurent, C.; Le Bras, F.; Haioun, C.; Fataccioli, V.; Amara, N.; Adélaïde, J.; Guille, A.; Schiano, J.-M.; et al. Gene Alterations in Epigenetic Modifiers and JAK-STAT Signaling Are Frequent in Breast Implant-Associated ALCL. *Blood* **2020**, *135*, 360–370. [CrossRef] [PubMed]
12. Los-de Vries, G.T.; de Boer, M.; van Dijk, E.; Stathi, P.; Hijmering, N.J.; Roemer, M.G.M.; Mendeville, M.; Miedema, D.M.; de Boer, J.P.; Rakhorst, H.A.; et al. Chromosome 20 Loss Is Characteristic of Breast Implant-Associated Anaplastic Large Cell Lymphoma. *Blood* **2020**, *136*, 2927–2932. [CrossRef] [PubMed]

13. Quesada, A.E.; Zhang, Y.; Ptashkin, R.; Ho, C.; Horwitz, S.; Benayed, R.; Dogan, A.; Arcila, M.E. Next Generation Sequencing of Breast Implant-Associated Anaplastic Large Cell Lymphomas Reveals a Novel STAT3-JAK2 Fusion among Other Activating Genetic Alterations within the JAK-STAT Pathway. *Breast J.* **2021**, *27*, 314–321. [CrossRef]
14. Lechner, M.G.; Megiel, C.; Church, C.H.; Angell, T.E.; Russell, S.M.; Sevell, R.B.; Jang, J.K.; Brody, G.S.; Epstein, A.L. Survival Signals and Targets for Therapy in Breast Implant-Associated ALK--Anaplastic Large Cell Lymphoma. *Clin. Cancer Res.* **2012**, *18*, 4549–4559. [CrossRef]
15. Di Napoli, A.; De Cecco, L.; Piccaluga, P.P.; Navari, M.; Cancila, V.; Cippitelli, C.; Pepe, G.; Lopez, G.; Monardo, F.; Bianchi, A.; et al. Transcriptional Analysis Distinguishes Breast Implant-Associated Anaplastic Large Cell Lymphoma from Other Peripheral T-Cell Lymphomas. *Mod. Pathol.* **2019**, *32*, 216–230. [CrossRef]
16. *ISO 14607:2024*; Non-Active Surgical Implants—Mammary Implants—Particular Requirements. International Organization for Standardization: Geneva, Switzerland, 2024.
17. SCHEER—Scientific Committee on Health, Environmental and Emerging Risks. *Scientific Opinion on the Safety of Breast Implants in Relation to Anaplastic Large Cell Lymphoma*; European Commission: Brussels, Belgium, 2021; p. 9.
18. Doloff, J.C.; Veiseh, O.; de Mezerville, R.; Sforza, M.; Perry, T.A.; Haupt, J.; Jamiel, M.; Chambers, C.; Nash, A.; Aghlara-Fotovat, S.; et al. The Surface Topography of Silicone Breast Implants Mediates the Foreign Body Response in Mice, Rabbits and Humans. *Nat. Biomed. Eng.* **2021**, *5*, 1115–1130. [CrossRef] [PubMed]
19. Santanelli di Pompeo, F.; Sorotos, M.; Canese, R.; Valeri, M.; Roberto, C.; Giorgia, S.; Firmani, G.; di Napoli, A. Study of the Effect of Different Breast Implant Surfaces on Capsule Formation and Host Inflammatory Response in an Animal Model. *Aesthet. Surg. J.* **2023**, *43*, 506–515. [CrossRef]
20. Hu, H.; Jacombs, A.; Vickery, K.; Merten, S.L.; Pennington, D.G.; Deva, A.K. Chronic Biofilm Infection in Breast Implants Is Associated with an Increased T-Cell Lymphocytic Infiltrate: Implications for Breast Implant-Associated Lymphoma. *Plast. Reconstr. Surg.* **2015**, *135*, 319–329. [CrossRef]
21. Mempin, M.; Hu, H.; Chowdhury, D.; Deva, A.; Vickery, K. The A, B and C's of Silicone Breast Implants: Anaplastic Large Cell Lymphoma, Biofilm and Capsular Contracture. *Materials* **2018**, *11*, 2393. [CrossRef] [PubMed]
22. James, G.A.; Boegli, L.; Hancock, J.; Bowersock, L.; Parker, A.; Kinney, B.M. Bacterial Adhesion and Biofilm Formation on Textured Breast Implant Shell Materials. *Aesthetic Plast. Surg.* **2019**, *43*, 490–497. [CrossRef]
23. Rezende-Pereira, G.; Albuquerque, J.P.; Souza, M.C.; Nogueira, B.A.; Silva, M.G.; Hirata, R.; Mattos-Guaraldi, A.L.; Duarte, R.S.; Neves, F.P.G. Biofilm Formation on Breast Implant Surfaces by Major Gram-Positive Bacterial Pathogens. *Aesthet. Surg. J.* **2021**, *41*, 1144–1151. [CrossRef]
24. Jones, P.; Mempin, M.; Hu, H.; Chowdhury, D.; Foley, M.; Cooter, R.; Adams, W.P.; Vickery, K.; Deva, A.K. The Functional Influence of Breast Implant Outer Shell Morphology on Bacterial Attachment and Growth. *Plast. Reconstr. Surg.* **2018**, *142*, 837–849. [CrossRef]
25. Chessa, D.; Ganau, G.; Spiga, L.; Bulla, A.; Mazzarello, V.; Campus, G.V.; Rubino, S. Staphylococcus Aureus and Staphylococcus Epidermidis Virulence Strains as Causative Agents of Persistent Infections in Breast Implants. *PLoS ONE* **2016**, *11*, e0146668. [CrossRef]
26. Mempin, M.; Hu, H.; Vickery, K.; Kadin, M.E.; Prince, H.M.; Kouttab, N.; Morgan, J.W.; Adams, W.P.; Deva, A.K. Gram-Negative Bacterial Lipopolysaccharide Promotes Tumor Cell Proliferation in Breast Implant-Associated Anaplastic Large-Cell Lymphoma. *Cancers* **2021**, *13*, 5298. [CrossRef]
27. Hu, H.; Johani, K.; Almatroudi, A.; Vickery, K.; Van Natta, B.; Kadin, M.E.; Brody, G.; Clemens, M.; Cheah, C.Y.; Lade, S.; et al. Bacterial Biofilm Infection Detected in Breast Implant-Associated Anaplastic Large-Cell Lymphoma. *Plast. Reconstr. Surg.* **2016**, *137*, 1659–1669. [CrossRef] [PubMed]
28. Walker, J.N.; Hanson, B.M.; Pinkner, C.L.; Simar, S.R.; Pinkner, J.S.; Parikh, R.; Clemens, M.W.; Hultgren, S.J.; Myckatyn, T.M. Insights into the Microbiome of Breast Implants and Periprosthetic Tissue in Breast Implant-Associated Anaplastic Large Cell Lymphoma. *Sci. Rep.* **2019**, *9*, 10393. [CrossRef]
29. Clemens, M.W.; Medeiros, L.J.; Butler, C.E.; Hunt, K.K.; Fanale, M.A.; Horwitz, S.; Weisenburger, D.D.; Liu, J.; Morgan, E.A.; Kanagal-Shamanna, R.; et al. Complete Surgical Excision Is Essential for the Management of Patients With Breast Implant-Associated Anaplastic Large-Cell Lymphoma. *J. Clin. Oncol.* **2016**, *34*, 160–168. [CrossRef] [PubMed]
30. Nanoporetech Oxford Nanopore Technologies. Available online: https://nanoporetech.com/document/experiment-companion-minknow (accessed on 26 August 2020).
31. EPI2ME. Available online: https://epi2me.nanoporetech.com/about/ (accessed on 26 August 2020).
32. RefSeq: NCBI Reference Sequence Database. Available online: https://www.ncbi.nlm.nih.gov/refseq (accessed on 26 August 2020).
33. Solcova, M.; Demnerova, K.; Purkrtova, S. Application of Nanopore Sequencing (MinION) for the Analysis of Bacteriome and Resistome of Bean Sprouts. *Microorganisms* **2021**, *9*, 937. [CrossRef]

34. Mechan Llontop, M.E.; Sharma, P.; Aguilera Flores, M.; Yang, S.; Pollok, J.; Tian, L.; Huang, C.; Rideout, S.; Heath, L.S.; Li, S.; et al. Strain-Level Identification of Bacterial Tomato Pathogens Directly from Metagenomic Sequences. *Phytopathology* **2020**, *110*, 768–779. [CrossRef] [PubMed]
35. Maguire, M.; Kase, J.A.; Roberson, D.; Muruvanda, T.; Brown, E.W.; Allard, M.; Musser, S.M.; González-Escalona, N. Precision Long-Read Metagenomics Sequencing for Food Safety by Detection and Assembly of Shiga Toxin-Producing Escherichia Coli in Irrigation Water. *PLoS ONE* **2021**, *16*, e0245172. [CrossRef]
36. Chamberlain, S.A.; Szöcs, E. Taxize: Taxonomic Search and Retrieval in R. *F1000Research* **2013**, *2*, 191. [CrossRef]
37. Conti, L.; Borro, M.; Milani, C.; Simmaco, M.; Esposito, G.; Canali, G.; Pilozzi, E.; Ventura, M.; Annibale, B.; Lahner, E. Gastric Microbiota Composition in Patients with Corpus Atrophic Gastritis. *Dig. Liver Dis.* **2021**, *53*, 1580–1587. [CrossRef]
38. Lin, H.; Peddada, S.D. Analysis of Compositions of Microbiomes with Bias Correction. *Nat. Commun.* **2020**, *11*, 3514. [CrossRef]
39. Simpson, E. Measurement of Diversity. *Nature* **1949**, *163*, 688. [CrossRef]
40. Spellerberg, I.F.; Fedor, P.J. A Tribute to Claude Shannon (1916–2001) and a Plea for More Rigorous Use of Species Richness, Species Diversity and the "Shannon–Wiener" Index. *Glob. Ecol. Biogeogr.* **2003**, *12*, 177–179. [CrossRef]
41. *R Statistical Software*; v4.5.0. The R Project for Statistical Computing: Vienna, Austria. Available online: http://www.R-project.org (accessed on 25 November 2025).
42. Schober, I.; Koblitz, J.; Sardà Carbasse, J.; Ebeling, C.; Schmidt, M.L.; Podstawka, A.; Gupta, R.; Ilangovan, V.; Chamanara, J.; Overmann, J.; et al. BacDive in 2025: The Core Database for Prokaryotic Strain Data. *Nucleic Acids Res.* **2025**, *53*, D748–D756. [CrossRef] [PubMed]
43. UK Health Security Agency, Culture Collection. Available online: https://www.culturecollections.org.uk (accessed on 26 August 2020).
44. Cébron, A.; Zeghal, E.; Usseglio-Polatera, P.; Meyer, A.; Bauda, P.; Lemmel, F.; Leyval, C.; Maunoury-Danger, F. BactoTraits—A Functional Trait Database to Evaluate How Natural and Man-Induced Changes Influence the Assembly of Bacterial Communities. *Ecol. Indic.* **2021**, *130*, 108047. [CrossRef]
45. Tessler, M.; Neumann, J.S.; Afshinnekoo, E.; Pineda, M.; Hersch, R.; Velho, L.F.M.; Segovia, B.T.; Lansac-Toha, F.A.; Lemke, M.; DeSalle, R.; et al. Large-Scale Differences in Microbial Biodiversity Discovery between 16S Amplicon and Shotgun Sequencing. *Sci. Rep.* **2017**, *7*, 6589. [CrossRef] [PubMed]
46. Smith, R.H.; Glendinning, L.; Walker, A.W.; Watson, M. Investigating the Impact of Database Choice on the Accuracy of Metagenomic Read Classification for the Rumen Microbiome. *Anim. Microbiome* **2022**, *4*, 57. [CrossRef]
47. Sczyrba, A.; Hofmann, P.; Belmann, P.; Koslicki, D.; Janssen, S.; Dröge, J.; Gregor, I.; Majda, S.; Fiedler, J.; Dahms, E.; et al. Critical Assessment of Metagenome Interpretation-a Benchmark of Metagenomics Software. *Nat. Methods* **2017**, *14*, 1063–1071. [CrossRef]

MDPI AG
Grosspeteranlage 5
4052 Basel
Switzerland
Tel.: +41 61 683 77 34

Cancers Editorial Office
E-mail: cancers@mdpi.com
www.mdpi.com/journal/cancers

www.ingramcontent.com/pod-product-compliance
Lightning Source LLC
Chambersburg PA
CBHW061451210926
51850CB00008B/1685